AF510671

Methods in Bioengineering

Alternative Technologies to Animal Testing

The Artech House Methods in Bioengineering Series

Series Editors-in-Chief

Martin L. Yarmush, M.D., Ph.D.
Robert S. Langer, Sc.D.

Methods in Bioengineering: Alternative Technologies to Animal Testing,
Tim Maguire and Eric Novik, editors

Methods in Bioengineering: Biomicrofabrication and Biomicrofluidics,
Jeffrey D. Zahn, editor

Methods in Bioengineering: Microdevices in Biology and Medicine,
Yaakov Nahmias and Sangeeta N. Bhatia, editors

Methods in Bioengineering: Nanoscale Bioengineering and Nanomedicine,
Kaushal Rege and Igor Medintz, editors

Methods in Bioengineering: Stem Cell Bioengineering,
Biju Parekkadan and Martin L. Yarmush, editors

Methods in Bioengineering: Systems Analysis of Biological Networks,
Arul Jayaraman and Juergen Hahn, editors

Methods in Bioengineering

Alternative Technologies to Animal Testing

Tim Maguire
Rutgers University

Eric Novik
Hurel Corporation

Editors

ARTECH HOUSE
BOSTON | LONDON
artechhouse.com

Library of Congress Cataloging-in-Publication Data
A catalog record for this book is available from the U. S. Library of Congress.

British Library Cataloguing in Publication Data
A catalogue record for this book is available from the British Library.

ISBN-13: 978-1-60807-011-4

Cover design by Vicki Kane

© 2010 Artech House. All rights reserved.

Printed and bound in the United States of America. No part of this book may be reproduced or utilized in any form or by any means, electronic or mechanical, including photocopying, recording, or by any information storage and retrieval system, without permission in writing from the publisher.

All terms mentioned in this book that are known to be trademarks or service marks have been appropriately capitalized. Artech House cannot attest to the accuracy of this information. Use of a term in this book should not be regarded as affecting the validity of any trademark or service mark.

10 9 8 7 6 5 4 3 2 1

Contents

Preface xiii

CHAPTER 1

Current Methods for Prediction of Human Hepatic Clearance Using In Vitro Intrinsic Clearance 1

1.1 Introduction 2

1.2 Materials 3

1.3 Methods 3

1.3.1 Thawing the hepatocytes 3

1.3.2 Clearance study using a hepatocyte suspension 3

1.3.3 Clearance study using a plated hepatocyte culture 4

1.3.4 Clearance study using a plated hepatocyte culture under a flow condition 4

1.3.5 Sampling for the clearance study 6

1.3.6 Sample analysis using LC-MS/MS 6

1.4 Data Acquisition, Anticipated Results, and Interpretation 7

1.4.1 Hepatocyte suspension and plated hepatocyte system 7

1.4.2 Physiologically based microfluidic systems 8

1.5 Discussion and Commentary 8

1.5.1 Hepatocyte suspension system 8

1.5.2 Plated hepatocyte system 11

1.5.3 Physiologically based microfluidic systems 12

1.6 Summary 13

References 15

CHAPTER 2

Use of Permeability from Cultured Cell Lines and PAMPA System and Absorption from Experimental Animals for the Prediction of Absorption in Humans 19

2.1 Introduction 20

2.2 Materials 21

2.3 Methods 21

2.3.1 Cultured cell system 21

2.3.2 PAMPA system 24

2.3.3 In vivo absorption measurements 25

2.4 Data Acquisition, Anticipated Results, and Interpretation 25
2.4.1 Data analysis 25
2.4.2 Results and interpretation 26
2.5 Discussion and Commentary 30
2.5.1 Cell culture and PAMPA systems 30
2.5.2 Absorption in experimental animals 35
2.5.3 Rats 35
2.5.4 Dogs 36
2.5.5 Monkeys 37
2.6 Summary 38
References 38

CHAPTER 3
Aggregating Brain Cell Cultures for Neurotoxicity Tests 41
3.1 Introduction 42
3.2 Experimental Design 43
3.3 Materials 44
3.3.1 Animals 44
3.3.2 Special equipment 45
3.3.3 Reagents 46
3.3.4 Preparation of solutions and media 47
3.4 Methods 49
3.4.1 Washing and sterilizing the glassware 49
3.4.2 Cell isolation and culture preparation 49
3.4.3 Maintenance of aggregating brain cell cultures (media replenishment and subdivision) 51
3.4.4 Preparation and treatment of replicate cultures 52
3.4.5 Harvest of replicate cultures for various analytical procedures 53
3.4.6 Examples of sample preparation and use for various analytical procedures 53
3.4.7 Data Analysis 54
3.5 Anticipated Results 55
3.6 Discussion and Commentary 57
3.7 Application Notes 57
3.8 Summary Points 58
Acknowledgments 59
References 59

CHAPTER 4
Approaches Towards a Multiscale Model of Systemic Inflammation in Humans 61
4.1 Introduction 62
4.2 Materials 64
4.2.1 Human endotoxin model and data collection 64

4.3 Methods 65
4.3.1 Transcriptional dynamics and intrinsic responses 65
4.3.2 Modeling inflammation at the cellular level 67
4.3.3 Modeling inflammation at the systemic level 73
4.4 Results 77
4.4.1 Elements of the multiscale host response model of human inflammation 77
4.4.2 Estimation of relevant model parameters 77
4.4.3 Qualitative assessment of the model 80
4.5 Conclusions 91
Acknowledgments 91
References 92

CHAPTER 5
A Liposome Assay for Evaluating the Ocular Toxicity of Chemicals 99
5.1 Introduction 100
5.2 Experimental Design 101
5.3 Materials 102
5.4 Methods 103
5.4.1 Preparation of calcein-loaded liposomes 103
5.4.2 Separation of bulk calcein from loaded liposomes with Sephadex 104
5.4.3 Ocular toxicity experiments using dye-loaded liposomes 106
5.5 Data Acquisition, Anticipated Results, and Interpretation 108
5.6 Discussion and Commentary 109
5.7 Application Notes 111
5.8 Summary Points 112
References 113

CHAPTER 6
Prediction of Potential Drug Myelotoxicity by In Vitro Assays on Hematopoietic Progenitors 115
6.1 Introduction 116
6.2 Experimental Design 116
6.3 Materials 117
6.3.1 Reagents 118
6.4 Methods 118
6.4.1 Preparation of methylcellulose stocks 118
6.4.2 Source of murine hematopoietic progenitors 118
6.4.3 Source of human hematopoietic progenitors 119
6.4.4 Technical procedure for GM-CFU test 121
6.4.5 Passing from screening phase to IC determination phase 122
6.4.6 Incubator humidity test 122

6.4.7 Scoring the colonies 123
6.4.8 Criteria for colony counting 123
6.5 Data Acquisition, Anticipated Results, and Interpretation 124
6.5.1 Statistical guidelines 125
6.6 Discussion and Commentary 126
6.7 Application Notes 128
6.8 Summary Points 128
Acknowledgments 130
References 130

CHAPTER 7

Epigenetically Stabilized Primary Hepatocyte Cultures: A Potential Sensitive Screening Tool for Nongenotoxic Carcinogenicity 133

7.1 Introduction 134
7.2 Experimental Design 135
7.3 Materials 135
7.3.1 Reagents 135
7.3.2 Facilities/Equipment 139
7.4 Methods 139
7.4.1 Isolation of hepatocytes from rat liver 139
7.4.2 Cultivation of primary rat hepatocytes (Troubleshooting Table) 141
7.5 Data Acquisition 141
7.6 Anticipated Results and Interpretation 141
7.7 Discussion and Commentary 143
7.8 Application Notes 144
7.9 Summary Points 144
Acknowledgements 145
References 145

CHAPTER 8

A Statistical Method to Reduce In Vivo Product Testing Using Related In Vitro Tests and ROC Analysis 147

8.1 Introduction 148
8.2 Experimental Design 149
8.3 Materials 149
8.4 Methods 150
8.4.1 Step-by-step protocol for the analysis of data using Analyse-It 152
8.5 Results 154
8.6 Discussion and Commentary 154
8.6.1 Selecting the proper secondary test 154
8.6.2 Determining the sample size for calibration and recalibration 155
8.6.3 Regulatory concerns 156
8.6.4 Determining the frequency of recalibration 156

8.6.5 Determining the need for confirmatory testing 157
8.6.6 Statistical analysis 157
8.7 Summary Points 158
Acknowledgments 158
References 158

CHAPTER 9
Application of the Benchmark Approach in the Correlation of In Vitro and In Vivo Data in Developmental Toxicity 159
9.1 Introduction 160
9.2 Materials and Methods 162
9.2.1 Derivation of in vitro BMC and BMD values 163
9.2.2 In vitro–in vivo correlation 164
9.3 Discussion and Commentary 167
References 168

CHAPTER 10
Three-Dimensional Cell Culture of Canine Uterine Glands 171
10.1 Introduction 172
10.2 Materials 173
10.2.1 Cell culture 173
10.2.2 Histological preparation for light microscopy 173
10.2.3 Histological preparation for electron microscopy 174
10.3 Methods 174
10.3.1 Cell culture 174
10.3.2 Histological preparation for light microscopy 176
10.3.3 Histological preparation for electron microscopy 177
10.3.4 Imaging 178
10.4 Anticipated Results 178
10.5 Discussion and Commentary 178
10.6 Application Notes 180
10.7 Summary Points 181
References 181

CHAPTER 11
Markers for an In Vitro Skin Substitute 183
11.1 Introduction 184
11.2 Experimental Design 185
11.3 Materials 185
11.3.1 Human tissue-engineered skin substitute reconstructed by the self-assembly approach 185
11.4 Methods 188

11.4.1 Preparation of solutions and materials for the in vitro fabrication of human skin substitutes by the self-assembly approach 188
11.4.2 In vitro fabrication of human skin substitutes by the self-assembly approach 189
11.4.3 Tissue preservation and sectioning 191
11.4.4 Preparation of solutions and materials for immunofluorescence 192
11.4.5 Immunofluorescent labeling of human skin substitutes 192
11.4.6 Histological analysis 193
11.4.7 Transmission electron microscopy 193
11.4.8 Statistical analysis 193
11.5 Anticipated Results 194
11.6 Discussion and Commentary 198
11.7 Application Notes 199
11.8 Summary Points 200
Acknowledgments 201
References 201

CHAPTER 12

3D Culture of Primary Chondrocytes, Cartilage, and Bone/Cartilage Explants in Simulated Microgravity 205

12.1 Introduction 206
12.2 Experimental Design 208
12.2.1 Culture models 208
12.2.2 The RCCS bioreactor and its operational conditions 208
12.2.3 Animals 209
12.3 Materials 211
12.3.1 Equipment for cell/tissue culture and preparation of samples 211
12.3.2 Chemicals 211
12.4 Methods 212
12.4.1 Preparation of tissue explants 212
12.4.2 Isolation of chondrocytes 212
12.4.3 2D culture of isolated chondrocytes (traditional monolayer in static fluid conditions) 214
12.4.4 3D culture of isolated chondrocytes (homotypic aggregates) 214
12.4.5 3D culture of fragments of articular cartilage explants 216
12.4.6 3D culture of undissected, complete proximal tibial epiphyses 217
12.4.7 Histomorphological study of chondrocytes and cartilage tissue 217
12.5 Anticipated Results 218
12.6 Discussion 220
12.6.1 Discussion of pitfalls 220
12.6.2 General discussion and commentary 221
12.7 Application Notes 222

12.8 Summary Points 223
Acknowledgments 224
References 224

CHAPTER 13
Alternatives for Absorption Testing 227
13.1 Introduction 228
13.2 Materials 229
13.2.1 Franz diffusion cell 229
13.2.2 Consumables 229
13.2.3 Chemicals and solutions 229
13.2.4 Technical equipment 230
13.3 Methods 230
13.3.1 Skin preparation 230
13.3.2 Determination of skin penetration using the Franz cell setup 230
13.3.3 Determination of skin permeation using the Franz cell setup 231
13.3.4 Skin absorption studies with commercially available 3D skin models 231
13.3.5 Quality control 231
13.3.6 Data evaluation 232
13.3.7 Biostatistics 233
13.4 Results and Discussion 234
13.5 Discussion of Pitfalls and Troubleshooting 236
13.6 Summary 236
References 237

CHAPTER 14
A 3D Model of the Human Epithelial Airway Barrier 239
14.1 Introduction 240
14.2 Experimental Design 241
14.3 Materials 241
14.3.1 General materials 241
14.3.2 Epithelial cell cultures—thawing 241
14.3.3 Epithelial cell cultures—culturing 241
14.3.4 Isolation of monocyte-derived macrophages (MDM) and dendritic cells (MDDC) 242
14.3.5 Triple cell coculture 242
14.3.6 Transepithelial electrical resistance (TEER) measurements 242
14.3.7 Staining for laser scanning microscopy (LSM) 243
14.3.8 Embedding for transmission electron microscopy (TEM) 243
14.4 Methods 243
14.4.1 Epithelial cells 243

14.4.2 Isolation of monocyte-derived macrophages (MDM) and dendritic cells (MDDC) 246
14.4.3 TEER measurements 249
14.4.4 Staining for LSM 250
14.4.5 Fixation and embedding of cells for transmission electron microscopy (TEM) 251
14.5 Anticipated Results 253
14.6 Discussion and Commentary 257
14.7 Application Notes 257
14.8 Summary Points 257
Acknowledgments 257
References 258

CHAPTER 15
Experimental Wear Assessment of Tibial Inserts for Total Knee Replacement 261
15.1 Introduction 262
15.2 Experimental Design 263
15.3 Materials 263
15.4 Methods 265
15.4.1 Management of the specimens 265
15.4.2 Wear test procedure 266
15.4.3 Examination of worn tibial inserts surfaces 267
15.5 Anticipated Results 268
15.6 Discussion and Commentary 269
15.7 Application Notes 269
15.8 Summary Points 270
Acknowledgments 271
References 271

About the Editors 273
Index 275

Preface

The USDA estimated that in 2007 more than 1 million animals were used in research, experiments, testing, and teaching. With today's rapidly advancing technological world, science has made giant leaps in many promising directions such as tissue and molecular engineering, in silico modeling, and medical devices. These advances have enabled many alternatives to the use of live animals in all forms of research and teaching.

In conjunction with the technological advancements, we are also witnessing a strong geopolitical movement, especially in the European Union, pushing for the development and implementation of alternatives to animal testing. Integral to making these changes a reality is the establishment of a repository of best practices and reproducible methods that avoid the use of animals. These methods must be agreed upon by the scientific community and vetted in a manner that validates the technologies as true alternatives for their animal-based assay counterparts.

With these facts in mind, we have compiled a book of alternatives to animal testing. The goal of this book is to bring forth techniques and methods at the forefront of scientific research that have the potential to replace certain whole animal tests. Although not encompassing of all techniques in the scientific community, this book provides a platform where other widely accepted techniques and scientific advancements can be collated into a concise set of methods that can then be implemented within both academic and industrial communities. We hope to one day have a database of methods and practices that replace animal testing where possible and provide data that not only match animal-based assays, but also offer even greater insight into true human in vivo conditions.

We would like to thank all the contributing authors who represent a number of countries and cultures and reflect a global perspective on the advancement of alternatives to animal testing.

Tim Maguire
Eric Novik
Editors
May 2010

CHAPTER

1

Current Methods for Prediction of Human Hepatic Clearance Using In Vitro Intrinsic Clearance

Piyun Chao,[1] Timothy J. Maguire,[1] Eric Novik,[1] Yi Han,[2] Annette S. Uss,[2] and K.-C. Cheng[2]

[1]Hurel Corporation, Beverly Hills, CA, [2]Merck and Co., Kenilworth, NJ,
Corresponding author: K.-C. Cheng, Building K15-D209, Merck Research Laboratories, Kenilworth, NJ, 07033, e-mail: kuo-chi.cheng@spcorp.com, phone: 908-740-4056, fax: 908-740-2916

Abstract

The use of intrinsic metabolic stability/clearance and other in vitro pharmacokinetic data for the selection of drug candidates for clinical evaluation during discovery lead optimization has become one of the primary focuses of research organizations involved in new drug discovery. Using intrinsic clearance determined from the human hepatocyte to predict human clearance has become more acceptable. This chapter focuses on the current methods for determining intrinsic clearance and scaling to predict human hepatic clearance and novel, physiologically based models for the improvement of human hepatic clearance prediction. In-house hepatocyte clearance data were compared with published in vivo human hepatic clearance data. Various scaling models and the effect of protein binding were examined. The use of a novel microfluidic model and other physiologically based models are presented. Metabolic clearance obtained using hepatocytes may work well in combination with the well-stirred model. Novel models incorporating flow and protein binding in the system may be the most complete models for prediction of human in vivo metabolism.

Key terms
metabolic stability
intrinsic clearance
hepatocyte metabolism
physiologically based model
prediction of human hepatic clearance

1.1 Introduction

One of the most challenging tasks in drug discovery is predicting a new chemical entity's (NCE) pharmacokinetic behavior in humans using data derived from in vitro model systems [1, 2]. For an orally administered drug, the primary goal during discovery lead optimization is to improve oral bioavailability and systemic half-life. This can be achieved by reducing the first-pass effect and/or metabolic clearance. Since the liver is one of the most important organs responsible for systemic drug metabolism, hepatic clearance has been a primary focus for lead optimization.

Many approaches have been developed to predict hepatic clearance using in vitro methodologies. For example, human liver microsomes have been used extensively for obtaining metabolic clearance and $T_{1/2}$ data have been used for the prediction of human clearance [3, 4]. It was suggested that the presence of plasma in the microsomal mixture may lead to better prediction of in vivo clearance [5, 6]. Additionally, the metabolic clearance determined from microsomal incubation may be mathematically corrected by factoring in the free fraction of the test article in plasma and/or the blood-plasma concentration ratio [7]. While these corrections may provide a more accurate clearance prediction, they still do not take into account other factors such as the uptake by the hepatocytes, which may also play a rate-limiting role [8].

With the improvement of human hepatocyte isolation and cryopreservation technology, it has become feasible to use cryopreserved primary hepatocytes for routine screening and estimation of metabolic clearance. The most commonly used procedure is based on first-order kinetics and is performed by monitoring the loss of the parent drug during incubation with hepatocyte suspensions [9, 10]. The resulting intrinsic clearance may be scaled up by a direct method [9] or the well-stirred method [11] to predict the in vivo hepatic clearance. There are two potential pitfalls with using the hepatocyte suspensions for measuring intrinsic clearance. First, hepatocytes in suspension have short life spans and therefore may not be sufficient for the evaluation of compounds with low clearance (long half-life). Second, the rate-limiting step(s) in the metabolism of some compounds may be plasma protein binding and/or cell uptake. For compounds with very high plasma protein binding, the available free fraction in the plasma may limit the hepatocyte uptake and, hence, the intracellular level of the compound. Additionally, compounds with a low membrane permeability may also have difficulty in achieving optimal levels in the liver for metabolism. In order to provide a better prediction, inclusion of plasma in the incubation or using the free fraction for mathematical correction has often been carried out [12, 13].

The recent introduction of microfluidic devices may provide a vast improvement over the previously mentioned systems for the prediction of in vivo metabolic clearance. A microfluidic cell culture analog (CCA) system for culturing cellular materials and evaluating the pharmacokinetic behavior of NCEs was used under conditions of perfusion (flow). The system described in a recent study comprises a biochip on which reside one or more discrete but microfluidically interconnected compartments. The different compartments can house various cell types, thereby simulating, through microfluidic intermediation, the metabolic interaction between different human organs. The specific chamber geometry is a physical analog to the concept of a physiologically based pharmacokinetic (PBPK) model—a mathematical model that represents the body as interconnected compartments specific for a particular organ [14–17].

1.2 Materials

The compounds with clinical data used in these studies were randomly selected from the literature and obtained from Sigma-Aldrich (St. Louis, Missouri). Methanol, acetonitrile, and water were purchased from Fisher Scientific (Rochester, New York). Plateable cryopreserved human hepatocytes, InVitroGRO HI (incubation) medium, InVitroGRO CP (plating) medium, custom albumin-free incubation medium, and Torpedo Antibiotic were acquired from Celsis In Vitro Technologies (Baltimore, Maryland). BD Biocoat collagen I 96-well microplates were obtained from BD Biosciences (Franklin Lakes, New Jersey). All the microfluidic parts and biochips were obtained from HμREL Corporation (Beverly Hills, California).

1.3 Methods

1.3.1 Thawing the hepatocytes

1. Prewarm the plating medium supplemented with 2% Torpedo antibiotics to 37°C in a water bath.
2. Transfer 5 mL of the supplemented plating medium to a sterile 50-mL conical tube.
3. Take one vial of cryopreserved human hepatocytes from liquid nitrogen and thaw the vial quickly in a water bath at 37°C for 2–3 minutes.
4. Transfer the cells to the 50-mL conical tube containing a 5-mL warm supplemented plating medium.
5. Rinse the vial with a ~1-mL warm plating medium from the conical tube.
6. Centrifuge the cells at 45×g (Beckman Coulter, TJ-25, Fullerton, California) for 5 minutes at room temperature.
7. Aspirate the supernatant carefully without touching the cells.

1.3.2 Clearance study using a hepatocyte suspension

The drug stock solution is prepared in dimethyl sulfoxide (DMSO, Sigma-Aldrich) or methanol with a concentration greater than 1,000 times higher than the working concentration, usually in the mM range.

1. Prewarm the incubation medium supplemented with 2% Torpedo antibiotics to 37°C.
2. Prepare the drug solution in the incubation medium from the stock solution to a final concentration of 2 μM. Prewarm the drug solution to 37°C.
3. Add 1 mL of the warm incubation medium to the cells.
4. Resuspend the cells by gently shake the tube.
5. Check the cell viability and number using trypan blue exclusion method with a hemacytometer.
6. Dilute the cell suspension to 2×10^6 cells/mL.
7. Add equal amount (25 μL) of the hepatocyte suspension and drug solution to wells of the sterile tissue culture treated flat bottom 96-well plates (Corning Inc., Corning, New York) in triplicates.
8. For control samples, add equal amount (25 μL) of the incubation medium without cells and drug solution.

9. Incubate the plate(s) at 37°C, 5% CO_2, under humidified condition on an orbital shaker with a shaking speed of 67 rpm.

1.3.3 Clearance study using a plated hepatocyte culture

1. After removing the supernatant, resuspend the cells in the plating medium with a cell density of 1×10^6 cells/mL.
2. Add 50 μL of the prewarmed plating medium to the wells of the Biocoat Collagen I coated 96-well plates (Becton Dickinson Labware, Bedford, Massachusetts)
3. Add 50 μL of the cell suspension to the wells containing the plating medium. By doing this, a homogenous distribution of the cells to the bottom of the plate can be achieved.
4. Allow the cells to settle down and attach to the bottom surface of the 96-well plate in the 37°C humidified CO_2 incubator for 2 hours.
5. Prepare drug solutions in the incubation medium with a final concentration of 1 μM containing <0.1% DMSO or methanol.
6. Prewarm the drug solutions to 37°C.
7. Carefully remove the culture medium of the plated hepatocytes by aspiration.
8. To each well, add 50 μL of the drug solution in triplicates. For control samples, add 50 μL of the drug solution to empty wells.
9. Incubate the plates at 37°C, 5% CO_2, under a humidified condition on an orbital shaker with a shaking speed of 67 rpm.

1.3.4 Clearance study using a plated hepatocyte culture under a flow condition

1.3.4.1 Prepare a biochip for cell seeding

1. Sterilize the biochips by soaking in 70% isopropanol for 1 hour followed by rinsing with sterile distilled water under sterile conditions.
2. Place the biochips to sterile rectangular 8-well plates (Nunc, ThermoFisher Scientific).
3. Allow the biochips to dry and subsequently treat the surface of the biochips with air plasma using a high-frequency generator (Electro-Technic Products Inc., Chicago, Illinois) for 3–5 seconds to modify the surface properties of the biochips to be more hydrophilic and to increase the collagen-coating efficiency.
4. Coat the "liver chamber" of the biochips with rat tail type I collagen (Becton Dickinson Labware, Bedford, Massachusetts) and store at 4°C under sterile conditions. The chips are stable for at least 3 months.
5. Rinse the chips with the plating medium three times before cell seeding.

1.3.4.2 Prepare a microfluidic pump for cell culture

The details of the microfluidic instrument assembly are similar to those described previously by Sin et al. [17], with some modifications. The complete setup is shown in Figure 1.1. In brief, a set of polycarbonate plates consisting of a housing bottom, a housing top,

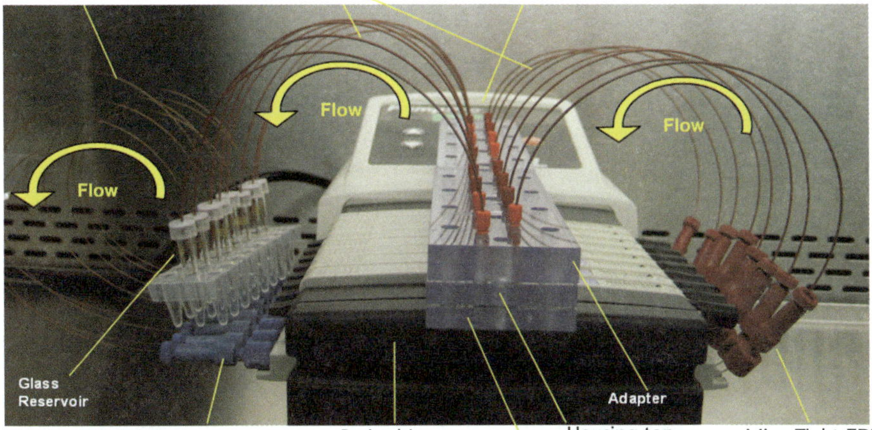

Figure 1.1 The complete setup of a HμREL prototype instrument. A peristaltic pump was used to generate the flow of culture medium. Two HμREL housing sets each containing four biochips with cells cultured upon them (for a total of eight microfluidic circuits configured to operate in parallel) were connected to the inlet and outlet PEEKsil tubing. The inlet PEEKsil tubing was connected to HμREL tubing, which was specially modified to be fitted inside the peristaltic pump. The other end of the HμREL tubing was connected to the fused silica tubing, which was inserted inside the reservoirs. The outlet PEEKsil tubing was inserted in the reservoirs to complete recirculation loops. All the connections between different tubing were MicroTight ZDV adapters. The glass vials were served as reservoirs and debubblers.

and an adapter is used to enclose the open features of four biochips after cells had been incubated upon them, thereby forming a microfluidically sealed, linear path for the recirculation of the culture medium through each of the four devices under positive pressure actuated by an off-board pump. The setup provides four microfluidic circuits, each of which utilizes one of the four biochips operating in parallel separately from the other three; hence, in this configuration of the instrument four separate microfluidic experiments can be run in parallel at one time. The adapter connects to four sets of PEEKsil tubing [100-μm inner diameter (I.D.) and 1/32-inch outer diameter (O.D.)], which serve as inlets to and outlets from the respective biochips through headless one-piece fittings. The inlet PEEKsil tubing in turn connects through MicroTight adapters to 1/16-inch high purity HμREL tubing (250-μm I.D.) with absorption characteristics composed to minimize the nonspecific binding of test substrates; the HμREL tubing conveys culture medium into the interior of a peristaltic pump (205U, 8-channel; Watson-Marlow, Wilmington, Massachusetts). Inside the pump, the HμREL tubing is inserted into the pump's cassette adapter, which holds the tubing in position for exposure to the pump's peristaltic actuation rollers. The other end of the HμREL tubing connects to fused silica tubing (100-μm I.D., 360-μm O.D.) through MicroTight adapters. The fused silica tubing is inserted into glass vials (Sun-sri 8-mm glass vials for microsampling, ThermoFisher Scientific, Pittsburgh, Pennsylvania), which serve as reservoirs capped with pre-slit snap caps (Sun-sri snap cap, 8-mm, preslit, ThermoFisher Scientific, Pittsburgh, Pennsylvania), to provide culture medium to the cells, and to which the outlet PEEKsil tubing is inserted to complete the four closed microfluidic circuits that afford the recirculation of the culture medium.

1.3.4.3 Cell culture for a microfluidic device

1. Suspend the hepatocytes in the plating medium with a cell density of 3×10^6 cells/mL.
2. Seed the cells are seeded in the "liver chamber"—one of the two cell culture compartments of the biochips—with a seeding density of 30,000 cells per chamber (10 μL).
3. Allow the cells to attach to the biochips in a humidified CO_2 incubator at 37°C for 4 hours before assembling the biochips to the HμREL housing sets.
4. Prepare the drug solution as described in the previous section (step 3 in Section 1.3.3). Prewarm the drug solution to 37°C.
5. Sterile the tubing by perfusing the tubing with 70% isopropanol, followed by sterile water for 30 minutes each.
6. Perfuse the tubing with the drug solution ~15 minutes before the assembling of the HμREL housing to the pump.
7. Enclose the biochips in the housing sets and connect the housing sets to the tubing and the pump under sterile conditions.
8. Transfer the microfluidic device to a humidified CO_2 incubator at 37°C for 10 minutes to equilibrate the system.
9. Add a 100-μL drug solution to glass microsample vials (ThermoFisher), and close the vials using snap caps (ThermoFisher).
10. Warm the drug solution in a humidified CO_2 incubator at 37°C.
11. Replace the reservoir to the drug solution vials to start the clearance study.

1.3.5 Sampling for the clearance study

1. Prepare the internal standard solution by adding a loperamide stock solution (40 μg/mL in methanol) to methanol with a final concentration of 10 ng/mL. The internal standard solution is stable under room temperature for at least 3 months if capped properly.
2. Add 100 μL of the internal standard solution to the wells of the 96-well microplate (ABgene Thermo-fast 96 skirted, low profile, 0.2 mL, Rochester, New York) immediately before sampling.
3. Take 5-μL aliquots from the hepatocyte suspension culture, plated hepatocyte culture, and/or reservoirs of the microfluidic device at predetermined time points (0, 0.5, 1, 2, and 4 hours post adding the drug solution).
4. Add the aliquots to the internal standard solution in the microplate.
5. Seal the plates containing samples using aluminum sealing tapes (Nunc, ThermoFisher Scientific, Rochester, New York) to minimize evaporation during and after the experiment.
6. Store the plates at –20°C until analyzed by LC-MS/MS.

1.3.6 Sample analysis using LC-MS/MS

1. Centrifuge the plates containing samples at 1,000 × g for 10 minutes.

2. Carefully remove the aluminum sealing tapes, reseal the plates using thin Teflon film for round opening 96-well plates (Analytical Sales & Products, Inc. Pompton Plains, New Jersey).
3. The supernatant of the samples is ready for analysis using LC-MS/MS:
    a. An aliquot (10 μL) of the supernatant was analyzed by LC-MS/MS. The LC-MS/MS system comprised a Shimadzu LC-10ADvp pump (Shimadzu, Columbia, Maryland), HTS PAL CTC autosampler (Leap Technologies, Carboro, North Carolina), and an API-4000 mass spectrometer with a Turbo Ion Spray probe (Applied Biosystems/MDS SCIEX, Ontario, Canada).
    b. The separation of compounds was achieved using a reversed phased stationary phase (Synergi Hydro-RP, 30 × 2.0 mm, Phenomenex, Torrance, California).
    c. The mobile phase was a gradient with 0.1% formic acid in water (A) and 0.1% formic acid in acetonitrile (B) with a flow rate of 0.8 mL/min.
    d. The time for each sample analysis is 1.5 minutes: the initial composition of the mobile phase was 5% of B for 0.1 minute, followed by a linear gradient to 90% of B over 1.1 minutes, held at 90% B for 0.2 minute, and back to 5% B in 0.1 minute.
    e. All the samples were detected using the multiple reaction monitoring (MRM) mode.
    f. The area ratio of the analytes to the internal standard was calculated using the Analyst software v. 1.4.1 (Applied Biosystems, Ontario, Canada).

1.4 Data Acquisition, Anticipated Results, and Interpretation

1.4.1 Hepatocyte suspension and plated hepatocyte system

The in vitro human hepatocyte intrinsic clearances (CL_{int}) are calculated from the substrate concentration profile in the hepatocyte incubation medium for static culture systems using (1.1) [9].

$$CL_{\text{int}} = \frac{C_0 - C_t}{AUC_{0-t}} \times \frac{V}{N} \tag{1.1}$$

C_0 and C_t are concentrations (μM) of compound at time 0 and t (minutes), respectively, AUC_{0-t} is the area under the concentration-time curve from 0 to t with a unit of min × μM, V is the volume of incubation solution, and N is the number of hepatocytes. The CL_{int} has a unit of μL/min/10^6 hepatocytes.

Two well-defined and widely used models, well-stirred and parallel tube models [11], can be used to scale up the in vitro intrinsic clearance obtained from the static culture conditions to the estimated hepatic clearance (CL_H), assuming the drug is totally unbound in the serum-free culture medium:

$$CL_H = Q_H \times \frac{CL_{\text{int}} \times SF}{Q_H + CL_{\text{int}} \times SF} \quad \text{well-stirred model} \tag{1.2}$$

$$CL_H = Q_H\left[1 - e^{\frac{-CL_{\text{int}} \times SF}{Q_H}}\right] \text{parallel tube model} \tag{1.3}$$

where Q_H is the human hepatic blood flow (20.7 mL/min/kg) and *SF* is the scaling factor (2.3) from in vitro to in vivo based on the consideration of the hepatocyte content in the liver (108×10^6 hepatocytes per g of liver) and the liver weight (20 grams liver weight per kilograms of body weight). The in vivo hepatic clearance (CL_H) has a unit of mL/min/kg.

1.4.2 Physiologically based microfluidic systems

The extraction ratio concept is used to scale up the clearance data obtained using the flow device to the estimated human hepatic clearance as shown here:

$$CL_h = \frac{(C_0 - C_t) \cdot V}{AUC_{0-t}} \tag{1.4}$$

where CL_h (μL/min/chip) is the clearance of the microfluidic system, C_0 and C_t are concentrations (μM) of compound in the reservoir at time 0 and *t*, respectively, AUC_{0-t} is the area under the concentration-time curve from 0 to *t*, and *V* is the volume of incubation. The extraction ratio of microfluidic system, E_h, is

$$E_h = \frac{CL_h}{Q_h} \tag{1.5}$$

The predicted human hepatic clearance from the flow device can be obtained by upscaling the extraction ratio:

$$CL_H = Q_H \times E_h \tag{1.6}$$

1.5 Discussion and Commentary

1.5.1 Hepatocyte suspension system

The use of isolated hepatocytes as an in vitro system for the prediction of in vivo metabolic clearance has recently been accepted for several reasons. First, it provides a more complete physiological system than other liver preparations, such as microsomes because hepatocytes contain both phase I and phase II enzymes and the reducing cofactor (NADPH) is continuously been produced. Second, hepatocytes contain functional enzymes enclosed by a layer of biomembrane. The permeability of certain compounds may be limited, which, in turn, limits the availability of the substrate for metabolic clearance. Third, isolated human hepatocytes, whether freshly prepared or cryopreserved, are widely available from multiple donor sources in high quality for metabolic evaluation of compounds under a high-throughput screening mode.

Although it is convenient to use freshly prepared hepatocytes from animal species, such as rat, dog, and monkey, to evaluate metabolic clearance, it is also feasible to use cryopreserved human hepatocytes for obtaining consistent and reproducible data for comparison. Human hepatocytes in suspension can survive up to 5 hours, sufficient for most clearance evaluations. The most commonly used method for determining intrinsic clearance is to monitor a compound loss according to first-order enzyme kinetics. The clearance may be derived by V_{max} and K_m based on the conversion of the parent compound to metabolite(s). Due to the presence of both phase I and phase II enzymes, monitoring the metabolite formation may be incomplete and result in an underestimation of total clearance.

Table 1.1 shows the intrinsic clearance of 26 marketed compounds using pooled cryopreserved hepatocytes from five individual donors. Scaling liver clearance to total clearance without the correction of plasma free fraction was performed according to (1.2) using the well-stirred model. A linear regression analysis of the correlation between the observed human hepatic clearance (the total clearance subtracting the renal clearance) and the scaled clearance from hepatocytes is shown in Figure 1.2. The correlation coefficient between the observed human in vivo clearance and the predicted CL based on the well-stirred CL_{int} for these 23 compounds was reasonable ($r^2 = 0.74$). More importantly, the regression line was close to unity (slope = 1), indicating that neither overprediction nor underprediction was achieved by this method. To test the need for using the free fraction (f_u) in plasma/blood for correction, we used (1.7) for scaling the total body clearance:

Table 1.1 Observed Hepatocyte Intrinsic Clearance, Scaled In Vivo Clearance, Predicted In Vivo Clearance (Well-Stirred Model), Literature Plasma Protein Binding, and Literature Hepatic Clearance of 23 Marketed Compounds

Compound	CL_{int} [a] (μL/min/10^6 cells)	Scaled $CL_{in\ vivo}$ [b] (mL/min/kg)	Well-Stirred $CL_{h,sys}$ [c] (mL/min/kg)	Plasma Protein Binding (%)	Literature Human Hepatic CL (mL/min/kg)
Acyclovir	0.22	0.5	0.5	15	0.9
Alprozolam	0.28	0.6	0.6	71	0.6
Fluconazole	0.46	1.1	1.0	75	0.3
Carbamazepine	0.46	1.1	1.0	74	1.3
Doxorubicin	0.9	2.1	1.9	75	5.4
Caffeine	0.93	2.1	1.9	36	1.4
Pindolol	0.98	2.3	2.1	40	3.8
Timolol	0.99	2.3	2.1	10	7.7
Diazepam	1.9	4.4	3.7	98.7	0.4
Gancyclovir	1.9	4.4	3.7	91	3.4
Phenytoin	1.95	4.5	3.7	89	5
Desipramine	2.2	5.1	4.1	15	10.3
Lorazepam	2.5	5.8	4.6	91	1.1
Clozapine	3.2	7.4	5.5	95	6.1
Omeprazole	3.4	7.8	5.8	95	7.5
Chlorpromazine	4.2	9.7	6.7	97	8.6
Diltiazem	10.0	23.0	11.2	78	11.8
Verapamil	20.5	47.2	15.0	90	15
Felodipine	26.2	60.3	16.1	99.6	12
Lovastatin	31.0	71.3	16.8	95	11.3
Naloxone	44.3	101.9	18.1	54	22
Nicardipine	55.5	127.7	18.8	95	13.4
Isradipine	61.2	140.8	19.0	95	12

Source: [50, 51].
a: hepatocyte intrinsic clearance, b: clearance in vivo (scaled), c: predicted in vivo clearance, well-stirred model, (1.1)

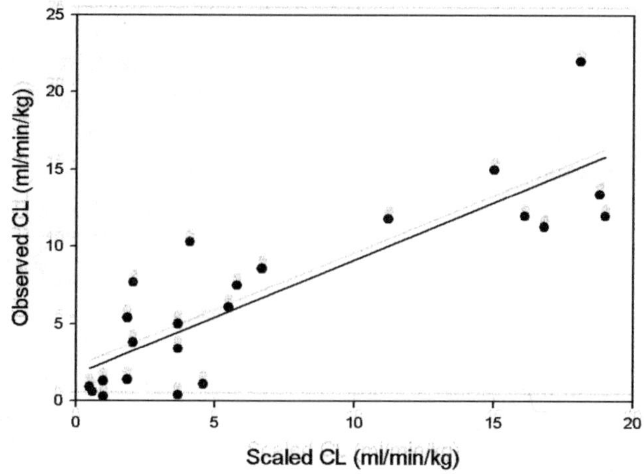

Figure 1.2 Correlation between the predicted CL using a well-stirred model from intrinsic clearance ($r^2 = 0.74$).

$$CL_H = Q_H \times \frac{f_u \times CL_{\text{int}} \times SF}{Q_H + f_u \times CL_{\text{int}} \times SF} \quad (1.7)$$

A linear regression analysis was performed to compare the observed CL and the free fraction corrected predicted human CL. As shown in Figure 1.3, using the plasma free fraction for correction resulted in not only diminishing the correlation ($r^2 = 0.46$), but also vastly underestimating the in vivo clearance. Here, it clearly demonstrated that (1.7) proposed by Pang and Rowland [11] with a plasma free fraction correction may not work properly when hepatocytes are used for predicting human in vivo clearance.

The hepatocyte system is a fully enclosed environment in which the unbound free compound within the hepatocyte may equal that in the plasma under a steady state condition. However, some have argued that the incorporation of plasma protein binding for low clearance compounds may provide a better prediction [18], perhaps due to the fact that lower clearance may allow the pharmacokinetic behavior of a compound to be

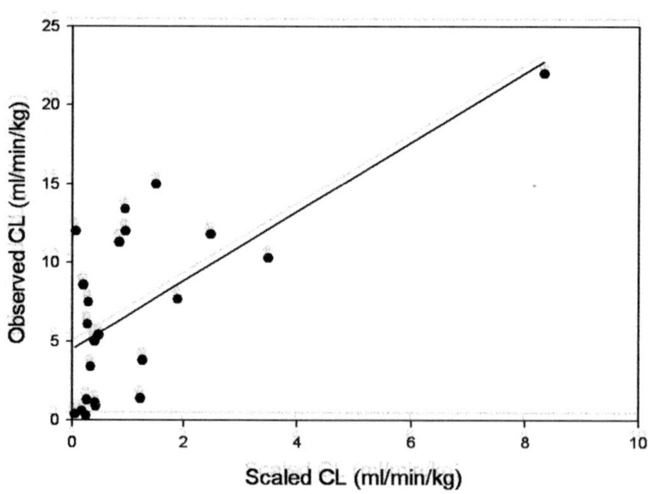

Figure 1.3 Correlation between the observed CL and predicted CL using the parallel tube model with plasma free fraction correction ($r^2 = 0.46$).

closer to the steady state, assumed by the well-stirred model. Several approaches have been proposed to address the potential issues of using the plasma free fraction to correct the hepatocyte clearance for the prediction of human in vivo clearance. For example, Lu et al. [8] proposed using the corrected free fraction based on the intracellular free fraction in the hepatocyte incubations for a model compound, midazolam. The practical usefulness of the plasma free fraction and intracellular free fraction in hepatocyte incubations according to (1.8) is still debatable:

$$CL = \frac{f_{up}/f_{uh} \times CL_{\text{int}} \times Q}{f_{up}/f_{uh} \times CL_{\text{int}} + Q} \tag{1.8}$$

where f_{up} is the free fraction in plasma and f_{uh} is the free fraction in hepatocytes. The measurement of an intracellular free fraction using hepatocyte homogenates ignores the compartmentalization of cells, and that drug distributes in the endoplasmic reticulum and may have direct access to the binding site of cytochrome P-450, which is a highly hydrophobic membrane-bound protein. But the central issue is whether the metabolic clearance process is governed by a steady-state (equilibrium) process. An alternative point of view is that the metabolic clearance is a kinetic process, where the k_{off} rate of the drug-plasma protein complex may represent a rate-limiting step. This could be especially true when a drug is highly plasma-protein bound. When entering the liver, the metabolic clearance could be represented by (1.9), which involves the k_{off} rate (or the dissociation rate constant):

$$CL_{app} = CL_{\text{int}} k_{off} \tag{1.9}$$

where CL_{app} is the apparent CL in the presence of plasma protein.

1.5.2 Plated hepatocyte system

A new way to assess metabolic clearance is using plated hepatocytes [19]. With the improvement of plating technology and the availability of platable human hepatocytes, it has recently become feasible to use plated hepatocytes for evaluation of hepatic clearance. When using plated hepatocytes, the drug is added in the medium and aliquots of samples are withdrawn from the incubation medium for analysis of the drug and/or the metabolite(s). This plated hepatocyte model is more physiologically based than hepatocyte suspensions for several reasons. Hepatocyte suspension models usually do not differentiate the drug inside or outside of the hepatocyte. A drug with a limited membrane permeability will display low clearance in both suspension and plated models. However, a drug with a high permeability and/or transporter-mediated uptake may show higher clearance in the plated model than in the suspension model. This is especially true for drugs that are transporter-mediated; their liver/plasma ratio could be very high after oral administration due to the first-pass uptake process. Plated hepatocytes can be cultured for weeks, and metabolic activity can also be maintained using coculture technology [20, 21]. This long life span could be most advantageous for evaluating drugs with better metabolic stability and potential long half-lives in vivo.

1.5.3 Physiologically based microfluidic systems

As discussed above, human in vivo clearance estimations based on hepatocytes are often limited by the lack of human physiological conditions, particularly with respect to geometrical and flow properties. For years, great efforts have been made to incorporate the flow parameter to the in vitro cell culture systems for various applications [15, 16, 22–28]. Among all the different in vitro systems studied, cultured hepatocytes seem to draw the most attention for mimicking the liver [29, 30] and provide a better understanding of drug clearance.

There is a body of evidence demonstrating the beneficial effects of applying hydrodynamics to various cultural systems, including: (1) maintaining stronger enzymatic activities for the long-term culture of primary hepatocytes [31, 32] and of cell lines [33]; (2) preserving the viability and morphology of liver tissue slices [34]; and (3) the ability to perform cellular/genomic analysis of toxicity of organ tissues due to the metabolism of xenobiotics by hepatocytes [15, 16, 35, 36]. It was reported that the drug biotransformation capacity was maintained for at least 14 days by primary rat hepatocytes cultured in a small-scale flat-membrane bioreactor by Bader et al. [37]. A similar study utilizing porcine hepatocytes showed that cells cultured in a flat membrane bioreactor maintained their phase I and phase II activities and responded to inducing drugs over a 3-week period [38]. Studies done with hepatocellular cell lines also have shown similar results indicating that this model system is useful in studying drug metabolism [39].

Despite the interest in microfluidic research, relatively few studies have focused on drug metabolism and pharmacokinetic applications [40]. One of the major obstacles in using the microfluidic device is the nonspecific binding of the testing articles to the device and the tubing [41]. This has been resolved by using materials that provide minimal nonspecific binding throughout the flow devices, incorporated in a recently developed microfluidic system [26]. The system was constructed as an analog of the PBPK modeling approach to drug metabolism and pharmacokinetics. The system comprises microchips, designed based on physiological geometries and flow-through housing, which interfaces with a peristaltic pump to drive the flow through the chips.

Figure 1.4 illustrates the relationship between the clearances of drugs by the liver chamber of the microfluidic device and the drug concentrations in the reservoir. The liver chamber in the system functions as the sole eliminating compartment and is con-

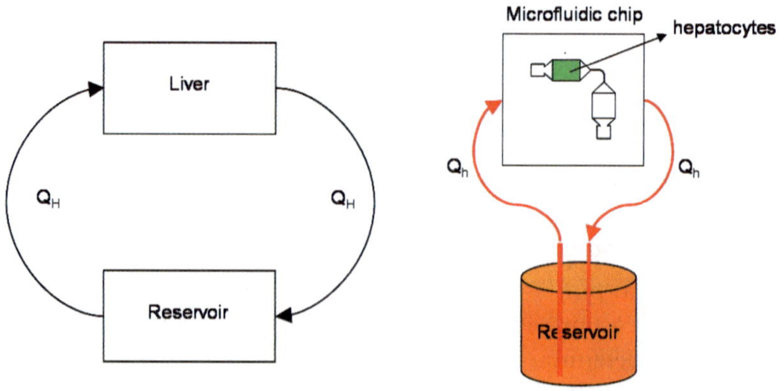

Figure 1.4 Schematic diagram of the microfluidic device.

nected to a noneliminating compartment, the reservoir. The liver compartment and the reservoir are connected via the medium flow with a flow rate of Q_h in a recirculation loop. Unlike traditional static culture systems, where the predictive hepatic clearance may be obtained by using the experimental intrinsic clearance with the well-stirred or parallel tube model to incorporate the hepatic flow parameter, the microfluidic device itself has a flow component. Therefore, the device itself represents a well-stirred model. The clearance data obtained using the flow device can be scaled up directly to predict the hepatic clearance.

The predicted data are comparable to the conventional intrinsic clearance calculation [9] with the model compounds tested, indicating that human hepatocytes cultured under a flow condition are at least as metabolically active as the ones cultured under a traditional static surrogate system. In addition, the device provides the apparent benefit of avoiding mathematical modeling to predict in vivo parameter value(s).

1.6 Summary

Drug metabolism is one of the most important pharmacokinetic processes affecting the extent and the duration of exposure of a drug in humans. Drug candidates must have adequate exposure and duration of action in order to achieve efficacy. Therefore, lead optimization based on metabolic behavior plays a major role in modern drug discovery. The liver plays a major role in the metabolism of drug candidate in humans due to the high enrichment of drug metabolism enzymes. Hence, using microsomes and/or hepatocytes derived from human donors for metabolic evaluation of drug candidates has been highly utilized by the pharmaceutical industry. Furthermore, because of major species difference in the metabolic enzymes, scaling from animal metabolism to human metabolism is not as reliable.

Cryopreserved hepatocytes derived from human donors are common tools for clearance evaluation during lead optimization in drug discovery. As the availability and quality of cryopreserved hepatocytes have improved, they have become the gold standard since cryopreserved hepatocytes are believed to have better metabolic competency and reproducibility [9, 42]. Most of the intrinsic clearance determinations using hepatocyte suspensions are based on the parent loss method. According to this approach, the parent compound, in the medium as well as inside the cells, was quantified based on the assumption that equilibrium is reached across the cell membrane. The intrinsic clearance obtained by using the hepatocyte suspension was reported by some studies [43] to provide much better prediction of human in vivo clearance when it is implemented with the well-stirred model [11]. It becomes clear that high clearance compounds may require scaling according to the human hepatic blood flow for better prediction of in vivo clearance in humans. Furthermore, refinement of the prediction includes the addition of human plasma in the hepatocyte incubation. This method takes direct measurement of the "plasma shift" of the intrinsic clearance due to the extent of plasma protein binding of the compound as well as the blood-plasma concentration ratio [44, 45]. Alternatively, it was proposed that an independent plasma protein binding determination may allow the estimation of a free fraction for the mathematical correction of the intrinsic clearance determined in the absence of the serum. These methods rely on the equilibrium hypothesis in which only unbound compound could reach the target enzyme(s) that carry out metabolism.

Cell uptake may play a rate-limiting role in metabolic clearance. This is consistent with our recent study that membrane permeability may limit the uptake of a compound and thus its ability to reach the target [46]. These studies tend to suggest that compounds with a low membrane permeability may display a low intrinsic clearance. Conversely, a highly permeable compound may display slow intrinsic clearance in hepatocyte suspensions due to a rapid uptake followed by slow metabolic turnover. However, this type of compound may hypothetically exhibit moderate to high clearance in vivo since high-clearance compounds are removed from the systemic circulation due to rapid liver uptake (high membrane permeability). In order to address this uptake/metabolism phenomenon, a physiologically based model is of the utmost importance for assessing the intrinsic clearance. We have recently established and tested a physiologically based model using plated human hepatocytes [19]. Clearance data obtained from this physiologically based in vitro model displayed a reasonable correlation with the observed in vivo clearance. Especially when well-stirred and parallel-tube models were used, there was no significant underestimation or overestimation, and the regression almost overlapped with the line of unity. Hence, it demonstrated the utility of using this physiologically based model for predicting the metabolic clearance of NCEs in humans. It is possible for some compounds to show lack of correlation due to the following reasons: high protein binding, nonlinear Michaelis-Menton kinetics, and bioanalytical variations.

As a general practice, the free fraction (f_u) is incorporated into the commonly used well-stirred model to correct the CL_{int} for the prediction of in vivo hepatic clearance [47, 48]. However, the role and the definition of f_u have long been a subject for discussion without clear consensus [8, 49]. We tested the hypothesis that only the "free fraction" of the compounds can be metabolized and the "free fraction" of compounds is under equilibrium condition during the clearance study (Figure 1.5). The results suggested that only about half of the compounds followed the equilibrium model, while the other half did not [19].

The microfluidic system may provide the best physiologically based model for the evaluation of intrinsic clearance. This system includes a hepatocyte culture that may last for a longer duration. In addition, intrinsic clearance is flow-rate limited and could use media containing plasma protein for the real-time evaluation of protein-bind-

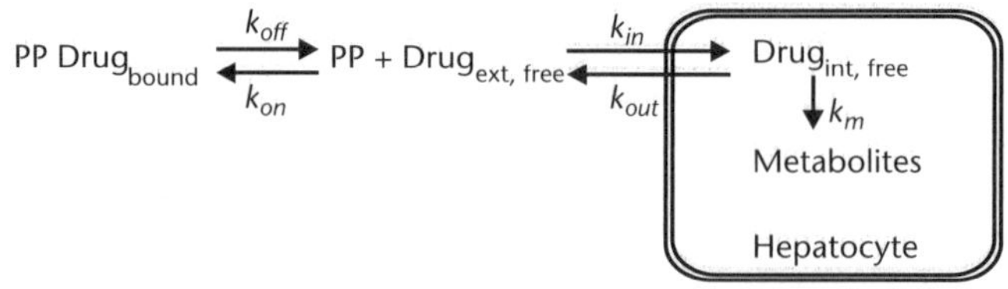

Figure 1.5 The schematic diagram describing the dynamics of the physiologically based model of drug clearance. The percentage of the unbound drug ($Drug_{\text{free}}$) can be influenced by the protein-bind release rate constant (k_{off}), protein-binding rate constant (k_{on}), uptake rate constant (k_{in}), and rate constant for $Drug_{\text{free}}$ to travel from inside the cells to outside of the cells (k_{out}). The k_{in} and k_{out} can be further affected by the rate of metabolism of the hepatocytes. The uptake of drug molecules to the hepatocytes can be either passive perfusion or active update, or both. The rate constant for intracellular unmetabolized drug molecules to travel extracellularly can be either passive perfusion or active efflux, or both. (PP: plasma protein, $Drug_{\text{bound}}$: protein bound drug.)

ing/uptake/metabolism interplay. Although the current system may not be suitable for high-throughput screening during lead optimization as part of the new drug discovery, it may be useful for late-stage compounds to provide a more precise prediction of hepatic clearance in humans.

In conclusion, intrinsic clearance determination has progressed from liver microsomal systems to more sophisticated microfluidic systems. In the near future, we may see continued use of the simple in vitro microsomal and hepatocyte systems for screening purposes during lead optimization in the pharmaceutical industry. As more progress is made with the microfluidic system and better prediction of human clearance is demonstrated, we foresee a system incorporating flow and plasma protein binding in an integrated manner that will gain more acceptance for performing routine evaluation of lead compounds in the drug discovery process.

References

[1] Chaturvedi, P. R., C. J. Decker, and A. Odinecs, "Prediction of Pharmacokinetic Properties Using Experimental Approaches During Early Drug Discovery," *Curr. Opin. Chem. Biol.*, Vol. 5, No. 4, 2001, pp. 452–463.

[2] Korfmacher, W. A., "Lead Optimization Strategies as Part of a Drug Metabolism Environment," *Curr. Opin. Drug Discov. Devel.*, Vol. 6, No. 4, 2003, pp. 481–485.

[3] Obach, R. S., et al., "The Prediction of Human Pharmacokinetic Parameters from Preclinical and In Vitro Metabolism Data," *J. Pharmacol. Exp. Ther.*, Vol. 283, No. 1, 1997, pp. 46–58.

[4] Obach, R. S., "Prediction of Human Clearance of Twenty-Nine Drugs from Hepatic Microsomal Intrinsic Clearance Data: An Examination of In Vitro Half-Life Approach and Nonspecific Binding to Microsomes," *Drug Metab. Dispos.*, Vol. 27, No. 11, 1999, pp. 1350–1359.

[5] Shibata, Y., H. Takahashi, and Y. Ishii, "A Convenient In Vitro Screening Method for Predicting In Vivo Drug Metabolic Clearance Using Isolated Hepatocytes Suspended in Serum," *Drug Metab. Dispos.*, Vol. 28, No. 12, 2000, pp. 1518–1523.

[6] Blanchard, N., et al., "Impact of Serum on Clearance Predictions Obtained from Suspensions and Primary Cultures of Rat Hepatocytes," *Eur. J. Pharm. Sci.*, Vol. 23, No. 2, 2004, pp. 189–199.

[7] Berezhkovskiy, L. M., "Some Features of the Kinetics and Equilibrium of Drug Binding to Plasma Proteins," *Expert Opin. Drug Metab. Toxicol.*, Vol. 4, No. 12, 2008, pp. 1479–1498.

[8] Lu, C., et al., "Comparison of Intrinsic Clearance in Liver Microsomes and Hepatocytes from Rats and Humans: Evaluation of Free Fraction and Uptake in Hepatocytes," *Drug Metab. Dispos.*, Vol. 34, No. 9, 2006, pp. 1600–1605.

[9] Lau, Y. Y., et al., "Development of a Novel In Vitro Model to Predict Hepatic Clearance Using Fresh, Cryopreserved, and Sandwich-Cultured Hepatocytes," *Drug Metab. Dispos.*, Vol. 30, No. 12, 2002, pp. 1446–1454.

[10] Naritomi, Y., et al., "Utility of Hepatocytes in Predicting Drug Metabolism: Comparison of Hepatic Intrinsic Clearance in Rats and Humans In Vivo and In Vitro," *Drug Metab. Dispos.*, Vol. 31, No. 5, 2003, pp. 580–588.

[11] Pang, K. S., and M. Rowland, "Hepatic Clearance of Drugs. I. Theoretical Considerations of a 'Well-Stirred' Model and a 'Parallel Tube' Model. Influence of Hepatic Blood Flow, Plasma and Blood Cell Binding, and the Hepatocellular Enzymatic Activity on Hepatic Drug Clearance," *J. Pharmacokinet. Biopharm.*, Vol. 5, No. 6, 1977, pp. 625–653.

[12] Shibata, Y., et al., "Prediction of Hepatic Clearance and Availability by Cryopreserved Human Hepatocytes: An Application of Serum Incubation Method," *Drug Metab. Dispos.*, Vol. 30, No. 8, 2002, pp. 892–896.

[13] Stringer, R., P. L. Nicklin, and J. B. Houston, "Reliability of Human Cryopreserved Hepatocytes and Liver Microsomes as In Vitro Systems to Predict Metabolic Clearance," *Xenobiotica*, Vol. 38, No. 10, 2008, pp. 1313–1329.

[14] Oh, T. I., et al., "Real-Time Fluorescence Detection of Multiple Microscale Cell Culture Analog Devices In Situ," *Cytometry A*, Vol. 71, No. 10, 2007, pp. 857–865.

[15] Viravaidya, K., A. Sin, and M. L. Shuler, "Development of a Microscale Cell Culture Analog to Probe Naphthalene Toxicity," *Biotechnol. Prog.*, Vol. 20, No. 1, 2004, pp. 316–323.

[16] Viravaidya, K., and M. L. Shuler, "Incorporation of 3T3-L1 Cells to Mimic Bioaccumulation in a Microscale Cell Culture Analog Device for Toxicity Studies," *Biotechnol. Prog.*, Vol. 20, No. 2, 2004, pp. 590–597.

[17] Sin, A., et al., "The Design and Fabrication of Three-Chamber Microscale Cell Culture Analog Devices with Integrated Dissolved Oxygen Sensors," *Biotechnol. Prog.*, Vol. 20, No. 1, 2004, pp. 338–345.

[18] Griffin, S. J., and J. B. Houston, "Comparison of Fresh and Cryopreserved Rat Hepatocyte Suspensions for the Prediction of In Vitro Intrinsic Clearance," *Drug Metab. Dispos.*, Vol. 32, No. 5, 2004, pp. 552–528.

[19] Chao, P., et al., "Prediction of Human Hepatic Clearance Using an In Vitro Plated Hepatocyte Clearance Model," *Drug Metab. Lett.*, Vol. 3, No. 4, 2009, pp. 296–307.

[20] Nishikawa, M., et al., "Enhanced Maintenance and Functions of Rat Hepatocytes Induced by Combination of On-Site Oxygenation and Coculture with Fibroblasts," *J. Biotechnol.*, Vol. 133, No. 2, 2008, pp. 253–260.

[21] Khetani, S. R., and S. N. Bhatia, "Microscale Culture of Human Liver Cells for Drug Development," *Nat. Biotechnol.*, Vol. 26, No. 1, 2008, pp. 120–126.

[22] Shackman, J. G., et al., "Perfusion and Chemical Monitoring of Living Cells on a Microfluidic Chip," *Lab Chip*, Vol. 5, No. 1, 2005, pp. 56–63.

[23] Kimura, H., et al., "An Integrated Microfluidic System for Long-Term Perfusion Culture and On-Line Monitoring of Intestinal Tissue Models," *Lab Chip*, Vol. 8, No. 5, 2008, pp. 741–746.

[24] Toh, Y. C., et al., "A Microfluidic 3D Hepatocyte Chip for Drug Toxicity Testing," *Lab Chip*, Vol. 9, No. 14, 2009, pp. 2026–2035.

[25] Leclerc, E., et al., "Study of Osteoblastic Cells in a Microfluidic Environment," *Biomaterials*, Vol. 27, No. 4, 2006, pp. 586–595.

[26] Chao, P., et al., "Evaluation of a Microfluidic Based Cell Culture Platform with Primary Human Hepatocytes for the Prediction of Hepatic Clearance in Human," *Biochem. Pharmacol.*, Vol. 78, No. 6, 2009, pp. 625–632.

[27] Tatosian, D. A., and M. L. Shuler, "A Novel System for Evaluation of Drug Mixtures for Potential Efficacy in Treating Multidrug Resistant Cancers," *Biotechnol. Bioeng.*, 2009. 103, No. 1, 2009, pp. 187–198.

[28] Kim, M., J. Yeon, and J. -K. Park, "A Microfluidic Platform for 3-Dimensional Cell Culture and Cell-Based Assays," *Biomedical Microdevices*, Vol. 9, No. 1, 2007, pp. 25–34.

[29] Park, J., et al., "Microfabricated Grooved Substrates as Platforms for Bioartificial Liver Reactors," *Biotechnol. Bioeng.*, Vol. 90, No. 5, 2005, pp. 632–644.

[30] Sivaraman, A., et al., "A Microscale In Vitro Physiological Model of the Liver: Predictive Screens for Drug Metabolism and Enzyme Induction," *Curr. Drug Metab.*, Vol. 6, No. 6, 2005, pp. 569–591.

[31] Schmitmeier, S., et al., "Development and Characterization of a Small-Scale Bioreactor Based on a Bioartificial Hepatic Culture Model for Predictive Pharmacological In Vitro Screenings," *Biotechnol. Bioeng.*, Vol. 95, No. 6, 2006, pp. 1198–1206.

[32] Zeilinger, K., et al., "Three-Dimensional Co-Culture of Primary Human Liver Cells in Bioreactors for In Vitro Drug Studies: Effects of the Initial Cell Quality on the Long-Term Maintenance of Hepatocyte-Specific Functions," *Altern. Lab Anim.*, 2002. 30, No. 5, 2002, pp. 525–538.

[33] Gebhardt, R., et al., "Improved Determination of Drug Metabolism by Perifusion of Recombinant V79 Cells Carrying Human CYP3A4," *Toxicology In Vitro*, Vol. 13, No. 4-5, 1999, pp. 639–643.

[34] Schumacher, K., et al., "Perfusion Culture Improves the Maintenance of Cultured Liver Tissue Slices," *Tissue Eng.*, Vol. 13, No. 1, 2007, pp. 197–205.

[35] Kane, B. J., et al., "Liver-Specific Functional Studies in a Microfluidic Array of Primary Mammalian Hepatocytes," *Anal. Chem.*, Vol. 78, No. 13, 2006, pp. 4291–4298.

[36] Gebhardt, R., H. Wegner, and J. Alber, "Perifusion of Co-Cultured Hepatocytes: Optimization of Studies on Drug Metabolism and Cytotoxicity In Vitro," *Cell Biol. Toxicol.*, Vol. 12, No. 2, 1996, pp. 57–68.

[37] Bader, A., et al., "Development of a Small-Scale Bioreactor for Drug Metabolism Studies Maintaining Hepatospecific Functions," *Xenobiotica*, Vol. 28, No. 9, 1998, pp. 815–825.

[38] Langsch, A., and A. Bader, "Long Term Stability of Phase I and Phase II Enzymes of Porcine Liver Cells in Flat Membrane Bioreactors," *Biotechnol. Bioeng.*, Vol. 76, No. 2, 2001, pp. 115–125.

[39] Iwahori, T., et al., "CYP3A4 Inducible Model for In Vitro Analysis of Human Drug Metabolism Using a Bioartificial Liver," *Hepatology*, Vol. 37, No. 3, 2003, pp. 665–673.

[40] El-Ali, J., P. K. Sorger, and K. F. Jensen, "Cells on Chips," *Nature*, Vol. 442, No. 7101, 2006, pp. 403–411.

[41] Toepke, M. W., and D. J. Beebe, "PDMS Absorption of Small Molecules and Consequences in Microfluidic Applications," *Lab Chip*, Vol. 6, No. 12, 2006, pp. 1484–1486.

[42] Lau, Y. Y., et al., "Ex Situ Inhibition of Hepatic Uptake and Efflux Significantly Changes Metabolism: Hepatic Enzyme-Transporter Interplay," *J. Pharmacol. Exp. Ther.*, Vol. 308, No. 3, 2004, pp. 1040–1045.

[43] Brown, H. S., M. Griffin, and J. B. Houston, "Evaluation of Cryopreserved Human Hepatocytes as an Alternative In Vitro System to Microsomes for the Prediction of Metabolic Clearance," *Drug Metab. Dispos.*, 2007. 35, No. 2, 2007, pp. 293–301.

[44] Berezhkovskiy, L. M., "Determination of Drug Binding to Plasma Proteins Using Competitive Equilibrium Binding to Dextran-Coated Charcoal," *J. Pharmacokinet. Pharmacodyn.*, Vol. 33, No. 5, 2006, pp. 595–608.

[45] Berezhkovskiy, L. M., "On the Calculation of the Concentration Dependence of Drug Binding to Plasma Proteins with Multiple Binding Sites of Different Affinities: Determination of the Possible Variation of the Unbound Drug Fraction and Calculation of the Number of Binding Sites of the Protein," *J. Pharm. Sci.*, Vol. 96, No. 2, 2007, pp. 249–257.

[46] Li, C., et al., "Correlation Between PAMPA Permeability and Cellular Activities of Hepatitis C Virus Protease Inhibitors," *Biochem. Pharmacol.*, Vol. 75, No. 5, 2008, pp. 1186–1197.

[47] Naritomi, Y., et al., "Prediction of Human Hepatic Clearance from In Vivo Animal Experiments and In Vitro Metabolic Studies with Liver Microsomes from Animals and Humans," *Drug Metab. Dispos.*, Vol. 29, No. 10, 2001, pp. 1316–1324.

[48] Ito, K., and J. B. Houston, "Comparison of the Use of Liver Models for Predicting Drug Clearance Using In Vitro Kinetic Data from Hepatic Microsomes and Isolated Hepatocytes," *Pharm. Res.*, Vol. 21, No. 5, 2004, pp. 785–792.

[49] Yang, J., et al., "Misuse of the Well-Stirred Model of Hepatic Drug Clearance," *Drug Metab. Dispos.*, Vol. 35, No. 3, 2007, pp. 501–502.

[50] Brunton, L. L., J. S. Lazo, and K. L. Parker, *Goodman & Gilman's The Pharmacological Basis of Therapeutics*, 11th ed., New York: McGraw-Hill, 2006.

[51] Li, C., et al., "Development of In Vitro Pharmacokinetic Screens Using Caco-2, Human Hepatocyte, and Caco-2/Human Hepatocyte Hybrid Systems for the Prediction of Oral Bioavailability in Humans," *J. Biomol. Screen.*, Vol. 12, No. 8, 2007, pp. 1084–1091.

CHAPTER

2

Use of Permeability from Cultured Cell Lines and PAMPA System and Absorption from Experimental Animals for the Prediction of Absorption in Humans

Cheng Li, Annette S. Uss, and K.-C. Cheng

Merck and Co., Kenilworth, NJ,
Corresponding author: K.-C. Cheng, Building K15-D209, Merck Research Laboratories, Kenilworth, NJ, 07033, e-mail: kuo-chi.cheng@spcorp.com, phone: 908-740-4056, fax: 908-740-2916

Abstract

Over the past decade, the use of drug metabolism and pharmacokinetic (DMPK) parameters to optimize the selection of candidates for drug development has become one of the primary focuses of research organizations involved in new drug discovery. This chapter will describe a cultured Caco-2 cell system, a parallel artificial membrane permeability assay (PAMPA) system, an in vivo absorption study using nonclinical animals, and the feasibility of using the data to predicting human absorption. In-house Caco-2 permeability and PAMPA data were compared with human absorption data from the literature. Animal absorption data obtained from current literature were also compared with human data. Since it is well recognized that human intestinal absorption cannot be precisely predicted by a single screening assay, it is important to utilize various in vitro and in vivo nonclinical studies during lead optimization in drug discovery.

Key terms
animal absorption
Caco-2 permeability assay
parallel artificial membrane permeability assay
PAMPA
prediction of human absorption

2.1 Introduction

Lead optimization of new chemical entities (NCEs) based on pharmacokinetic behavior plays a major role in modern drug discovery. Despite advancement of drug delivery methods, the oral route remains the most frequent route of administration for approved new drugs. Therefore, during lead optimization it is essential to identify NCEs with sufficient predicted oral absorption using a variety of in vitro and in vivo assays. Recent advancements in chemistry, such as parallel and combinatorial synthesis, have resulted in a multifold increase in the number of compounds that are available for evaluation in new drug discovery. Furthermore, other improvements in structural chemistry tools, such as X-ray crystallography, structural modeling, and ligand/substrate docking algorithms, provide new drug discovery scientists with an unprecedented level of structure-based designs to further guide the synthesis of new chemotypes as potential drug leads. Along with the advancement of chemistry and biology, new automated screening tools have became commercially available which can carry out complex, programmable, and adaptable robotic operations to test hundreds of thousands of compounds in a speedy and precise manner. As a result, these new forces have come together to increase our ability to create NCEs that exhibit targeted pharmacological activity. Hence, the task of screening compounds for their biopharmaceutical properties, such as solubility, permeability, and metabolic stability, has become a major challenge in drug discovery. This change, in turn, has compelled the invention and implementation of high-throughput screening methods that predict oral absorption in humans.

It is well recognized that in order for an NCE to achieve reasonable oral absorption, it must have adequate aqueous solubility and intestinal permeability [1]. Drug absorption from the gastrointestinal (GI) tract following oral administration is a rather complex and dynamic process. Passive diffusion can occur through the cell membranes of enterocytes (transcellular) or the tight junctions between the enterocytes (paracellular) [2–4]. Influx and efflux through various drug transporters also contribute to drug absorption. Since many processes are occurring simultaneously, it is often impossible for a single in vitro model to simulate the entire in vivo process. Two of the most commonly used models that have been developed and are used by most major pharmaceutical companies in their drug discovery efforts over the last decade are the Caco-2 cell permeability assay system and the parallel artificial membrane permeability assay (PAMPA) system [5].

The prediction of human pharmacokinetic behavior of a compound in clinical development is often based upon in vivo nonclinical pharmacokinetic data in animals. Typically, for a lead candidate, pharmacokinetic data are obtained in at least two or three nonclinical species, often including rat, dog, and/or monkey, prior to commencing clinical studies. However, a clear understanding of the relative ability of each nonclinical animal species to estimate human oral exposure is very crucial. Since species-dependent differences in physiology and the metabolic enzyme/transporter composition have been noted, investigators must understand the limitations [6].

In this chapter, we will describe the Caco-2 cell permeability assay system and the PAMPA system, their pros and cons, and how to use them for lead optimization in the discovery process. We will also discuss through the pros and cons of the three most commonly used nonclinical species, namely, rats, dogs, and monkeys, with respect to their suitability for the prediction of oral absorption in humans.

2.2 Materials

The Caco-2 cell line was obtained from the American Tissue Culture Collection (ATCC, Rockville, Maryland). Hank's Balanced Salt Solution (HBSS), Phosphate Buffered Saline (PBS), Dulbecco's Modified Eagle's Medium (DMEM), L-glutamine, penicillin-streptomycin (PEST), and nonessential amino acids (NEAA) were obtained from Lonza Walkerwille (Walkerwille, Maryland). The fetal bovine serum (FBS) was obtained from Invitrogen (Carlsbad, California). All the reagents were stored following conditions that were specified by the vendors. Transport studies were performed using the HTS Multiwell insert system with PET membranes (24 wells, 1.0-μm pore size, 0.33 cm^2 area) and bottom plates (Becton Dickinson Labware, Franklin Lakes, New Jersey). The PAMPA lipid was obtained from pION Inc. (Woburn, Massachusetts; GIT-0 Lipid Solution) and was stored at –20°C until use. A concentrated system solution and the acceptor sink buffer from pION Inc. were used as directed for dosing and receiving sides. A multiscreen 96-well PVDF filter plate (pore size = 0.45 μm, surface area = 0.24 cm^2, porosity = 0.76) and a transport receiver plate from Millipore Corporation (Billerica, Massachusetts) were used for PAMPA "sandwich" preparation. All other chemicals were obtained from Sigma.

2.3 Methods

2.3.1 Cultured cell system

There are several cell monolayer models that are frequently used for the evaluation of drug permeability and absorption potential (Table 2.1). Caco-2 cells (adenocarcinoma cells derived from colon) are the most extensively characterized and frequently used of the available cell lines [7–11]. A unique feature of Caco-2 cells is that they undergo spontaneous enterocyte differentiation in cell culture when grown on a porous filter membrane. Unlike intestinal enterocytes, Caco-2 cells are immortalized and replicate rapidly into confluent monolayers. When the cells reach confluency during culture on a semiporous membrane, they start to polarize and form tight junctions, creating an ideal system for permeability and transport studies. During the past decade, the use of Caco-2 cells in high-throughput screening for NCEs in the pharmaceutical industry has increased tremendously. The utility of this monolayer model to either rank order or filter out compounds has been exemplified by many previous studies [12, 13]. The following sections describe the detailed procedures for Caco-2 cell culture and permeability measurement using the Caco-2 cell monolayers.

Table 2.1 Commonly Used Monolayer Cell Models for Estimating Permeability and Transcellular Flux

Cells	Species/Tissue Origin	Cell Type
Caco-2	Human/colon	Epithelial
2/4/A1	Rat/intestine	Epithelial
MDCK	Canine/kidney	Epithelial
LLC-PK1	Porcine/kidney	Epithelial

2.3.1.1 Caco-2 cell culture

Prepare cell culture medium To make 1L of Caco-2 cell culture medium, mix 870 mL of Dulbecco's Modified Eagle Medium (DMEM), 100 mL of the fetal bovine serum (FBS), 10 mL of the nonessential amino acid NEAA (100×), and 10 mL of a 200-mM L-glutamine (i.e., 2 mM) well; adjust the pH to 7.4 with 1N NaOH or HCl. Sterilize the medium by filtering through the 0.2-μm filter membrane (Thermo Scientific Nalgene MF75 Series Sterile Disposable Tissue Culture Filter Units—Part # 09-741-02) and store at 4°C before use. Use the medium within 6 months.

Seed and subculture cells

1. Thaw one or two frozen vials by rapid agitation in a 37°C water bath.
2. Immediately after the cell suspension is melted, remove it from the water bath and transfer the cell suspension from the vial to a 225 cm^2 culture flask, which contains a 40-mL culture medium using a serological pipette.
3. Incubate the flask at 37°C in a CO_2 incubator with an atmosphere of 5% CO_2 and 90% relative humidity. Change the culture medium three times weekly and observe growth using a light microscope. Subculture when it gets to ~85% confluence using the next steps.
4. Aspirate the culture medium from the culture flask using a serological pipette, which is connected to a vacuum flask to the waste container (or it can be manually aspirated out using the pipette if vacuum setup is not available).
5. Wash the flask twice with 15 mL of cold phosphated buffered saline (PBS).
6. Rinse the flask with ~5 mL of cold 0.25% trypsin and incubate at room temperature and/or at 37°C until the cells start to round up and detach.
7. Remove the trypsin solution from the flask and add 10 mL of the cold cell culture medium immediately to the flask and rapidly break up the cell lumps by pipetting.
8. Split the cell suspensions 1:2 into two new flasks with a 40-mL medium each.
9. The cells are then maintained in the CO_2 incubator as described above.

Seed and culture Caco-2 cell monolayers

1. Prepare the cell suspension as described above.
2. Transfer cell suspension from each flask to a BD Falcon 50-mL Conical Centrifuge Tube (sterile) and centrifuge the tube at 800 rpm for 10 minutes at 10°C IEC-8R centrifuge.
3. Decant the supernatant and resuspend the cell pellets in the tube with the culture medium to a final volume of ~10 mL. Count the cell number as shown in the cell counting section. Based on the cell count, adjust the cell concentration with culture medium so that the cells can be seeded with the targeted insert volume (0.5 mL for the BD 24-well insert plate) at a density of 50,000–60,000 cells/cm^2. (Note: The seeding density may vary depending on each person's technique. The cell passage number could also be important; the higher the passage number, the lower seeding density it is likely to need.)
4. Add a 20-mL warm culture medium to the bottom common well of the 24-well plate and seed a 0.5-mL cell suspension to each insert.

5. Place the cell plates in the CO_2 incubator and change the culture medium every other day prior to being used for the transport study (usually 21–25 days postseeding).

Count cells

1. Mix 0.2 mL of the cell suspension with 0.3 mL of PBS in a 1-mL tube.
2. Add 0.5 mL of trypan blue dye (0.4%) to the solution from Step 1 (total volume 1.0 mL, dilution factor 5).
3. Gently mix the cells and load the resulting cell suspension in both chambers of a hemocytometer with a cover slip in place. Count both viable and nonviable cells in all four 1-mm squares of each chamber under the microscope.
4. Calculate the cell concentration (cell/ml) and cell viability (%) using:

$$\text{Cell density (\# of cells/mL)} = \text{Avg. count per square} \times \text{dilution factor} \times 10^4$$

$$\%\ \text{cell viability} = \left[(\text{unstained cells})/(\text{stained} + \text{unstained cells})\right] \times 100\%$$

2.3.1.2 Caco-2 permeability measurement

Transport medium (TM) for dosing solution Hank's Balanced Salt Solution (HBSS) with 10-mM HEPES, 25 mM D-glucose, adjust pH to 7.4 with 1N NaOH or 1N HCl. (If pH 6.5 TM is preferred, replace 10-mM HEPES with 10-mM MES.)

Receiver Buffer: TM_pH7.4.

Experimental procedures

1. Remove the culture medium from both the apical (AP) and basolateral (BL) compartments and wash the wells twice with warmed TM. Measure the transepithelial electrical resistance (TEER) across each monolayer.
2. Dump the TM and gently dry the insert with a cotton swab to minimize the remaining buffer. Place the cell insert into the bottom assay plate.
3. Add a 1-mL receiver buffer to the receiver side. [Note: A TM with protein (adjust pH to 7.4), such as a bovine serum albumin, could be added to the receiver buffer for lipophilic compounds to improve permeability measurement. In this lab, we routinely use 4% BSA for our permeability assay.]
4. Add 0.42 mL of the dosing solution at a targeted concentration by diluting a DMSO or MeOH stock to the AP side. Immediately take 20-μL donor samples for zero time dosing concentration measurement. (Note: The dosing concentration could range from one to several hundred micromolars, depending on the compound solubility and analytical methods to be used; the final organic solvent concentration in the dosing solution needs to be less than 1% to maintain the cell monolayer integrity.)
5. Incubate the Caco-2 cell plates for 2 hours at 37°C with 5% CO_2 and 90% relative humidity with shaking using a DPC Micromix 5 shaker.
6. Collect receiver samples (100 μL) for up to 2 hours with a replacement and a donor sample at the end of the incubation. Measure the TEER again. (Note: It would be normal for the TEER value to be lower by 20–30% after a 2-hour incubation.)
7. Quantify the test article concentration in each sample using LC-UV, HPLC, or LC-MS methods.

8. Calculate the apparent permeability and mass balance of the test compound using the equation as shown in Section 2.4.1.

For using other cell line systems, similar cell culture and transport methods can be applied to obtain the permeability. Proper validation using standard compounds should be done to establish the cell membrane integrity and/or transporter expression criteria for the new cell line.

2.3.2 PAMPA system

PAMPA assesses passive diffusion of a molecule, the most common pathway for drug absorption and uptake by target tissues [5, 14]. PAMPA was first demonstrated by Kansy et al. [15] as a high-throughput screening tool for predicting absorption. This model utilizes an artificial lipid membrane created by coating a hydrophobic filter with a mixture of lecithin/phospholipids dissolved in an inert organic solvent such as dodecane. Since then, several variants of PAMPA have been developed using different types of lipid barriers. The first commercially available lipid barrier was based on 2% w/v dioleyoylphosphotidylcholine (DOPC) dissolved in n-dodecane [16]. An improved preparation was subsequently introduced [17] based on a 20% w/v mixture of negatively charged phospholipids also dissolved in n-dodecane in a variant called double-sink PAMPA (DS-PAMPA). The following sections describe the detailed procedures for PAMPA assay for absorption prediction purpose.

2.4.2.1 Transport buffer for dosing solution

Dilute the system buffer from pION: 250-μL concentrate to 100-mL DDwater (pH~3) (0.25%v/v). Adjust the pH to 7.4 with a 1-mM NaOH (expires within 1 week) to be used for compound dosing solution preparation.

Receiver buffer: pION Acceptor Sink Buffer (pH7.4)

2.4.2.2 Experimental procedures

1. Prepare test compound dosing solutions ranging from 10 μM to 100 μM by diluting 1–10 mM DMSO or MeOH compound stock solution using the diluted system buffer with a final organic concentration of 1%. For compounds with solubility issues, up to 5% DMSO could be used.
2. Prepare the PAMPA sandwich. Thaw pION GIT-0 lipid solution at room temperature. Apply a 5-μL lipid solution to each filter membrane (Millipore Multiscreen 96-well PVDF filter plate).
3. Add a dosing solution (200 μL) to the top of each well of the filter plate. The filter plate is then put onto a receiver plate with 300 μL preloaded with an acceptor sink buffer.
4. Take a donor sample from the dosing solution for an initial compound concentration analysis.
5. Incubate the sandwich for 4–5 hours or overnight (up to 16 hours for a low permeability compound) at room temperature.

6. Take samples from both receiver and donor sides at the end of the incubation and analyze using an UV spectrometer (SpectraMax 190, Molecular Devices, Sunnyvale, California) and/or LC-MS/MS.

On a technical note, the pION buffers are perfect for the UV detection of the compound because of its minimal background UV absorbance. If more specific analytical methods will be used, such as HPLC, LC-MS/MS, the phosphate buffer (10–20 mM) with or without protein will generate similar results.

2.3.3 In vivo absorption measurements

This section describes a general method that was used within this study group to assess the oral absorption in animal species. All animal study protocols were preapproved by the company's Animal Care and Use Committee (ACUC). Absolute absorption in animals was measured by dosing the ^{3}H-labeled compound by both intravenous injection (IV) and oral (PO) administration. Briefly, 100 μCi ^{3}H-labeled compound solution at 3 mg/kg (or lower depending on compound solubility) prepared in 0.4% aqueous hydroxypropyl methylcellulose or 20% hydroxypropyl-β-cyclodextrin (from Sigma-Aldrich) was administered to rats, dogs, or monkeys intravenously or orally. Plasma samples were taken at 0.25, 0.5, 1, 2, 4, 6, 8, 12, 24, and 48 hours postdose. The samples were lyophilized, scintillation cocktailed was added to each sample, and total radioactivity at each time point was counted using a Packard 2250A Tri-Carb Scintillation Counter.

2.4 Data Acquisition, Anticipated Results, and Interpretation

2.4.1 Data analysis

The apparent permeability and aqueous recovery from both Caco-2 and PAMPA assay were calculated using the following equations:

$$P_{\text{app}} = \frac{(dC_R/dt) * V_R}{S * C_{D,0\text{hr}}}$$

$$\text{AqueousRecovery}(\%) = \frac{C_{D,\text{Final}}}{D_{D,0\text{hr}}} \times 100 + \frac{\text{Receiver Accumulated Amount}}{C_{D,0\text{hr}} * V_D} \times 100$$

where dC_R/dt is the slope of the cumulative concentration in the receiver compartment versus time; S is the surface area of the cell monolayer or lipid membrane; $C_{D,0\text{hr}}$ and $C_{D,\text{Final}}$ are the donor compound concentrations immediately after dosing and at the end of incubation, respectively; and V_D and V_R are the volumes of the donor and receiver compartments, respectively.

The extent of absorption (Fa) in animals was calculated by dividing the area under the total plasma radioactivity versus the time curve (AUC) calculated from PO administration by that from the IV administration normalized with the dose:

$$Fa(\%) = \frac{AUC_{\_PO(\text{total radioactivity})} / Dose_{\_PO}}{AUC_{\_IV(\text{total radioactivity})} / Dose_{\_IV}} * 100$$

2.4.2 Results and interpretation

The Caco-2 cell monolayer integrity can be monitored by measuring the TEER value across each monolayer before and after experiment and/or by measuring the permeability of mannitol (0.5 μCi/mL) or Lucifer Yellow (1 mg/mL in TM), which are transported through a paracellular pathway. A TEER value greater than 300–600 $\Omega \bullet cm^2$ and/or mannitol and Lucifer Yellow permeability less than 20 or 4 nm/s, respectively, could be used as criteria for the cell monolayer's integrity. Reported permeability coefficients, however, may vary significantly from laboratory to laboratory, particularly for compounds that pass through the Caco-2 monolayer via the paracellular route. According to Walter and Kissel [18], the permeability coefficients of mannitol in the literature range from 2 to 221 nm/s. Even within the same laboratory, several factors, such as cell source, age of cell culture, and culturing conditions, may affect the reproducibility of permeability results. The cell passage number may specifically affect the tight junctions and the expression level of drug transporters, such as P-glycoprotein (P-gp). Nevertheless, it is not difficult to differentiate a highly permeable drug from a poorly permeable drug if the conditions of the assay are carefully maintained. This is particularly true for compounds that are absorbed transcellularly.

Table 2.2 summarizes the Caco-2 cell validation results for three different passage cells using five standards compounds measured in our laboratory. The results demonstrated that, with proper cell culture techniques, the permeability variation could be controlled fairly well in the same laboratory. The high and low permeability standards, such as ^{14}C-mannitol and ^{3}H-propranolol, each had shown low and high permeability in each batch, while the P-gp substrate (^{3}H-digxoin) had consistently shown a greater than

Table 2.2 Caco-2 Cell System Validation at Different Cell Passage Number with Five Standard Compounds

	Caco-2 Cell Passage #	P_{app}_AP to BL (nm/s)*	Recovery _AP to BL (%)	P_{app}_BL to AP (nm/s)	Recovery _BL to AP (%)	Efflux Ratio
^{14}C-Mannitol	50	2.9 ± 0.2	96 ± 1.1	3.5 ± 0.3	95 ± 8.5	1.2
	60	8.8 ± 1.3	101 ± 0.6	7.0 ± 0.3	100 ± 10	0.8
	70	2.9 ± 0.6	98 ± 3.9	2.7 ± 0.2	99 ± 0.4	0.9
^{14}C-Urea	50	52 ± 4.1	96 ± 0.7	56 ± 1.7	91 ± 1.8	1.1
	60	46 ± 0.4	104 ± 3.1	45 ± 0.6	94 ± 1.3	1.0
	70	50 ± 1.0	103 ± 2.3	46 ± 2.2	101 ± 0.5	0.9
^{3}H-Propranolol	50	272 ± 24	90 ± 1.9	339 ± 13	91 ± 3.8	1.3
	60	366 ± 14	81 ± 2.4	435 ± 6	108 ± 0.1	1.2
	70	453 ± 15	86 ± 2.6	400 ± 34	97 ± 2.6	0.9
^{3}H-CSA	50	22 ± 1.6	56 ± 0.9	55 ± 8.9	62 ± 7.7	2.5
	60	43 ± 2.7	78 ± 4.4	72 ± 12	100 ± 0.8	1.7
	70	56 ± 2.8	80 ± 2.8	86 ± 9.6	100 ± 0.9	1.6
^{3}H-Digoxin	50	18 ± 1.2	95 ± 4.5	145 ± 16	92 ± 0.7	8.1
	60	17 ± 1.2	105 ± 2.9	128 ± 2	81 ± 2.4	7.4
	70	17 ± 0.6	98 ± 2.6	127 ± 7	86 ± 2.6	7.6

*: $n = 3$

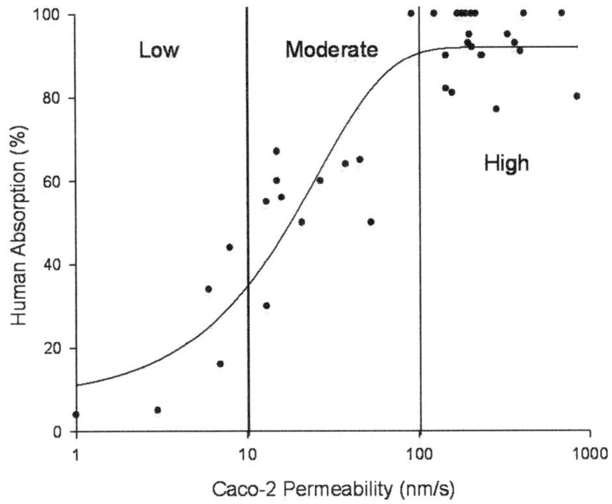

Figure 2.1 Flux of 37 commonly prescribed drugs across Caco-2 monolayers versus absorption in humans.

2.0 efflux ratio, indicating sufficient P-gp transporter expression in the cells. Table 2.3 shows Caco-2 permeability values obtained in this laboratory and human absorption reported in the literature for 37 marketed compounds [19–21] using the method described in Section 2.3.1. These compounds were randomly selected based on structural diversity. Figure 2.1 shows the correlation between permeability coefficients and the oral absorption of these 37 marketed compounds. As reported by many other laboratories, a sigmoidal relationship was observed between the fraction absorbed and the permeability coefficients in the Caco-2 model [22, 23]. Based on this sigmoidal relationship, one may design cutoff values for low, moderate, and high groups. The dynamic range of permeability values for highly absorbed compounds versus poorly absorbed compounds is usually close to two orders of magnitude in the Caco-2 model. The steepness of the slope may reflect the predictability of the model for compounds that are moderately absorbed. Figure 2.1 also demonstrates the difficulty in differentiating compounds in the midrange due to large variability in this range of the curve. During routine screening, a few compounds may be used repeatedly as benchmarks for calibration of day-to-day and batch-to-batch variations. For example, pindolol may be used for medium-to-high permeability cutoffs, whereas lovastatin may be used for medium-to-low permeability cutoffs. Metoprolol is a favored compound for the benchmarking of high permeability compounds.

Table 2.4 shows the PAMPA permeability and human absorption of 32 marketed compounds. As shown in Figure 2.2, there is a good correlation between the flux rate across the PAMPA membrane using DS-PAMPA and the absorption of these 32 structurally diverse compounds in humans. These results appear to be consistent with recent findings, which show that DS-PAMPA may yield the best correlation in comparison with other variant PAMPA systems [17].

Using the nonclinical animal species for the prediction of human absorption is commonly accepted in lead optimization during drug discovery. Table 2.5 lists the literature data for the fraction of absorption after oral administration in humans, rats, monkeys,

Table 2.3 Caco-2 Permeability and Human Absorption of 37 Marketed Drugs and Excipients

Compound	Human Absorption (%)	Caco-2 Permeability (nm/s)
PEG 4000	0	0
Gancyclovir	4	1
autoDoxorubicin	5	3
Mannitol	16	7
Acyclovir	30	13
Nadolol	34	6
Terbutaline	44	8
Chlorpromazine	50	53
Ranitidine	50	21
Furosemide	55	13
Atenolol	56	16
Cimetidine	60	27
Etoposide	60	15
Metolazone	64	38
Sulfasalazine	65	46
Hydrocholorothiazide	67	15
Omeprazole	77	285
Fluoxetine	80	860
Timolol	81	159
Methylprednisolone	82	146
Alprazolam	90	234
Phenytoin	90	146
Propranolol	90	394
Hydrocotisone	91	207
Isradipine	92	197
Alprenolol Hcl	93	366
Lorezapam	93	200
Metoprolol	95	332
Pindolol	95	236
Cisapride	100	192
Corticosterone	100	414
Desipramine	100	171
Dexamethasone	100	125
Diazepam	100	217
Felodipin	100	92
Fluconazole	100	700
Imipramine	100	182

and dogs for 34 marketed drugs. Linear regression analyses between individual animal species and human, as shown in Figures 2.3 through 2.5, clearly demonstrate similarity between human and rat or monkey. However, the lack of an overall correlation between human and dog suggests caution in using a dog as the sole animal species for the prediction of human absorption.

Table 2.4 PAMPA Permeability and Human Absorption of 32 Compounds

Compound ID	M.Wt.	Human Absorption (%)	PAMPA P_{app} (nm/s)	SD (n = 3)
Ganciclovir	255	5	0	0
Acyclovir	225	20	0	0
Lisinopril	406	25	0	0
Nadolol	309	34	0	0
Terbutaline	274	44	0	0
Ranitidine	351	50	3.2	4
Furosemide	331	55	0	0
Atenolol	266	56	0	0
Cimetidine	252	60	0	0
Etoposide	589	60	39	18
Metolazone	366	64	108	17
Sulfasalazine	398	65	0	0
Hydrocholorothiazide	298	67	3	4
Methylprednisolone	375	82	124	23
Propranolol	296	90	152	33
Pindolol	248	90	18	4
Hydrocortisone	363	91	123	23
Alprenolol	286	93	115	43
Metoprolol	685	95	23	2
Antipyrine	188	97	13	1
Caffeine	194	100	31	11
Carbamazepine	236	100	263	43
Dexamethasone	393	100	154	57
Ketoprofen	254	100	10	1
Naproxen	230	100	11	3
Theophylline	180	100	12	3
Corticosterone	347	100	233	23
Ibuprofen	206	100	111	10
Erythromycin	734	100	41	3
Azidothymidine	267	100	15	1
Levofloxacin	361	100	121	24

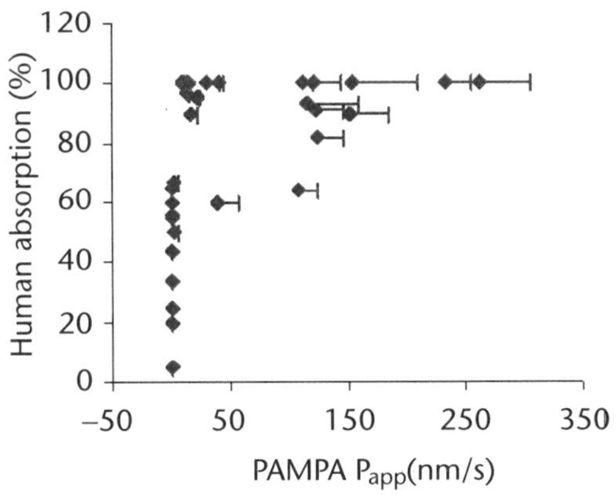

Figure 2.2 PAMPA permeability and human absorption.

Table 2.5 Summary of Fraction of Absorption After Oral Administration in Humans, Rats, Monkeys, and Dogs

	Absorption (% of Dose)			
Drug	Humans	Rats	Monkeys	Dogs
Acarbose	1.5	1.5		4
Acyclovir	30	21		100
Atenolol	56	50	45	100
Benazepril	30	50	32	39
Bepredil	99		83	
Bisprolol	100		100	98
Brompcriptin	30	36	35	
Captopril	75	71	79	
Coumarin	100		87	81
Doxazosin	100	100		
Droloxifene	100		100	
Etiodronate	10		16	
Flunisolide	100		100	
Fluvastatin	98	100	100	
Furosemide	55	60	60	54
Isoxepac	98	99		100
Ketanserin	100	100		100
Ovastatin	30	29	31	23
Menogaril	59		63	
Metoprolol	95		92	
Miglitol	100	100		100
Moxifloxacin	95		82	
Nadolol	34	18	23	98
Nimodipine	100	100		100
Nitredipine	88	90		75
Nisodipine	97		97	
Omeprazole	97	100		100
Pelrinone	98	71		96
Propranolol	90	99	100	
Ramipril	60	56		43
Remoxipride	100	100		99
Tamulosin	100	100		90
Xamoterol	9	16		39
Zomepirac	96		94	

Source: [3, 42, 43].

2.5 Discussion and Commentary

2.5.1 Cell culture and PAMPA systems

It is generally accepted that the Caco-2 cell system is used as an intestinal permeability screen during lead optimization in the pharmaceutical industry. This type of early screening would provide medicinal chemists with guidance in choosing more favorable chemotypes. Hence, in order to accommodate a large number of compounds, the trend over the last decade has been to fully automate or semiautomate the entire Caco-2 screening. A fully automated Caco-2 system allows a higher throughput and increases productivity without a proportional increase of resources. The components for a fully automated robotic system include an automated cell seeding and feeding system. This

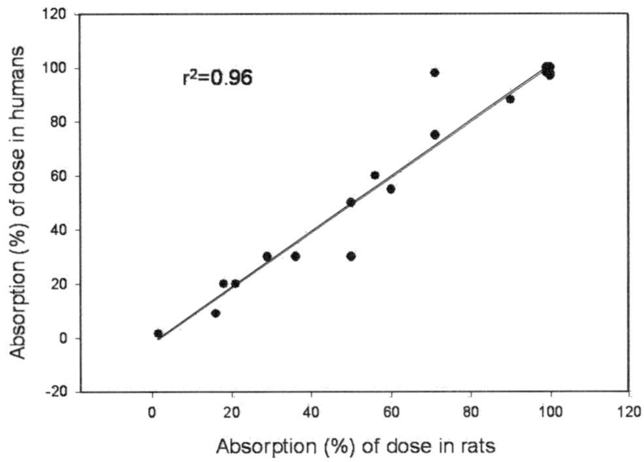

Figure 2.3 Correlation of percent of oral dose absorbed between humans and rats for 23 marketed drugs (data are taken from Table 2.5).

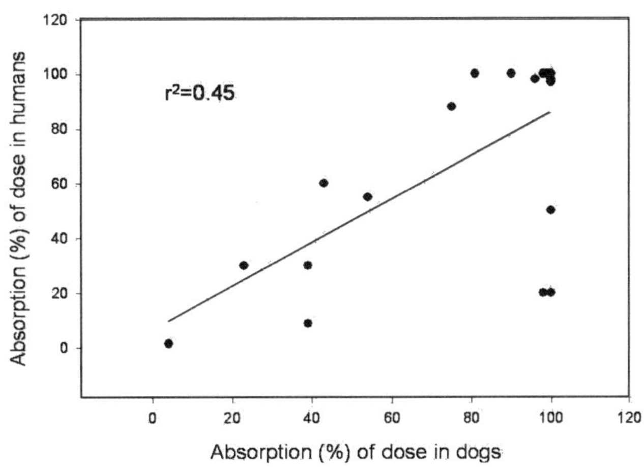

Figure 2.4 Correlation of percent of oral dose absorbed between humans and monkeys for 20 marketed drugs (data are taken from Table 2.5).

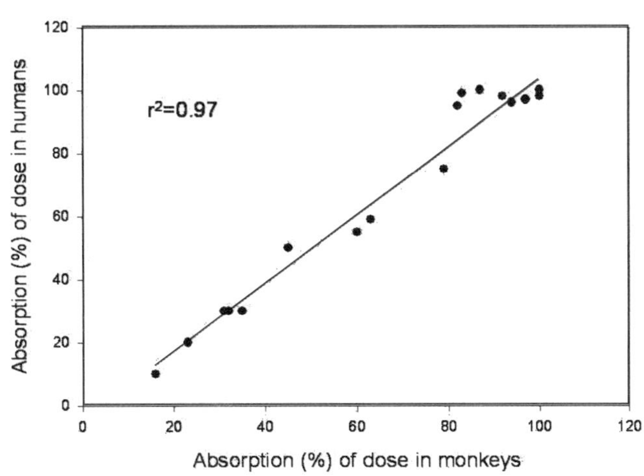

Figure 2.5 Correlation of percent of oral dose absorbed between humans and dogs for 20 marketed drugs (data are taken from Table 2.5).

system is often housed in a clean room to prevent the potential contamination of the cells. The second component comprises a robot that doses and aliquots samples from the Caco-2 plate. When the Caco-2 system is teamed up with a high-throughput LC/MS, a group of two to three scientists could screen 500–2,000 compounds per month. In addition, using a 24-well or even a 96-well plate system coupled with LC/MS significantly reduces the amount of compound required to perform permeability experiments. This is a remarkable feature that is useful when combinatory chemistry generates a large array of structurally related compounds. The Food and Drug Administration (FDA) issued guidelines in 2000 for the Biopharmaceutical Classification System (BCS), which utilizes permeability and solubility for the classification of compounds [24]. A Class 1 compound with high permeability and high solubility may obtain a waiver for modification of formulation in clinical trials, whereas a Class 4 compound with low permeability and low solubility may have difficulty in obtaining such a waiver. For passively transported compounds, Caco-2 cell permeability is considered an appropriate method for drug permeability classification for BCS.

During the differentiation process in a cell culture which leads to the formation of tight junctions and development of cell polarity, various drug transporters may also be expressed in Caco-2 cells in a polarized manner. It is well documented that the major efflux pumps, such as P-gp and breast cancer resistant protein (BCRP), are highly expressed in Caco-2 cells [25, 26]. Both drug transporters are located on the apical membrane when the Caco-2 cells establish cell polarity. In addition, other transporters, such as multidrug resistance proteins (MRPs), may also play a significant role in a Caco-2 efflux. Hence, the Caco-2 cell model has also been used for evaluation of efflux and identification of efflux substrates with a bidirectional transport study, in which the test compound is dosed in either the apical compartment or the basolateral compartment and the apical-to-basal and basal-to-apical permeability coefficients are measured and compared in order to calculate an efflux ratio ($P_{app\_BL\ to\ AP}/P_{app\_AP\ to\ BL}$). When the efflux ratio is greater than 2, the test compound is flagged as a potential efflux substrate. It is usually feasible to use specific inhibitors to identify the transporter responsible for the efflux. For example, the efflux ratio for digoxin ranges from 10 to 40 [27]. When codosed with cyclosporine (a P-gp inhibitor), the efflux ratio of digoxin decreases to about 1. Verapamil, cyclosporine A [28], and ketoconazole are commonly used as P-gp inhibitors for such experiments, while probenecid and indomethacin are commonly used to inhibit MRP2 [29].

There are several potential issues that may restrict the use of the Caco-2 model as the sole screening tool for evaluation of absorption potential. First, despite the fact that Caco-2 cells show sufficient levels of expression of P-gp, BCRP, and MRPs, many other transporters involved in the influx of substrates are underexpressed. These transporters include OATPs, OCT, and PepT1. For example, PepT1 substrates, such as beta-lactam antibiotics (cefadroxil and amoxicillin), are well absorbed, but show poor Caco-2 permeability [30]. The Caco-2 model is likely to generate false negatives with NCEs actively transported by underexpressed transporters. When Caco-2 permeability data differ greatly from animal absorption data, it may require additional studies to understand if the drug candidate is the substrate of active transporters. A comparison of human duodenum and Caco-2 gene expression profiles using GeneChip analysis showed a significant difference (greater than fivefold) in expression of only approximately 10% of the total genes [31]. The in vivo permeability of 26 drugs, including transporter substrates, determined in healthy human subjects correlates with that determined in Caco-2 cells. Thus, the Caco-2 system is still a reasonable model to study the permeability of transporter substrates.

A second issue limits the use of the Caco-2 model for compounds that diffuse paracellularly. Many compounds, such as atenolol and furosemide, show poor Caco-2 cell permeability despite adequate absorption (greater than 50%) in humans [32]. The plausible explanation for this discrepancy is that the Caco-2 cell tight junctions are smaller and thus less permeable than those in vivo. A perturbation of the tight junctions by treatment with calcium-chelating chemicals, such as EDTA, may be helpful in evaluating such compounds; however, the use of organic solvents is contraindicated. The integrity of tight junctions in the Caco-2 model is easily affected by commonly used organic solvents or cosolvents (e.g., methanol, ethanol, PEG, DMSO) at relatively low concentrations (<1% V/V). Additionally, NCEs that diffuse paracellularly and have a poor aqueous solubility may not be adequately evaluated by this model. It has become a common practice in the pharmaceutical industry to test the solubility of compounds before performing any other in vitro screens and to eliminate NCEs with poor aqueous solubility, thus preventing false negatives due to this issue.

A third issue arises with hydrophobic NCEs that bind to plastic surfaces. A significant loss due to this type of nonspecific binding to the Caco-2 device may lead to the underestimation of permeability. The addition of protein (e.g., bovine serum albumin) in the receiver compartment can sometimes prevent nonspecific binding [33].

Lastly, the preparation of fully differentiated Caco-2 cells generally requires a 3-week cell culture period with eight to ten cell feedings. A fully automated system for cell seeding and feeding can be used for industrial high-throughput screening purposes; however, for more limited use, this labor-intensive cell culture process could become a rate-limiting factor. The cell preparation time can be substantially reduced by using an available coating material and supplemented growth medium. The generation of a functional monolayer in a shorter cell culturing period not only increases the productivity, but also decreases the chance of having contamination.

Other potential cell models are Madin-Darby canine kidney cells (MDCK), 2/4/A1 cells, HT-29, and LLC-PK1 cells. The 2/4/A1 cell line originates from the fetal rat intestine and mimics the permeability of human small intestine with regards to both paracellular and transcellular drug transport [34]. The permeability of some poorly permeable compounds, such as mannitol and creatine, across a 2/4/A1 cell monolayer appears to be comparable to that in human jejunum suggesting that this cell line could be a better model for compounds that are absorbed via the paracellular route. MDCK cells differentiate into columnar epithelial cells and form tight junctions when cultured on semipermeable membranes. Cho et al. [35] compared the permeability values of 55 compounds in an MDCK model with human absorption values and found a good correlation between these two values. The major concern with using MDCK cells for permeability screening is that the cell line is derived from dog kidney cells, and it is very likely that the expression levels and substrate specificity of various transporters may be different from the in vivo situation in humans. On the other hand, the major advantage of MDCK cells is that they require only a 3-day culture period.

Despite the availability of other cell lines, Caco-2 cells remain the most widely used intestinal cell culture model at present. This model has provided valuable information necessary for lead optimization in the drug discovery process. However, it is important to understand that compounds with high permeability in this model are typically well absorbed, whereas compounds with low solubility and low permeability in this model may not necessarily be poorly absorbed in vivo. Although this type of positive selection limits the usefulness in providing a structure-permeability relationship, the Caco-2 model has the most effect in drug discovery when the screen is implemented early and in conjunction with other types of in vitro and in vivo permeability/absorption screens. Frequently, in silico screening based upon physicochemical properties, such as cLogP, LogD, molecular weight, and H-bonding, is used during lead optimization in combination with in vitro membrane permeability screening to identify potential lead compounds. The relationship between physicochemical properties and absorption of chemicals was summarized by Lipinski [36]. A recent publication from our lab [14] indicates that the Caco-2 cell system and/or the parallel artificial membrane permeability system (PAMPA) remain the best in vitro systems for the evaluation of membrane permeability.

For a PAMPA system, the use of 96-well microtiter plates coupled with the rapid quantification by a spectrophotometric plate reader allows this system to screen a large array of compounds in a relatively short period of time. Alternatively, the compound analysis can be performed with fast and sensitive LC/MS by pooling the samples together. Both the spectrophotometric and mass spectrometric methods can be used for low solubility compounds since the limit of quantitation can reach nanomolar concentrations. The major advantage of PAMPA is that it is much less labor intensive than cell culture or in vivo studies that may have a similar predictive power. There are several limitations, however, in using this model for prediction of human absorption. First, as mentioned previously, the composition of the phospholipid bilayers in different variants of PAMPA systems may cause some major variation in permeability coefficients. Second, PAMPA can significantly underestimate the absorption of compounds that are actively transported by influx drug transporters, such as PEPT1 and OATPs, or hydrophobic small molecules, which may be transported through the paracellular route. Despite these limitations, PAMPA has become a popular permeability screen during the early drug discovery process because of its high-throughput capability.

PAMPA and the Caco-2 cell model are often used in combination to evaluate permeability potential. While the Caco-2 cell model has been the in vitro gold standard for permeability screening, PAMPA has recently become a more efficient, economical, and higher-throughput screen. Both models show approximately two orders of magnitude in their dynamic range of permeability values. Both work well in categorizing compound permeability (high, medium, and low), but lack predictive accuracy. In order to efficiently combine these two models in drug discovery, one will need to understand the pros and cons. The sole purpose of using PAMPA for screening is to identify compounds that have good transcellular passive permeability across a lipid barrier. This type of evaluation of "intrinsic" passive permeability may provide a positive selection of compounds that will demonstrate good in vivo absorption. Since PAMPA is often placed as a tier-1 screen, compounds with good permeability may obtain a higher priority for subsequent screening. Obviously, a simple lipid barrier such as the PAMPA system lacks other biological characteristics of intestinal enterocyte cell membranes. Hence, PAMPA may not predict compounds that are transported by influx and efflux drug transporters, permeate through the tight junctions, or are absorbed by other types of cellular mechanisms. Similar to PAMPA, the Caco-2 cell model can also identify compounds with good passive permeability via the transcellular route. It is also possible to identify compounds such as small hydrophobic molecules, which may permeate via the tight junctions (paracellular) with proper perturbation of the tight junctions. The high levels of expression of certain drug transporters, such as p-GP, BCRP, and MRPS, certainly allow Caco-2 cells to retain many pharmacological characteristics of intestinal enterocytes. However, many influx transporters, such as PEPT1 and OATPs, are not well expressed in Caco-2 cells. Understanding these capacities and limitations, PAMPA and Caco-2 models can both be used to better predict absorption during the early stages of drug discovery.

2.5.2 Absorption in experimental animals

During the lead optimization phase, animal pharmacokinetic data often play a key role in the progression/selection of the lead candidate for clinical development. Very often the animal pharmacokinetic study will begin in a rodent species, such as rat or mouse, and progress further to large animal species, such as dog and monkey. Increasingly, the pharmacokinetic behavior is used as "go-no-go" selection criteria for further lead progression for several reasons [37]. First, an unfavorable pharmacokinetic performance may disqualify the animal from serving as the toxicological species. Second, it may indirectly imply that human pharmacokinetic performance may also be poor. When a lead candidate displays uniform interspecies pharmacokinetic behavior (either favorable or unfavorable), using these data for decision making is relatively straightforward. In practice, the nonclinical species may well disagree with each other, and hence, it often becomes a question regarding which nonclinical species may be best suited for the prediction of pharmacokinetics humans.

Despite the challenges, several comparative studies evaluating the oral exposure of marketed drugs may shed light on predictability using pharmacokinetic data derived from experimental animals [38–42]. In the following sections, we will examine the pros and cons of the three most common nonclinical species with respect to their suitability for the prediction of oral absorption in humans.

2.5.3 Rats

The rat is the most common nonclinical species because of its relatively low cost and ready availability comparing with using other animals, such as dogs and monkeys. Rats are convenient for surgical operation, including cannulation and removal of intestinal tissues for ex vivo studies. Hence, a lot of emphasis has been placed on using rats for in vivo oral absorption, in vivo upper-GI (small intestine) versus lower-GI (colon) absorption comparison, and ex vivo intestinal permeability studies for human absorption prediction and correlation. A recent review [38] on the comparative evaluation of absorption and metabolism mechanisms on drug oral bioavailability in rats and humans pointed out significant differences in the anatomy and physiology of the GI tract. These differences include: the thickness of the unstirred water layer in the lumen, which is approximately 25 μm in humans and 106 μm in rats; the small intestinal length, which is about 5m in humans and 0.8m in rats; and the small intestinal radii, which is about 10 times wider in humans than in rats. However, these differences may not be associated with significant differences in the rate or extent of drug absorption between rats and humans. The small intestinal pH in rats is one unit higher than that in humans [39]. This could also affect the correlation between these two species. The different pH values present at the site of absorption affects the degree of ionization, especially when weak basic compounds are under evaluation. Ionization can affect absorption of a compound by increasing its solubility and decreasing its permeability.

An early study comparing the fraction of oral dose absorbed of 64 drugs showed excellent linearity and correlation between humans and rats [40]. It should be pointed out that the following types of drugs were excluded: drugs that are unstable in the GI lumen, such as erythromycin and penicillin; drugs that are known to be affected by particle size, such as nitrofurantoin; and drugs with dose-dependent absorption. In a separate comparative evaluation of 56 drugs in nonclinical animals and humans, the rat predicted human exposure the worst [41]. It needs to be pointed out that oral exposure relies on two pharmacokinetic events: absorption and hepatic first-pass. The lack of good correlation in oral exposure between rat and human is most likely due to a large difference in the hepatic first-pass between these two species.

During lead optimization an obvious disadvantage of using animals to predict oral absorption in human is the cost to support this in vivo screening. Generally, however, a rapid in vivo screen in rats will delineate the oral exposure of a given compound, complementing in vitro Caco-2 permeability data and hepatocyte clearance data, to understand the lack of favorable pharmacokinetic performance. This will lead to a new cycle of chemical synthesis of compounds with improved absorption and/or metabolic stability profiles.

2.5.4 Dogs

The dog has also been commonly used as an animal model for oral absorption in drug discovery and development. Several reviews have focused on the similarities and differences in physiology and anatomy of the GI tract of humans and dogs and their potential effects on drug absorption [41–43]. Although fasted dogs and humans have similar stomach-emptying times for solutions, the transit time in the dog GI tract (2 hours) is much shorter than that in human (4 hours). This is reflected by a shorter plasma level peak time (T_{max}) after oral administration for many drugs. For example, the T_{max} for ranitidine, sumatriptan, and atenolol is much shorter in dogs than in humans. To compensate for the shorter transit time in dogs, the apparent intestinal first-order absorption rate constant of drugs needs to be higher in dogs than in humans in order to achieve similar oral absorption in both species. In addition, dogs have longer villi than humans [42] that may help to offset the shorter intestinal transit time. Furthermore, dogs have a higher bile salt secretion rate and a higher bile salt concentration in the intestine that may improve the absorption of poor water soluble compounds under fed condition [43]. A higher intestinal pH (about one unit) in fasted dogs than in humans may partially explain the better absorption of weak basic drug in dogs [44].

In a comparative evaluation of 43 drugs using the dog as the animal model, only a moderate correlation was observed between the fraction of oral dose absorbed in dogs and humans [45]. A closer examination showed 34 drugs (out of the 43) had similar oral absorption in both dogs and humans. Seven of the 43 drugs had significantly higher oral absorption in dogs than in humans. These seven drugs include many water-soluble neutral compounds, and the increased absorption in dogs may be due to the greater size and frequency of tight junctions for paracellular permeability. This study demonstrated that the use of a smaller set of compounds may yield more favorable correlation coefficients, and hence, the similarity/difference in oral absorption between dogs and humans is dependent on the drugs selected for evaluation. Certainly, a better understanding of the mechanism(s) responsible for efficient absorption in dogs will aid in predicting drug absorption in humans.

2.5.5 Monkeys

The monkey is the only primate among the frequently used animal species in nonclinical drug discovery and development. The cost and ethical concerns are some of the factors limiting the use of monkeys. Among the old-world monkeys, rhesus and cynomolgus macaques are most frequently used for drug metabolism and pharmacokinetic studies. Due to evolutionary proximity with humans, the monkey is the most similar to humans in terms of physiology and biochemistry of the GI tract. Monkeys and humans are generally similar with regard to the rate and extent of drug absorption [46]. When comparable doses were used, humans and monkeys showed similar T_{max}. When drugs (most notably ifetroben, rolipram, and venlafaxine) were administered orally to monkeys at doses far exceeding those to humans, T_{max} values tended to be prolonged, which is probably due to extended absorption in the lower GI tract, incomplete absorption in the upper GI tract, and solubility/dissolution issues. Hence, for prediction purposes, the dose administered to monkeys should be similar to that administered to humans. In a study by Chiou and Buchler [46], a comparison of oral absorption of 35 drugs in monkeys and humans showed excellent correlation, despite dosage form (solution, fast-release suspension, tablet, or capsule) differences [46]. A similar study by Ward et al. [41] comparing the oral exposure of 56 drugs in rats, dogs, monkeys, and humans also showed that monkeys are most similar to humans with the caveat that monkeys often demonstrated higher metabolic clearance and, hence, lower oral bioavailability of the drug. The lower oral bioavailability predicted using monkey pharmacokinetic data should not bias the investigator against using monkeys as a reasonable predictor for oral absorption.

One of the concerns with using monkeys in the lead optimization stage is the consecutive dosing of different drug candidates with short washout periods. It is not unusual that compounds elected for dosing in monkeys have very long half-lives (>14 days). These compounds will remain in the animal for a rather long period of time and, hence, may affect the metabolic activity in two opposite ways. If the compound has a CYP inhibition activity, it may boost the exposure of a subsequent compound. Conversely, if the compound is a CYP 3A inducer, it may cause a reduction in the exposure. In addition, compounds that are hepatotoxic may affect the overall metabolic activity of the animal even after a single oral dose. Other considerations include the genetic makeup and the origin of the monkeys used for the pharmacokinetic study. These factors may contribute to large individual variations in a given pharmacokinetic study and cause bias in using the data for human prediction. In addition, the use of nonhuman primates may need careful management to minimize resources and meet animal use regulations. However, comparison studies between monkeys and humans suggest that the monkey should continue to be an important animal species in drug discovery for use in predicting human pharmacokinetic data.

2.6 Summary

The Caco-2 cell system is the most widely used intestinal cell culture model at present. This model has provided a high-throughput screening for lead optimization in the drug discovery process. However, it is important to understand that compounds with high permeability in this model are typically well absorbed, whereas compounds with low solubility and low permeability in this model may not necessarily be poorly absorbed in vivo. Although this type of positive selection limits the usefulness in providing a structure-permeability relationship, the Caco-2 model has the most effect in drug discovery when the screen is implemented early and in conjunction with other types of in vitro and in vivo permeability/absorption screens. During lead optimization an obvious disadvantage of using preclinical animals to predict oral absorption in humans is the cost to support this in vivo screening. However, in vivo preclinical pharmacokinetic studies play a key role in the overall drug discovery paradigm.

Typically, for a given drug candidate, the in vivo pharmacokinetic evaluation will be performed in at least two or three preclinical species, which often include rat, dog, and/or monkey. Thus, the Caco-2 cell system, in conjunction with animal pharmacokinetic screens is the most frequently used hybrid in vitro–in vivo model for the screening of NCEs in the drug discovery.

References

[1] Chiou, W. L., "The Rate and Extent of Oral Bioavailability Versus the Rate and Extent of Oral Absorption: Clarification and Recommendation of Terminology," *J. Pharmacokinet. Biopharm.*, Vol. 28, 2001, pp. 3–6.

[2] Balimane, P. V., S. Chong, and R. A. Morrison, "Current Methodologies Used for Evaluation of Intestinal Permeability and Absorption," *J. Phamacol. Toxico. Methods,* Vol. 44, 2000, pp. 301–312.

[3] Balimane, P. V., Y. H. Han, and S. Chong, "Current Industrial Practices of Assessing Permeability and P-Glycoprotein Interaction," *AAPS J.,* Vol. 8, 2006, pp. E1–E13.

[4] Bergstrom, C. A. S., "In Silico Predictions of Drug Solubility and Permeability: Two Rate-Limiting Barriers to Oral Drug Absorption," *Basic Clin. Pharmacol. Toxicol.,* Vol. 96, 2005, pp. 156–161.

[5] Avdeef, A., *Absorption and Drug Development,* New York: Wiley-Interscience, 2003, pp. 116–246.

[6] De Zwart, L. L., et al., *Anatomical and Physiological Differences Between Various Species Used in Studies on the Pharmacokinetics and Toxicology of Xenobiotics,* RIVM Report 623860010, National Institutes of Public Health and the Environment, 1999.

[7] Artursson, P., "Cell Cultures as Models for Drug Absorption Across the Intestinal Mucosa," *Crit. Rev. Ther. Drug Carrier Syst.,* Vol. 8, 1991, pp. 305–330.

[8] Artursson, P., K. Palm, and K. Luthman, "Effect of Molecular Charge on Intestinal Epithelial Drug Transport: pH-Dependent Transport of Cationic Drugs," *Adv. Drug Deliv. Rev.,* Vol. 46, 2001, pp. 27–43.

[9] Borchardt, R., "Application of Cell Culture Systems in Drug Discovery and Development," *J. Drug Target,* Vol. 3, 1995, pp. 179–182.

[10] Hidalgo, I., T. Raub, and R. Borchardt, "Characterization of the Human Colon Carcinoma Cell Line (Caco-2) as a Model System for Intestinal Epithelial Permeability," *Gastroenterology,* Vol. 96, 1989, pp. 736–749.

[11] Rubas, W., N. Jezyk, and G. M. Grass, "Comparison of the Permeability Characteristics of a Human Colonic Epithelial (Caco-2) Cell Line to Colon of Rabbit, Monkey, and Dog Intestine and Human Drug Absorption," *Pharm. Res.,* Vol. 10, 1993, pp. 113–118.

[12] Artursson, P., and J. Karlsson, "Correlation Between Oral Drug Absorption in Humans and Apparent Drug Permeability Coefficients in Human Intestinal Epithelial (Caco-2) Cells," *Biochem. Biophys. Res. Comm.,* Vol. 175, 1991, pp. 880–885.

[13] Chong, S., S. Dando, and R. A. Morrison, "Evaluation of Biocoat Intestinal Epithelium Differentiation Environment (3-Day Cultured Caco-2 Cells) as an Absorption Screening Model with Improved Productivity," *Pharm. Res.,* Vol. 14, 1997, pp. 1835–1837.

[14] Li, C., et al., "Correlation Between PAMPA Permeability and Cellular Activities of Hepatitis C Virus Protease Inhibitors," *Biochem. Pharma.,* Vol. 75, 2008, pp. 1186–1197.

[15] Kansy, M., F. Senner, and K. Gubermator, "Physicochemical High Throughput Screening: Parallel Artificial Membrane Permeation Assay in the Description of Passive Absorption Processes," *J. Med. Chem.*, Vol. 41, 1998, pp. 1007–1010.

[16] Avdeef, A., et al., "Drug Absorption In Vitro Model: Filter-Immobilized Artificial Membranes. 2. Studies of the Permeability Properties of Lactones in Piper Methysticum Forst," *Eur. J. Pharmaceut. Sci.*, Vol. 14, 2001, pp. 271–280.

[17] Avdeef, A., and A. Tsinman, "PAMPA—A Drug Absorption In Vitro Model 13. Chemical Selectivity Due to Membrane Hydrogen Bonding: In Combo Comparisons of HDM-, DOPC-, and DS-PAMPA Models," *Eur. J. Pharmaceut. Sci.*., Vol. 28, 2006, pp. 43–50.

[18] Walter, E., and T. Kissel, "Transport of Peptidomimetic Renin Inhibitors Across Monolayers of a Human Intestinal Cell Line (Caco-2): Evidence for Self-Enhancement of Paracellular Transport Route," *J. Pharm. Sci.,* Vol. 3, 1995, pp. 215–230.

[19] Artursson, P., and R. T. Borchardt, "Intestinal Drug Absorption and Metabolism in Cell Cultures: Caco-2 and Beyond," *Pharm. Res.,* Vol. 14, 1997, pp. 1655–1658.

[20] Artursson, P., et al., "Effect of Chitosan on the Permeability of Monolayers of Intestinal Epithelial Cells (Caco-2)," *Pharm. Res.,* Vol. 11, 1994, pp. 1358–1361.

[21] Thummel, K. E., and D. D. Shen, "Design and Optimization of Dosage Regimens: Pharmacokinetic Data," in J. G. Hardman, L. E. Limbard, and A. G. Gilman, (eds.), *Goodman and Gilman's The Pharmacological Basis of Therapeutics*, 10th ed., New York: McGraw-Hill, 2001, pp. 1927–2023.

[22] Bailey, C. A., P. Bryla, and A. W. Malick, "The Use of the Intestinal Epithelial Cell Culture Model, Caco-2, in Pharmaceutical Development," *Adv. Drug Del. Rev.*, Vol. 22, 1996, pp. 85–103.

[23] Yamashita, S., et al., "Optimized Conditions for Prediction of Intestinal Drug Permeability Using Caco-2 Cells," *Eur. J. Pharm. Sci.,* Vol. 10, 2000, pp. 195–204.

[24] FDA, The Biopharmaceutics Classification System (BCS) Guidance, CDER, http://www.fda.gov/AboutFDA/CentersOffices/CDER/ucml28219.htm.

[25] Adachi, Y., H. Suzuki, and Y. Sugiyama, "Comparative Studies on In Vitro Methods for Evaluating In Vivo Function of MDR1 P-Glycoprotein," *Pharm Res.,* Vol. 18, 2001, pp. 1660–1668.

[26] Breedveld, P., J. H. Beijnen, and J. H. Schellens, "Use of P-Glycoprotein and BCRP Inhibitors to Improve Oral Bioavailability and CNS Penetration of Anticancer Drugs," *Trends Pharmacol. Sci.,* Vol. 27, 2006, pp. 17–24.

[27] Yao, H. M., and W. L. Chiou, "The Complexity of Intestinal Absorption and Exsorption of Digoxin in Rats," *Int. J. Pharm.,* Vol. 322, 2006, pp. 79–86.

[28] Loor, F., et al., "Cyclosporins: Structure-Activity Relationships for the Inhibition of the Human FPR1 Formylpeptide Receptor," *J. Med. Chem.,* Vol. 45, 2002, pp. 4598–4612.

[29] Mallants, R., et al. "Multidrug Resistance-Associated Protein 2 (MRP2) Affects Hepatobiliary Elimination But Not the Intestinal Disposition of Tenofovir Disoproxil Fumarate and Its Metabolites," *Xenobiotica,* Vol. 35, 2005, pp. 1055–1066.

[30] Chong, S., et al., "In Vitro Permeability Through Caco-2 Cells Is Not Quantitatively Predictive of In Vivo Absorption for Peptide-Like Drugs Absorbed Via the Dipeptide Transporter System," *Pharm. Res.,* Vol. 13, 1996, pp. 120–123.

[31] Sun, D., et al., "Comparison of Human Duodenum and Caco-2 Gene Expression Profiles for 12000 Gene Sequences Tags and Correlation with Permeability of 26 Drugs," *Pharm. Res.*, Vol. 19, 2002, pp. 1400–1416.

[32] Bourdet, D. L., G. M. Pollack, and D. R. Thakker, "Intestinal Absorptive Transport of the Hydrophilic Cation Ranitidine: A Kinetic Modeling Approach to Elucidate the Role of Uptake and Efflux Transporters and Paracellular Vs. Transcellular Transport in Caco-2 Cells," *Pharm Res.,* Vol. 23, 2006, pp. 1178–1187.

[33] Saha, P., and J. Kou, "Effect of Bovine Serum Albumin on Drug Permeability Estimation Across Caco-2 Monolayers," *Eur. J. Pharm. Biopharm.,* Vol. 54, 2002, pp. 319–324.

[34] Tavelin, S., et al., "A Conditionally Immortalized Epithelial Cell Line for Studies of Intestinal Drug Transport," *J. Pharmacol. Exper. Ther.*, Vol. 290, 1999, pp. 1212–1221.

[35] Cho, M., et al., "The Madin Darby Canine Kidney (MDCK) Epithelial Cell Monolayer as a Model Cellular Transport Barrier," *Pharm. Res.,* Vol. 6, 1989, pp. 71–77.

[36] Lipinski, C. A., et al., "Experimental and Computational Approaches to Estimate Solubility and Permeability in Drug Discovery and Development Settings," *Adv. Drug Deliver. Rev.,* Vol. 46, 2001, pp. 3–26.

[37] Cheng, K.-C., et al., "Lead Optimization in Discovery Drug Metabolism and Pharmacokinetics/ Case Study: The Hepatitis C Virus (HCV) Protease Inhibitor SCH 503034," *Perspective in Med. Chemistry*, Vol. 1, 2007, pp. 1–9.

[38] Hurst, S., et al., "Impact of Physiological, Physicochemical and Biopharmaceutical Factors in Absorption and Metabolism Mechanisms on the Drug Oral Bioavailability of Rats and Humans," *Expert Opin. Drug Metab. Toxicol.*, Vol. 3, 2007, pp. 469–489.

[39] Kararli, T. T., "Comparison of the Gastrointestinal Anatomy, Physiology and Biochemistry of Humans and Commonly Used Laboratory Animals," *Biopharm. Drug Dispos.*, Vol. 16, 1995, pp. 351–380.

[40] Chiou, W. L., and A. Barve, "Linear Correlation of the Fraction of Oral Dose Absorbed of 64 Drugs Between Humans and Rats," *Pharm. Res.*, Vol. 15, 1998, pp. 1792–1795.

[41] Ward, K. W., R. Nagilla, and L. J. Jolivette, "Comparative Evaluation of Oral Systemic Exposure of 56 Xenobiotics in Rat, Dog, Monkey and Human," *Xenobiotica,* Vol. 35, 2005, pp. 191–210.

[42] Dressman, J. B., "Comparison of Canine and Human Gastrointestinal Physiology," *Pharm. Res.,* Vol. 3, 1986, pp. 123–131.

[43] Kalantzi, L., et al., "Canine Intestinal Contents Vs Simulated Media for the Assessment of Solubility of Two Weak Bases in the Human Small Intestinal Contents," *Pharm. Res.*, Vol. 23, 2006, pp. 1373–1381.

[44] Lin, J. H., "Species Similarities and Differences in Pharmacokinetics," *Drug Metab. Dispos.*, Vol. 23, 1995, pp. 1008–1021.

[45] Chiou, W. L., et al., "Evaluation of Using Dog as an Animal Model to Study the Fraction of Oral Dose Absorbed of 43 Drugs in Humans," *Pharm. Res.,* Vol. 17, 2000, pp. 135–140.

[46] Chiou, W. L., and P. W. Buehler, "Comparison of Oral Absorption and Bioavailability of Drugs Between Monkey and Human," *Pharm. Res.,* Vol. 19, 2002, pp. 868–873.

CHAPTER

3

Aggregating Brain Cell Cultures for Neurotoxicity Tests

Paul Honegger, Florianne Monnet-Tschudi, and Marie-Gabrielle Zurich

Department of Physiology, University of Lausanne, Switzerland
Corresponding author: Marie-Gabrielle Zurich, e-mail: Marie-Gabrielle.ZurichFontanellaz@unil.ch, Departement de Physiologie, Universite de Lausanne, Rue du Bugnon 7, CH-1005, Lausanne, Switzerland, phone: +4121-692 5500, fax: +4121-692 5595

Abstract

The aggregating brain cell cultures described here are three-dimensional primary cell cultures derived from mechanically dissociated embryonic brain tissue and grown in a chemically defined medium. The initial cell suspension contains neural stem cells, neural progenitor cells, immature glial cells, and early postmitotic neurons. Under continuous gyratory agitation, aggregates form spontaneously and are maintained in a suspension culture for several weeks. Within the aggregates, the cells reproduce critical morphogenic events such as migration, proliferation, differentiation, synaptogenesis, and myelination, reconstituting a highly differentiated histotypic brain architecture. These cultures therefore offer a suitable model to study the adverse effects of agents during brain development and in the mature brain. The high number of aggregates available and their excellent reproducibility facilitate routine test procedures. This chapter presents a detailed description of culture preparation for their use in fundamental studies and toxicity screening. Examples for suitable endpoints are given, including developmental, structural, and functional criteria.

Key terms
aggregating cell cultures, brain
CNS, development, maturation
gliotoxicity, high content analysis
neurotoxicity, organ toxicity
three-dimensional cell culture

3.1 Introduction

The central nervous system is a frequent target of pharmaceuticals, drugs, pesticides, herbicides, solvents, and other industrial and environmental organic and inorganic chemicals. Therefore, risk assessments for any compound must include the evaluation of its potential neurotoxicity. There is now a general consensus that the largest part of risk assessment needs to be done by in vitro and in silico approaches rather than in animals. Furthermore, the evaluation of chemical adverse effects should focus on the interference with vital biological mechanisms rather than the dose-dependent cytotoxicity or lethality. The ideal in vitro model for neurotoxicity studies should mirror as close as possible the structural and functional features of the brain, and at the same time should permit efficient and reproducible testing. These requirements are fulfilled to a large extent by aggregating brain cell cultures, a three-dimensional suspension culture prepared from embryonic vertebrate brain cells. In a recent study aimed at the in vitro prediction of acute systemic toxicity (www.acutetox.org), the potential neurotoxicity of 57 reference chemicals was tested in five different in vitro models, that is, the human neuroblastoma cell line SH-SY5Y, rat primary cortical neurons, rat and mouse primary cerebellar granule cells, rat brain slices, and aggregating brain cell cultures. Neurotoxic alerts were defined as significant effects of compounds at concentrations sevenfold lower than those causing a general cytotoxicity in a mouse fibroblast (Balb/3T3) cell line. Of the five models, aggregating brain cell cultures identified the highest number of neurotoxic alerts [1]. They also proved to be the most complete, most versatile, and most efficient in vitro test system, particularly when compared with brain slice cultures, another three-dimensional culture system.

The rotation-mediated reaggregation of immature cells was first described by A. Moscona in 1961 [2]. The spontaneous reaggregation and rearrangement of cells, reconstituting a three-dimensional histotypic architecture, was shown to be an intrinsic capacity of immature cells originating from various organs, including the brain. The use of a mechanical dissociation method greatly increased the reproducibility and robustness of aggregating brain cell cultures [3]. Maintained under continuous gyratory agitation in a CO_2 incubator, dissociated embryonic brain cells form numerous aggregates, containing all cell types present in the brain. In this tissue-specific environment, the cells can interact metabolically as well as by signaling between adjacent cells and through soluble intrinsic messengers. This enables them to differentiate extensively and to develop histotypic features, including the natural extracellular matrix, mature synapses, functional neuronal networks, and myelinated axons. The aggregate cultures also offer unique advantages from a practical point of view, in particular, high yield, robustness, high reproducibility, and the possibility to use in permanence a serum-free, chemically defined culture medium [4].

In principle, aggregating brain cell cultures can be prepared from any part of a vertebrate brain, provided the presence of sufficiently immature brain cells. It is also conceivable to use multipotent neural stem cells, once the techniques to control cellular differentiation and cell expansion will be fully established. For the time being, primary brain cells are needed for culture preparation. For routine work, a 16-day embryonic rat brain proved to be the most convenient source. One embryonic rat brain (comprising telencephalon, mesencephalon, and rhombencephalon) yields enough cells for one initial flask of aggregate cultures, giving rise to about 1,200 aggregates. For neurotoxicity studies, replicate cultures are prepared (i.e., the aggregates of several flasks are pooled), and aliquots are redistributed into smaller culture vessels. The use of replicate cultures increases both the testing capacity of the cultures and the statistical power of the data. Routinely, for one test series, 42 replicate cultures are prepared from 7 original flasks, giving replicates of about 200 individual aggregates. Thus, from the embryos of one pregnant rat, an average of 84 replicate cultures can be obtained, enough for two test series. This high yield allows of a considerable throughput, while the high number of aggregates per replicate enables multidisciplinary (high content) analysis. Cytotoxic drug concentrations can be assessed by measuring an LDH release. Selective adverse effects are examined at subcytotoxic drug concentrations, using criteria representative for pathogenic events, and for functional and structural features of the different brain cell types. In principle, any analytical approach applicable to brain tissue can be used also for aggregating brain cell cultures. The maturational characteristics of aggregating brain cell cultures are particularly advantageous for work in the domain of developmental neurotoxicity, in both an acute setting and a chronic setting, and for toxicity screening as well as fundamental studies.

The use of aggregating brain cell cultures requires special equipment not always found in a standard cell culture laboratory. For most special equipment and procedures, substitutes are possible, provided that the following critical points are respected:

1. The mechanical dissociation of the embryonic brain tissue into a single cell fraction must be complete, but sufficiently gentle to provide a single cell suspension with a maximum of viable cells.
2. During the initial phase of cell aggregation, the culture vessels need to be equal in geometry to assure the formation of uniform aggregates and hence comparable cultures. The vortex in each culture vessel is critical for the aggregate formation, and depends on the dimensions of the culture vessel, the volume of the medium, and the speed of agitation. To find the optimal vortex, it is recommended to keep the shaking protocol unchanged, while adjusting the volume of medium in the culture vessels.
3. The culture vessels need to be equipped with a gas-permeable cap.
4. The developmental stage of the original tissue is critical, requiring the exact timing of pregnancy.
5. Culture conditions, medium composition, and osmolarity (340 mOsm) need to be kept constant. Extreme changes in the pH values of media should be avoided. The highest standards of purity have to be respected for solutions, media, and glassware.

3.2 Experimental Design

Aggregating brain cell cultures are prepared from a 16-day embryonic rat brain, counting the day following mating/conception as day 0. The dissected brain tissue, comprising the telencephalon, mesencephalon, and rhombencephalon, is dissociated mechanically into a single cell fraction by the sequential passage through nylon sieves of 200 μm and 100-μm pore sizes. The cells are then pelleted, washed in Puck's D1 solution by centrifugation, resuspended in a cold culture medium (modified DMEM), and aliquoted into the culture flasks (modified 25-ml Erlenmeyer flasks), which are placed on a rotating platform (68 rpm, 5-cm rotation diameter) in a CO_2 incubator (37°C, 10% CO_2/90% humidified air). Aggregation of the dissociated cells starts immediately after incubation and continues for several hours. Initially, all reaggregated cells are immature, as evidenced by low or undetectable levels of cell type-specific differentiation markers. While most of the neurons are already postmitotic, the majority of the macroglia (astrocytes and oligodendrocytes) arise from glial precursor cells and continue to proliferate during the first 2 weeks in vitro. Cellular maturation progresses during about 4 weeks, giving rise to highly differentiated histotypic cultures. Under appropriate culture conditions, they can be maintained at this high level of maturation for several months. Throughout the culture period, the media are replenished regularly. The high culture density is beneficial for the cellular stability and maturation, but it requires frequent media replenishment and occasional subdivision (split) of the cultures to assure the adequate supply of metabolic substrates.

For acute experimentation, for example, the assessment of adverse effects of chemicals and pharmaceuticals, replicate cultures are prepared, that is, aggregates are collected from several flasks, randomized, and aliquoted into culture vessels. For an individual test, up to 48 replicates are used, prepared with the randomized aggregates from 8 original culture flasks. In general, concentration-effect studies require three replicates for each drug concentration to be tested. At least three different concentrations should be tested per chemical, with a dilution factor of at least 2. Several types of controls may be included: (1) solvent control: cultures receive the same volume of solvent as added in the aliquots of the test chemicals; (2) negative control: a reference compound is dissolved in the same solvent as the test compounds and is added to the cultures at a nontoxic concentration; (3) positive control: a reference compound is dissolved in the same solvent as the test compounds and is added to the cultures at a concentration known to cause a defined adverse effect; and (4) total (100%) cytotoxicity control: sodium lauryl sulfate (2×10^{-4} M final concentration) known to cause 100% cytotoxicity.

For developmental studies, repetitive sampling from the same replicate cultures may be a suitable approach, which increases the statistical power of the data. For repetitive sampling, depending on the endpoints to be analyzed, it may be necessary to increase the size of the replicate cultures (e.g., taking twice the amount of aggregates in a total volume of 8 mL, using 50-mL culture vessels).

Given the high yield of the cultures, it is recommended to apply a high content analysis, by the assay of multiple endpoints. The combination of such a multiparametric endpoint strategy with the use of replicate cultures contributes to the high statistical power of the experimental data to be obtained.

In the routine testing approach, each data point is an average of the effects occurring in some 200 individual aggregates. To increase the number of replicates and thus the throughput, it would be possible to reduce the number of aggregates per replicate to about 50, without compromising the multiparametric approach and its high statistical power.

3.3 Materials

3.3.1 Animals

There are timed-pregnant rats, SPF, Sprague-Dawley, 16 days in gestation, counting the day following mating as embryonic day 0 (E0). Embryos of rats mated during the day are taken in the morning of E16 for culture preparation; embryos of rats mated overnight are taken in the afternoon of E16 for culture preparation. For more information about the staging and dissection of rat embryos, see [5]. For rat housing, handling, and euthanasia, the existing specific regulations and guidelines have to be followed.

3.3.2 Special equipment

- Incubator: 37°C ± 1°C; 90% ± 5% humidity; 10% ± 1% CO_2/air (e.g., Heraeus, which can be heat-sterilized, and accommodates the gyratory shaker listed next).
- Gyratory shaker with a 50-mm shaking diameter, that is, a deviation of 25 mm to each side (e.g., a Kühner gyratory shaker X-SL, driven by magnetic induction, tolerates high humidity and heat sterilization and creates only little heat in the incubator; http://www.kuhner.com/).
- Osmometer (e.g., Fiske 210 Micro-Sample Osmometer, Advanced Instruments Inc.).
- Dissecting tools (Figure 3.1).
- Dissecting block with wells for Petri-dishes and perfused with cooling liquid (optional).
- Culture vessels (Figure 3.1), for example, modified Erlenmeyer flasks, 25 ml and 50 ml. [The culture flasks are custom-made from Schott Duran Erlenmeyer flasks

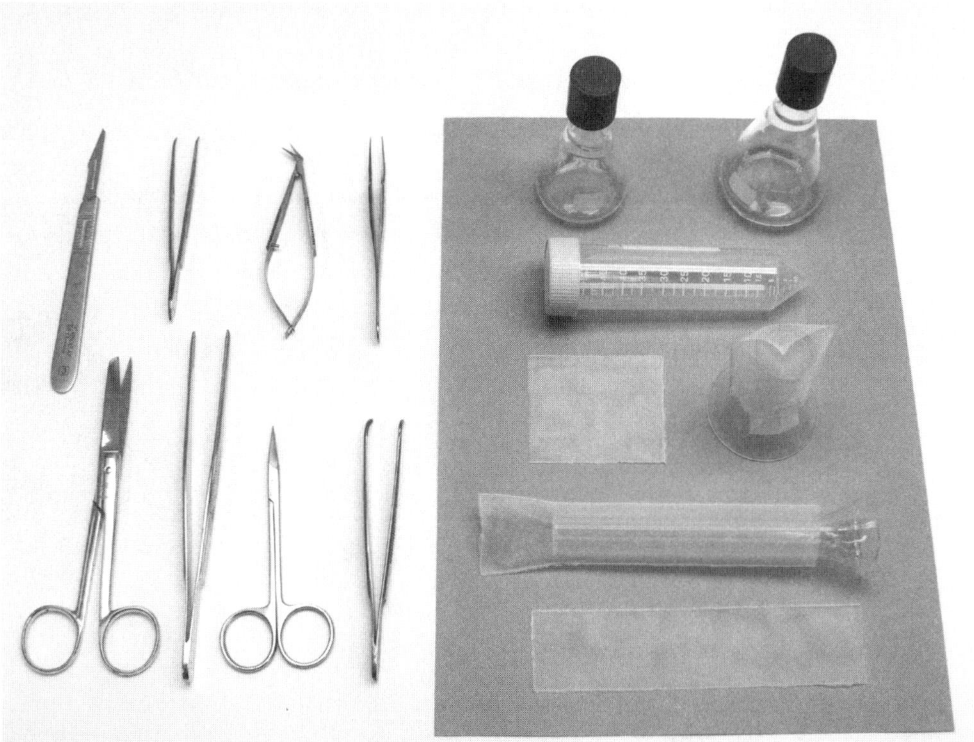

Figure 3.1 Appliance for culture preparation and maintenance. Left side: scissors (large, small pointed, and Vannas scissors); forceps (large, curved, and Dumont forceps); scalpel. Right side: nylon mesh filter bags 140 × 35 mm, 200-μm pores; 55 × 50 mm, 100-μm pores; appropriate glass funnels; 50-mL plastic tubes; culture vessels, small (25-ml modified Erlenmeyer) for culture initiation and culture replicates; large (50-mL modified Erlenmeyer) for culture maintenance.

(Verrerie de Carouge; http://www.vdc.ch/), of which the necks are cylindric to accommodate a gas-permeable plastic cap.]

- Plastic caps (autoclavable) for culture vessels (e.g., Bellco KAP-UTS cat no. 2007-18004; http://www.bellcoglass.com/; or Labocap 17/18, cat no. LD107-410, Verrerie de Carouge, http://www.vdc.ch/).
- Nylon-mesh filter bags (Figure 3.1), prepared from Nitex nylon monofilament [Sefar; http://www.sefar.com/] with either a 200-μm mesh opening [cat. no. 03-200/54] or 100-μm mesh openings [cat. no. 03-100/49]. The filter bags can be prepared from double sheets of nylon mesh by sealing and cutting them at once using a 40-W soldering iron. The 200-μm mesh sacs are 14-cm long and 3.5-cm wide; the 100-μm mesh sacs are 5.5-cm long and 5-cm large. Prior to their use, they are rinsed in ultrapure water, dried between clean absorbent paper, and autoclaved at 120°C, assembled with a filter bag.
- Appropriate glass funnels around which the filter bags are fitted (Figure 3.1): for 200-μm mesh sacs, the funnels have a stem of 12 cm and an outside diameter of 2.5 cm (top) and 1.6 cm (bottom); for 100-μm mesh bags, the funnels have a stem of 4 cm and an outside diameter of 4.5 cm (top) and 2.5 cm (bottom). Only the dimensions of the second, shorter funnel are critical. To this funnel, the bag is attached with adhesion tape (Figure 3.1) before the sterilization in the autoclave at 120°C.
- Glass rods for mechanical dissociation, 0.6 cm diameter, 20 cm long, with blunt, fire-polished ends. (Note: Nylon mesh filter bags, glass funnels, and glass rods are used for the mechanical dissociation of the brain tissue. Alternative mechanical dissociation methods are described by Cole and de Vellis [6]. However, in our hands, the method described here is the most gentle and most efficient.)
- Peristaltic pump for media filtration (optional, for the preparation of media, PBS, and Puck's solution D1).
- Pyrex glass jar (10L) (optional).
- Pyrex glass bottles 500 mL (optional) and 100 mL with air-tight closures for media.
- Filtration units for sterile filtration of small volumes (Millipore, 0.2 μm) and for large volumes (Microgon MediaKap-10 hollow fibers) (optional).
- Glass pipettes, 5 mL (optional).
- Cotton plugger for glass pipettes (optional, but convenient if recyclable glass pipettes are used for media replenishment, as cost-reducing measure).
- Serological plastic pipettes, 5 mL, 2 mL. (Note: For culture sampling and transfer, it is recommended to use plastic pipettes with a relatively wide orifice to avoid damage to the aggregate.)

3.3.3 Reagents

- DMEM powder mix for 10L (Gibco-BRL cat. no. 52100-039), containing high glucose (4.5 g/L) and glutamine, but no bicarbonate and no pyruvate
- Various trace elements
- Gentamicin sulfate (Sigma cat. no. G 1264)
- Vitamins BME 100-fold concentrated (Sigma cat. no. B 6891)
- L-Carnitine (Fluka, Buchs, Switzerland cat. no. 22018)

- Choline chloride (Sigma cat. no. C 7527)
- Hydrocortisone hemisuccinate (Sigma cat. no. H 2270)
- Insulin (Sigma cat. no. I 5500)
- Linoleic acid (Sigma cat. no. L 8134)
- Lipoic acid (Sigma cat. no. T 5625)
- Transferrin (human) (Sigma cat. no. T 2252)
- Triiodothyronine Sigma cat. no. T 2752)
- Vitamin A alcohol (Fluka cat. no. 95144)
- Vitamin B12 (Fluka, cat. no. 95190)
- Vitamin E (Sigma cat. no. T 3251)
- Phenol red (Fluka cat. no. 77660)

3.3.4 Preparation of solutions and media

Protocols are given for all solution preparations, although it may be possible to purchase custom-made solutions and media. All solutions are made with ultrapure water, which is an absolute requirement when working with serum-free culture media.

- 20× Concentrated salts for Puck's solution D1: Dissolve in ultrapure water 160 g/L NaCl, 8 g/L KCl, 1.5 g/L $Na_2HPO_4 \times 12\ H_2O$, 0.6 g/L KH_2PO_4, and 100 mg/L phenol red. Sterilize by filtration and store frozen in 100-mL aliquots.
- 20× Concentrated glucose/sucrose for Puck's solution D1: Dissolve in ultrapure water 22 g/L D-glucose monohydrate, and 400 g/L D-sucrose. Sterilize by filtration and store frozen in 100-mL aliquots. After thawing, shake vigorously before use.
- 500× Concentrated gentamicin stock solution: 12.5 mg/mL of gentamicin sulfate. Sterilize by filtration and store at 4°C.
- Puck's solution D1: To 450-mL ultrapure sterile water add 25 mL of 20× concentrated salts for Puck's solution D1 plus 25 mL of 20× concentrated glucose/sucrose for Puck's solution D1. Check osmolarity and adjust, if necessary, to 340 ± 5 mOsm. Store at 4°C. Shortly before use, adjust to pH 7.4 with sterile 0.2N NaOH (the color of solution should be "salmon" red, neither yellowish nor purplish) and add 1 mL of 500× concentrated gentamicin stock solution.
- Trace elements, 10^4-fold concentrated: Prepare the following nine aqueous solutions separately and store them at –20°C in 1-mL aliquots:
  - 2.5 mM $Na_2SiO_3 \times 5\ H_2O$
  - 0.15 mM Na_2SeO_3
  - 0.05 mM $CdSO_4 \times 8\ H_2O$
  - 0.1 mM $CuSO_4 \times 5\ H_2O$
  - 0.05 mM $MnCl_2 \times 4\ H_2O$
  - 0.005 mM $(NH_4)_6Mo_7O_{24} \times 4\ H_2O$
  - 0.0025 mM $NiSO_4 \times 6\ H_2O$
  - 0.0025 mM $SnCl_2 \times 2\ H_2O$
  - 0.5 mM $ZnSO_4 \times 7\ H_2O$

- Media supplements, 10^3-fold concentrated: Prepare the following 5 aqueous solutions separately, sterilize them by filtration, and store at –20°C in aliquots of 5 to 10 mL:
  - 0.02 mM hydrocortisone hemisuccinate
  - 0.03 mM triiodothyronine
  - 10 mM linoleic acid
  - 5 mg/mL insulin
  - 1 mg/mL transferrin
- Concentrated mixture of vitamins A and E: Dissolve 114 mg vitamin A in 0.2 mL absolute ethanol and mix this solution with 1.8 g of vitamin E. Store this mixture at –20°C protected from light.

Modified DMEM serum-free culture medium

- Fill a 10-L glass jar (Pyrex) with about 8L of ultrapure water and dissolve under gentle stirring (using a large magnetic stirring bar) DMEM powder mix for 10L (Gibco-BRL cat. no. 52100-039), containing high glucose (4.5 g/L) and glutamine, but no bicarbonate and no pyruvate.
- Add the following supplements: 1.35g choline chloride, 20-mg L-carnitine, 2-mg lipoic acid, and 13.6-mg vitamin B12.
- Add 1 mL of each of the 9 trace elements 10^4-fold concentrated.
- Add 37g $NaHCO_3$ and immediately adjust to pH 7.0–7.2 (color yellowish to salmon red) by bubbling the CO_2 gas into the medium (note that for storage, the medium is kept slightly below the optimal pH, since it increases somewhat by warming the medium up to 37°C prior to use).
- Adjust the osmolarity of the medium to 340 ± 2 mOsm by adding the required volume of ultrapure water.
- Sterilize the medium by filtration using MediaKap-10 hollow fiber filters and a peristaltic pump. Store the medium in 500-ml Pyrex bottles with gas-tight closures, at 4°C in the dark. (Note: The shelf life of this medium is only about 2 months because of high glutamine, forming ammonia by its degradation.)

Completion of the culture medium Shortly before use (maximum 1 week), complete the medium by adding the following sterile supplements per 500 mL of medium:

- 0.5 ml of each of the five 10^3-fold concentrated media supplements, that is, 0.02-mM hydrocortisone hemisuccinate, 0.03-mM triiodothyronine, 10-mM linoleic acid, 5-mg/mL insulin, and 1-mg/mL transferrin.
- 1 mL of 500× concentrated gentamin stock solution.
- 0.05 mL of concentrated mixture of vitamins A and E sonicated in 10 mL of medium and added to 500 mL of the medium by sterile filtration (Millex-GP 0.2 μm) (note that only a small fraction of the vitamins will pass the filter).
- 5 mL of vitamins BME a hundredfold concentrated.

Note: The sterility of the final culture medium is checked by the incubation of a 4-mL aliquot under normal culture conditions for at least 1 day.

3.4 Methods

3.4.1 Washing and sterilizing the glassware

Glass culture vessels and glass pipettes can be washed and sterilized for recycling. The detergents to be used for the washing of culture glassware should be devoid of organic additives. Used culture glassware should never be allowed to dry before it is cleaned, but should be stored immersed in a special detergent solution (e.g., Neodisher LM-10, 5% v/v). After washing with special detergent (e.g., Neodisher FT), the glassware is rinsed with water, neutralized with a solution containing citric acid or phosphoric acid, further rinsed with water, and finally rinsed with ultrapure water. After drying, the glassware is sterilized in the autoclave (except for cotton-plugged 5-mL glass pipettes used for media replenishment, which are sterilized in dry air for 2.5 hours at 220°C).

3.4.2 Cell isolation and culture preparation

Aggregating brain cell cultures are prepared from a 16-day embryonic (E16) rat brain comprising telencephalon, mesencephalon, and rhombencephalon. The duration of the dissection should not exceed 2 hours. All dissecting instruments and solutions should be sterile. From step 5 onwards, all work is done under strictly aseptic conditions. During steps 5 to 21, the tissue/cells remain at 0–4°C.

1. Sacrifice the pregnant rat by decapitation using a special guillotine for rats. Avoid all unnecessary stress to the animal and follow the ethical guidelines. Clean the place of decapitation for each animal to be sacrificed. (Note: When properly done, the decapitation of the conscient animal proved to create less stress than using anesthetics. Periodically, the brain of a sacrificed animal is taken for the preparation of standards for the diverse assay procedures.)
2. Lay the decapitated animal on its back on an adsorbent paper and rinse the abdomen with 95% alcohol.
3. Open the peritoneal cavity. Using the large forceps, grasp the skin about 1 cm above the genitalia. With the large scissors, cut through the skin and fascia, and extend the cut on both sides of the abdomen, until the entire peritoneal cavity lies open.
4. Remove the uterine horns containing the fetuses, using small curved forceps and small pointed scissors, and transfer them into 50-mL plastic tube containing 25–30 mL of ice-cold Puck's solution D1.
5. Transfer the uterine horns to one of the Petri dishes placed on the cold dissecting block (or on ice, if no dissection block is available) within a horizontal laminar flow bench.
6. Using the Vannas scissors and small forceps, separate the embryos from the uterus and from their amniotic sac and placenta and transfer them to a fresh Petri dish containing Puck's solution D1.
7. Dissect the brain of each embryo. Fix the head of the embryo in a lateral position with the aid of the Dumont forceps, and make an extended lateral incision at the base of the brain using the Vannas scissors. Lift the brain through this opening using the blunt side of the closed Vannas scissors.
8. Using a scalpel, divide the brain roughly into two halves to reduce the size of the brain pieces.

9. Transfer the dissected brain parts to a 50-mL plastic tube filled with 40 mL of Puck's solution D1.
10. Repeat steps 1 to 6 for each pregnant rat and steps 7 to 9 for each embryo. The brain tissue collected from 50 embryos is the maximum amount to be kept in one tube. If necessary, use several tubes filled with 40-mL Puck's solution D1 to collect the dissected tissue, and replace the liquid periodically to avoid a drop of the pH.
11. At the end of dissection, rinse each batch of tissue three times with about 40 mL of cold Puck's solution D1. Take these batches separately for the subsequent mechanical dissociation, using a fresh 200-μm mesh bag each time.
12. For the dissociation procedure, place the 200-μm mesh bag (with a glass funnel fitted inside) in a 50-mL conical plastic tube containing 25-mL cold Puck's solution D1. Pour the tissue of one batch into the bag. Remove the funnel and hold the upper ends of the closed bag against the outer side of the tube wall.
13. Using the glass rod, gently stroke downwards, from the immersed outside of the bag, to squeeze the tissue through the mesh into the surrounding solution. At the end of this first passage, remove the bag and close the tube.
14. Repeat steps 12 and 13 for every batch of dissected brain tissue.
15. In filtration of the cell suspension take the 100-μm mesh nylon bag attached with tape to the short glass funnel and place it on top of an empty 50-mL conical plastic tube. Using a 5-mL serological plastic pipette, transfer the cell/tissue suspension to the nylon bag and let the suspension pass through the filter by gravity flow into the 50-mL tube. Towards the end of the filtration, add Puck's solution D1 up to a final volume of 50 mL.
16. Repeat steps 15 for each batch of dissociated tissue.
17. Centrifuge the filtrate(s) containing the dissociated cells (300g, 15 minutes at 4°C, with slow acceleration and deceleration).
18. Of each tube, remove the supernatant, and resuspend the pellet with cold Puck's solution D1 (start with 2.5 ml) by gentle trituration (5 to 6 strokes up and down using a 5-mL plastic pipette without foaming). Bring the volume to 50 mL with cold Puck's solution D1.
19. Centrifuge the cell suspension(s) (300g, 15 minutes at 4°C).
20. Of each tube, remove the supernatant and resuspend the cells in a cold serum-free culture medium by gentle trituration. Transfer the resulting cell suspension to a plastic culture flask for further dilution.
21. Dilute the cell suspension with cold medium to the final volume (i.e., routinely, the number of brains used per batch of cell suspension times 4 ml of medium) to obtain an average of one flask of cultures per embryo with an initial cell number of 2.5 to 3.5 $\times 10^7$ cells per flask. [Note: For routine work, cell counting is not required. The method is sufficiently robust to permit considerable variations in cell number without affecting the fundamental properties of the cultures. Nevertheless, it might be useful (particularly for beginners) to determine cell number and viability. To this end, an aliquot (0.1 ml of 50 ml of the cell suspension) is taken after the completion of step 18 in the isolation procedure, diluted at least 10 times, and counted using either a hemocytometer or an electronic cell counter. Cell viability, determined by the exclusion of trypan blue, should be about 50%.]

22. Transfer 4-mL aliquots of the final cell suspension to 25-mL culture flasks, and place them onto the platform of a rotating (68 rpm) gyratory shaker in the CO_2 incubator (37°C, 10% CO_2/90% humidified air).
23. On the evening of the day of culture preparation (DIV 0), accelerate the gyratory agitation to 70 rpm, and the next day accelerate it to 74 rpm.
24. At DIV 2, transfer the cultures to 50-mL culture flasks. Pour the content of each flask along the glass wall of a fresh flask, and again in the same way with an additional 4 mL of fresh prewarmed medium, so that no aggregates remain attached on the wall of the fresh culture flask. Continue with the transfer of the other cultures, but do not exceed 30 minutes until the cultures are allowed to equilibrate (the pH increases dramatically during this procedure). At the end of the transfer, accelerate the gyratory agitation to 77 rpm. On the following days, further increase the agitation speed by 1 rpm/day until the maximal frequency of 80 rpm is reached at DIV 5.

3.4.3 Maintenance of aggregating brain cell cultures (media replenishment and subdivision)

The maintenance of aggregating brain cell cultures begins at DIV 5 and consists of regular replenishment of the culture media and the subdivision (split) of cultures at DIV 20. All this work is done in a vertical laminar flow clean bench.

Media are replenished every third day from DIV 5 to DIV 14, and every other day in the following. Media replenishment, that is, the replacement of 5 mL of media supernatant (of a total of 8 mL) with fresh prewarmed serum-free culture medium, is done according to the following protocol:

1. Fill the amount of fresh medium required for replenishment into a 150-mL sterile glass bottle and warm it in a water bath at 37°C for 10 minutes.
2. Put five flasks of cultures on a slanted support, and transfer them to the laminar flow bench. Using a 5-mL pipette, remove 5 mL of media supernatant and replace it by a slightly larger volume (5.2 mL, to compensate evaporation) of fresh medium. (Note: To avoid contamination, use always a fresh pipette with fresh medium. Furthermore, the rim of the caps should be shortly immersed in 100% ethanol, taking care that no ethanol enters when closing the flask.)
3. Return the flasks to the incubator and continue with another group of five cultures. (Note: The duration of media replenishment should not exceed 30 minutes, and the cultures should be allowed to equilibrate for at least 1.5 hours before a further series of the same incubator is replenished.)

The subdivision (split) of cultures at DIV 20 is necessary because of the greatly increased metabolic activity of the now more mature cells. The cultures are split into two as follows:

1. Add 4 mL of cold culture medium to the fresh series of flasks, and pre-equilibrate in the incubator under gyratory agitation for at least 1 hour.
2. Transfer half of one culture (4 ml) to one of the pre-equilibrated flask. Using a 5-mL plastic pipette, quickly resuspend the aggregates by short aspiration and brisk expulsion of the supernatant medium, then quickly take a 4-mL aliquot and transfer it to the fresh flask.

3. Replenish the original flask by adding 4 mL of fresh prewarmed medium.
4. Put the flasks back to the incubator and continue with the next series.

Note: The duration of subdivision should not exceed 30 minutes. Thereafter, the cultures should be allowed to equilibrate for at least 1.5 hours before a further series is put into the same incubator.

3.4.4 Preparation and treatment of replicate cultures

Due to their free-floating nature, aggregate cultures can be easily pooled from individual cultures and divided into aliquots for the preparation of replicate cultures. The original culture flask contains some 1,200 aggregates. In general, six replicate cultures are prepared for each original flask, each replicate receiving about 200 aggregates. For a single experiment, a maximum of 48 replicate cultures (from eight original flasks) are taken, according to the following protocol:

1. At the day of experimentation (e.g., DIV 18), put the required number of 25-ml culture flasks into the laminar flow hood and label them with a series of sequential numbers.
2. Add to each flask 3 ml of cold culture medium, and pre-equilibrate in the incubator under gyratory agitation for at least 1 hour.
3. Pool in a 50-ml plastic tube the aggregates from as many original cultures as necessary for the number of replicates to be prepared (e.g., six replicates are prepared per original culture). Adjust the final volume of the pool according to the number of replicates to be prepared (requiring 1 ml of suspension per replicate).
4. Take a first series of six pre-equilibrated 25-ml flasks from the incubator. Using a 2-ml plastic pipette, quickly resuspend the aggregates by short aspiration and brisk expulsion of the supernatant medium, then quickly take a 1-ml aliquot and transfer it to the first flask in the row. Continue this way with the next flasks, until all six flasks have received an aliquot of aggregates. Make sure that the sequence of numbering corresponds with that of aliquoting.
5. Put the flasks back to the incubator and continue the same way with a further series. Continue until all aliquots are distributed.
6. Equilibrate the replicate cultures for at least 90 minutes before exposing them to test chemicals.
7. For the treatments, take the equilibrated replicate cultures (groups of six at a time) from the incubator and transfer them to the laminar hood.
8. To each flask, add an aliquot of stock solution under slight agitation (usually 2 μl of stock). For solvent control, cultures receive an equal volume of solvent.
9. Return the flasks immediately to the shaker in the incubator.
10. Continue the same way with the other replicates.
11. Incubate the treated series under standard culture conditions (i.e., at 37°C in 10% CO_2/90% humidified air, and under 80 rpm gyratory agitation). If the test duration exceeds 3 days, media have to be replenished regularly.

3.4.5 Harvest of replicate cultures for various analytical procedures

Note: For analysis by qRT-PCR, always use sterilized materials and solutions for the harvest and wear disposable gloves to prevent RNase contamination. For biochemical analyses and qRT-PCR, the aggregates of individual flasks/replicates are washed in ice-cold PBS and stored frozen according to the following protocol:

1. Take one replicate at the time, transfer (pour) the aggregates into a 15-ml plastic tube placed in ice.
2. Let the aggregates settle by gravity sedimentation and remove the supernatant by suction.
3. Resuspend the aggregates with 5 ml of ice-cold PBS, let the aggregates settle, and remove again the supernatant.
4. Repeat step 3 once more (second wash). (A third wash is necessary for cultures subjected to metabolic radiolabeling.)
5. Store the pelleted aggregates at –80°C.

For immunocytochemistry, a pool of three to four replicates (at least 600 aggregates) are washed in prewarmed PBS (30–37°C) as described above (steps 1 to 4, but in prewarmed PBS). Thereafter, using a 1-mL plastic pipette, the aggregates are transferred to a gelatin capsule (Feton International, Brussels, no. 00). The supernatant is quickly removed with a Pasteur pipet, and the aggregates are covered with cryoform (Cryomatrix, Shandon Scientific, Runcorn, Cheshire, United Kingdom). The capsule is immediately immersed in isopentane cooled in liquid nitrogen. The frozen capsules are stored at –80°C.

3.4.6 Examples of sample preparation and use for various analytical procedures

For biochemical assays, the aggregates of each replicate are sonicated in 150–400 μl of 2-mM potassium phosphate buffer pH 6.9 (3 pulses of 3–4 seconds duration at 30W using a microtip). The homogenate is then divided into aliquots for the different assays and stored at –80°C for analysis. Several useful methods for the determination of biochemical parameters have been described, for example, for the measurement of enzymatic activities, protein, and DNA content [7], thymidine incorporation [8], and 2-deoxyglucose uptake [9].

For qRT-PCR, the total RNA is extracted of each replicate culture, using a kit from Qiagen. The total RNA content of each extract is determined. If the overall rate of RNA synthesis is determined, an aliquot of the extract is taken for liquid scintillation counting, to measure uridine incorporation. Reverse transcription (RT) is performed using the reverse transcription kit and protocols from Applied Biosystems (ABI). The RT is run with 200 ng of total RNA in a reaction volume of 50 μl. The cDNA obtained is then used for subsequent PCR reactions. The cDNA (3.2 ng) is added to the PCR mixture composed of primers (see Table 3.1), 1 × SYBR Green PCR master mix (ABI), and H_2O, in a final volume of 10 μl.

For immunocytochemistry, cryostat sections are prepared from unfixed aggregates. The frozen capsules are transferred from –80°C to –20°C the day before cutting on the cryostat. The gelatin capsule is removed with the use of a razor blade. The cryosections

Table 3.1 Sequences and Concentrations of PCR Primers

Gene	Sense Primer (5′-3′)	Antisense Primer (5′-3′)	nM
NF-H	CAGGACCTGCTCAACGTCAA	CTTCGCCTTCCAGGAGTTTTCT	300
TH	TGCTGTTCTCAACCTGCTCTTCT	CTAAGTGGTGAATTTTGGCTTCAA	300
GFAP	CCTTGACCTGCGACCTTGAG	GCGCATTTGCCTCTCCAA	150
MBP	GCACGCTTTCCAAAATCTTTAAG	AGGGAGGCTCTCAGCGTCTT	200
HO-1	AGGTGTCCAGGGAAGGCTTT	TCCAGGGCCGTATAGATATGGT	300

(10 μm) are fixed for 10 minutes in 4% paraformaldehyde/PBS at RT, washed in PBS, and processed the same day or the next day (in the latter case, sections are kept overnight at 4°C). To block nonspecific binding, the sections are incubated with horse serum (1:25 in PBS with 0.1% Triton-X-100) for monoclonal antibodies, and with normal donkey serum (1:25 in PBS with 0.1 % Triton-X-100) for polyclonal antibodies, prior to the exposure, overnight at 4°C, to the primary antibodies. Primary antibodies are diluted in PBS [antinestin, monoclonal, Millipore, 1:100; antimicrotubule associated protein-2 (MAP-2), polyclonal, Chemicon, 1:200; antiglial fibrillary acidic protein (GFAP), monoclonal, Sigma, 1:800; antimyelin basic protein (MBP), monoclonal, Chemicon, 1:40]. The sections are then exposed to the secondary antibodies (biotinylated horse antimouse IgG (Vector, 1:400), followed by FITC-avidine (Vector, 1:100) for monoclonal antibodies; and Cy3 donkey antirabbit IgG (Jackson Laboratories, 1:400) for polyclonal antibodies, respectively. The sections are mounted in ProLong Gold antifade reagent (Invitrogen).

More methodological and practical details for endpoint analyses in aggregating brain cell can be found in [7–11].

3.4.7 Data Analysis

For biochemical measurements (e.g., mitotic activity, enzymatic activity, glucose consumption), the data are calculated relative to the protein content (or, alternatively, relative to the DNA content), and normalized by expressing them as percentage of untreated control cultures. Effect concentration (EC) values are determined, with EC_{20} and EC_{50} reflecting the concentrations causing an increase or decrease by 20% and 50%, respectively, compared to the control values. The collected data are subjected to conventional statistical analysis.

For qRT-PCR measurements, the data are calculated with the μCt method. Results are expressed as fold change relative to control cultures receiving only the solvent of the stock solutions (equal to 1). The lowest observed effective concentration (LOEC) is determined, defined as the lowest drug concentration causing a change in gene expression compared to the controls by at least a factor of 1.7 (i.e., a decrease to = 0.59-fold the control value or an increase to = 1.7-fold the control value).

For immunocytochemistry, the stained surface and staining intensity are quantified by Image J, evaluating 20 different exposures taken from 4 different sections, and cut at different levels. The entire procedure is performed twice, for two different culture samples. Values are expressed as percentage of untreated cultures and subjected to statistical analysis.

For all analytical procedures, criteria for the acceptance of experimental results should be defined, for example, for biochemical and qRT-PCR data, a given experiment would be acceptable if: (1) the values of duplicates do not deviate from their mean value by more than 15%; (2) the control values are in the (normal) range of previously established values; and (3) effective compounds show a concentration-effect relationship. Furthermore, for gene expression, actin analysis is performed first. If acceptable according to criterion 1, the other genes are analyzed in the following. If the actin data are not acceptable, reverse transcription and qPCR analysis are repeated.

3.5 Anticipated Results

Typical maturation-dependent changes of biochemical parameters and mRNA expression data in aggregate cultures derived from rat embryonic whole brain are summarized in Table 3.1. The spontaneous mitotic activity is high in very early cultures, and decreases rapidly with cellular maturation. However, a small population of mitotically active cells persists even in highly differentiated cultures. The mitotic activity in immature cultures is greatly stimulated by bFGF, EGF, and $PDGF_{AA}$. These growth factors are mitogenic also in cultures at more advanced developmental stages, and the effects of the growth factors are additive (not shown), suggesting the existence of distinct responsive subpopulations, presumably progenitors of astrocytes and oligodendrocytes, respectively. The increase in cell type-specific enzymatic activities reflects the maturation of GABAergic neurons (GAD), cholinergic neurons (ChAT), astrocytes (GS), and oligodendrocytes (CNP). Maximal values for the enzymatic activities were found at DIV 21 for CNP, and at DIV 28 for GAD, ChAT, and GS (not shown). The mRNA expression data for genes encoding cell type-specific structural proteins are in line with the progressive maturation of neurons (MAP2, NF-H), astrocytes (GFAP) and oligodendrocytes (MBP) in these cultures. The drop of MBP mRNA in highly differentiated cultures is found also in vivo and may reflect its localization close to myelinated axons and the relatively slow turnover of myelin. The immunocytochemical analysis of these cultures (Figure 3.2) shows high immunoreactivity for MAP2 and a maturation-related increase in GFAP and MBP staining, in good agreement with the developmental changes observed for the biochemical parameters and the gene expression data. Furthermore, it shows a reproducible maturation-dependent cellular organization in each aggregate. In particular, nestin-positive (i.e., immature) cells were found dispersed throughout the aggregates of early cultures, whereas they occurred in two distinct locations, that is, at the periphery and in the center of the aggregates at a more mature stage. In accordance with the measurements of the mitotic activity, these findings indicate that immature precursor cells and perhaps radial glia persist in these highly differentiated cultures, while mature neurons, extensive neuronal networks, and myelinated axons are organized in a broad layer between the center and the periphery of the aggregates [12]. The presence of mature synapses and compact myelin at an advanced developmental stage was shown previously [13, 14]. The average diameter of the mature aggregates was around 600 μm.

Typical results from neurotoxicity tests with aggregate cultures at an advanced stage of maturation are shown in the Troubleshooting Table. Parathion, as it can be expected from known in vivo data, was highly toxic, decreasing the activity of the neuron-specific enzymes GAD and ChAT at micromolar concentrations. Similar to in vivo, this organophosphorous insecticide greatly decreased the activity of AChE. Since AChE is the direct target of paraoxon, the toxic metabolite of parathion, these results show that

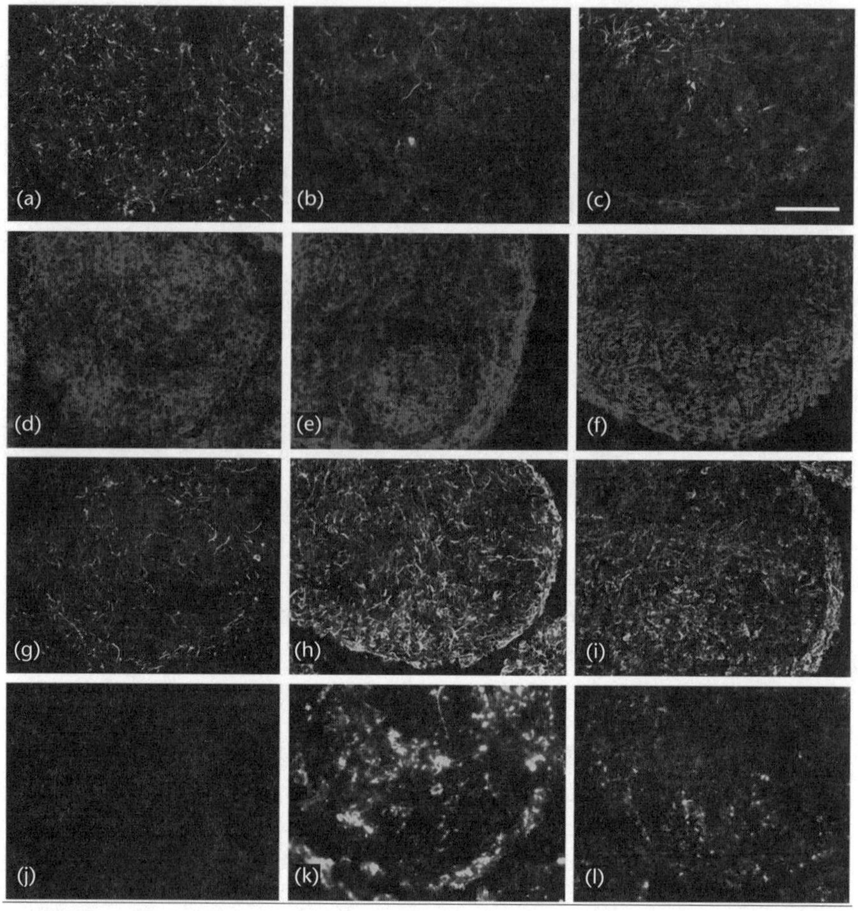

Figure 3.2 Cell type-specific immunostaining of aggregating brain cell cultures at different stages of maturation. Representative examples of the developmental expression of nestin (A,B,C), MAP-2 (D,E,F), GFAP (G,H,I) and MBP (J,K,L), as markers of precursor cells, neurons, astrocytes, and oligodendrocytes, respectively. Cultures were harvested at DIV 7 (A,D,G,J), DIV 21 (B,E,H,K), and DIV 35 (C,F,I,L) and processed for immunocytochemistry. Bar = 100 μm.

the metabolic conversion of the parent compound occurs in these cultures [15]. Of the genes tested, the NF-H expression was most affected, in agreement with the known neurotoxicity of parathion. Glufosinate, a herbicide, greatly affected GS activity, in agreement with its known role as an irreversible inhibitor of GS. However, it strongly decreased also GAD activity and the expression of NF-H, indicating that this glia-specific toxicant also affected neurons at a relatively low concentration. The fact that in vivo glufosinate exhibits relatively low neurotoxicity can be explained by the low permeability of the blood-brain barrier for this compound. Lindane, an organochlorine insecticide, is known as a central nervous system toxicant causing hyperexcitability and oxidative stress. In aggregate cultures, glucose consumption was increased at a relatively low concentration of lindane, in accordance with an increase in neuronal activity and earlier findings that a significant proportion of glucose consumption in differentiated aggregate cultures is linked to neuronal activity [9]. The neuron-specific enzymes and the mRNA expression of the four selected genes were affected at comparable concentration levels.

3.6 Discussion and Commentary

The preparation of aggregating brain cell cultures is relatively easy. However, their maintenance is time-consuming, because of the high culture density requiring frequent media replenishments. The costs involved in such repetitive work could be reduced by automation, for example, by the use of a flow-through system by which a continuous flow of media through the culture vessel maintains the substrate levels constant, without changing the media volume and vortex. However, so far, such a system is not commercially available for aggregate cultures. Another obstacle is the need for special culture vessels. It is proposed here to use modified Erlenmeyer flasks as culture vessels, prepared by a glass-working company. Similar culture vessels, called DeLong flasks, can be obtained from Bellco. However, we have found batch-to-batch differences in geometry that were incompatible with the preparation of reproducible cultures. For acute experimentation, the geometry of the culture vessel is less critical. Also, simpler disposable culture vessels could be used. Furthermore, the size of the replicate cultures could be reduced to about 50 aggregates per replicate, the presumed limit of reproducibility. In this case, 6-well plates (with wells of a diameter of 1 cm) could be suitable, requiring a volume of 0.8 mL/well to assure a sufficient vortex.

The experimental design and the choice of representative endpoints will depend on the goal of the study, whether it addresses developmental neurotoxicity, pathways of neurotoxic adverse effects, or routine neurotoxicity testing. For routine testing, the use of a multiparametric assay strategy proved to be advantageous. Changes in gene expression at the mRNA level for a series of representative genes were determined together with changes in glucose consumption (measuring the glucose concentrations in media supernatants) and the rate of overall RNA synthesis (incorporation of radiolabeled uridine). The upregulation of heme oxidase or heat shock protein-32 (HO-1/HSP32), an indicator of oxidative stress, was found to be a sensitive marker of adverse effects for a great variety of neurotoxicants. Also, changes in the mRNA expression of genes coding for cell type-specific proteins such as NF-H, GFAP, and MBP, were found to be useful. While a decrease in GFAP expression often indicates astrocyte loss, GFAP upregulation is a convenient marker for astrogliosis. These few examples may illustrate the suitability of this approach; however, it will be important to find additional genomic endpoints representing predictive biomarkers for drug targets and/or common pathways of adverse effects. From a practical point of view, gene expression data are relatively unaffected by the variability of aggregate sampling, since equal amounts of RNA are taken for the measurements. In addition, qRT-PCR measurements allow to increase the throughput and efficacity of testing (e.g., by the use of 384-well plate for analysis), combined with a Laboratory Automation Workstation Biomek 3000 (Beckman Coulter) for automatic pipetting. Multiple endpoints increase the likelihood to detect adverse effects at concentrations relevant to the in vivo situation. Concentration-dependent effects observed on several endpoints increase the significance of the data and reduce the number of false positive findings.

3.7 Application Notes

Aggregating cell cultures derived from embryonic rat brain are used in our laboratory for more than 30 years. They proved to be a unique system for neurotoxicological and

Troubleshooting Table

Problem	Possible Cause	Potential Solution
Irregular aggregate size	Culture vessels not uniform or vortex too weak. Irregular shaking. Initial cell density too low.	Replace culture vessels. Reduce initial volume. Repair or replace orbital shaker. Reduce stress on cells during preparation. Increase initial density of viable cells.
Too large aggregates	Initial volume too big for diameter of vessels.	Reduce initial volume.
Too small aggregates	Initial volume too small for diameter of vessels. Initial cell density too high.	Increase initial volume. Reduce initial cell density.
A few big clumps beside regular aggregates	Presence of fibers (DNA) in initial cell suspension.	Replace plastic pipette several times while aliquoting the cell suspension.
Low control values, small aggregates	Incomplete medium. Too low osmolality. Toxic contaminants.	Use complete medium. Check osmolarity. Be aware that commercial media are often at = 300 mOsm. Always use the highest grade for water and chemicals. Apply highest standards for glassware cleaning (see protocol).
Variable amounts of aggregates in original cultures	Loss during manipulation.	Be aware of possible loss during replenishment (aspiration) and during transfer. Aggregates, particularly immature ones, stick readily to glass and plastic.
Variable amounts of aggregates in replicates	Uneven aliquots taken for replicate preparation. Loss during pipetting	Be aware that the aggregates sediment rapidly, and that they have to be resuspended quickly, each time before taking an aliquot. They may also stick to the plastic pipette used for the transfer.

developmental studies. The use of serum-free culture medium enabled the study of the role of growth factors and hormones in the developing brain. The cultures were also found to be useful for the investigation of developmental neurotoxicants [15, 16], the detection and evaluation of acute neurotoxicity [1], demyelination [14], hypoglycemic and ischemic adverse effects [7], and neuroinflammatory reactions [17]. Three-dimensional cell cultures most closely reproduce the morphogenic events occurring during brain development that depend on direct cell-to-cell signalling and on endogenous soluble signaling factors. Therefore, the interference of chemicals with such critical developmental events can be detected in this system. Furthermore, the histotypic architecture and high degree of maturation afford a maximum of potential targets (structural as well as functional) for neurotoxic compounds. The high yield, excellent reproducibility, and robustness of these cultures qualify them also for routine toxicity screening and for the application of large-scale profiling technologies such as metabolomics [18].

3.8 Summary Points

1. A detailed protocol describes the preparation, maintenance, and use of serum-free aggregating brain cell cultures derived from embyonic rat brain.

2. The characteristics of these three-dimensional cell cultures are described with respect to their developmental features, and examples are included for their use in neurotoxicity test procedures.
3. Examples are given for the choice of suitable endpoints in neurotoxicity tests.
4. Possible ways are suggested to increase predictivity, efficiency, and throughput for routine neurotoxicity screening.

Acknowledgments

The authors are grateful to Denise Tavel and Brigitte Delacuisine for excellent technical assistance and to M. E. Bernardi for the artwork. This work was supported by the EU-FP6 grant (FP6-LIFESCIHEALTH-2004-512051) and by the 3R Research Foundation Switzerland.

References

[1] Forsby, A., et al., "Neuronal In Vitro Models for the Estimation of Acute Systemic Toxicity," *Toxicology In Vitro*, Vol. 23, No. 8, 2009, pp. 1564–1569.

[2] Moscona, A. A., "Rotation-Mediated Histogenetic Aggregation of Dissociated Cells: A Quantifiable Approach to Cell Interactions In Vitro," *Exp. Cell Res.*, Vol. 22, 1961, pp. 455–475.

[3] Honegger, P., and E. Richelson, "Biochemical Differentiation of Mechanically Dissociated Mammalian Brain in Aggregating Cell Culture," *Brain Res.*, Vol. 109, 1976, pp. 335–354.

[4] Honegger, P., D. Lenoir, and P. Favrod, "Growth and Differentiation of Aggregating Fetal Brain Cells in a Serum-Free Defined Medium," *Nature*, Vol. 282, 1979, pp. 305–308.

[5] Dunnett, S. B., and A. Björklund, "Staging and Dissection of Rat Embryos," in S. B. Dunnett and A. Björklund, (eds.), *Neural Transplantation: A Practical Approach*, Oxford, U.K.: Oxford University Press, 1992, pp. 1–19.

[6] Cole, R., and J. de Vellis, "Preparation of Astrocytes, Oligodendrocytes, and Microglia Cultures from Primary Rat Cerebral Cultures," in S. Fedoroff and A. Richardson, (eds.), *Protocols for Neural Cell Culture*, 3rd ed., Totowa, NJ: Humana Press, 2001, pp. 117–127.

[7] Pardo, B., and P. Honegger, "Selective Neurodegeneration Induced in Rotation-Mediated Aggregate Cell Cultures by a Transient Switch to Stationary Culture Conditions: A Potential Model to Study Ischemia-Related Pathogenic Mechanisms," *Brain Res.*, Vol. 818, 1999, pp. 84–95.

[8] Honegger, P., and M. Tenot-Sparti, "Developmental Effect of Basic Fibroblast Growth Factor and Platelet-Derived Growth Factor on Glial Cells in a Three-Dimensional Cell Culture System," *J. Neuroimmunol.*, Vol. 40, 1992, pp. 295–304.

[9] Honegger, P., and B. Pardo, "Separate Neuronal and Glial Na+,K+-Atpase Isoforms Regulate Glucose Utilization in Response to Membrane Depolarization and Elevated Potassium," *J. Cerebral Bloods Flow and Metabol.*, Vol. 19, 1999, pp. 1051–1059.

[10] Honegger, P., and F. Monnet-Tschudi, "Aggregating Neural Cell Cultures," in S. Fedoroff and A. Richardson, (eds.), *Protocols for Neural Cell Culture*, 3rd ed., Totowa, NJ: Humana Press, 2001, pp. 199–218.

[11] Zurich, M.-G., et al., "Unusual Astrocyte Reactivity Caused by the Food Mycotoxin Ochratoxin A in Aggregating Rat Brain Cell Cultures," *Neuroscience,* Vol. 134, 2005, pp. 771–782.

[12] Braissant, O., et al., "Ammonium-Induced Impairment of Axonal Growth Is Prevented Through Glial Creatine," *J. Neurosci.*, Vol. 22, 2002, pp. 9810–9820.

[13] Monnet-Tschudi, F., et al., "Microglial Responsiveness as a Sensitive Marker for Trimethyltin (TMT) Neurotoxicity," *Brain Res.*, Vol. 690, 1995, pp. 8–14.

[14] Defaux A., et al., "Effects of the PPAR-Beta Agonist GW501516 in an In Vitro Model of Brain Inflammation and Antibody-Induced Demyelination," *J. Neuroinflammation,* Vol. 6, 2009, pp. 15–27.

[15] Monnet-Tschudi, F., et al., "Maturation-Dependent Effects of Chlorpyrifos and Parathion and Their Oxygen Analogs on Acetylcholinesterase and Neuronal and Glial Markers in Aggregating Brain Cell Cultures," *Toxicol. Appl. Pharmacol.*, Vol. 165, 2000, pp. 175–183.

[16] Monnet-Tschudi, F., M.-G. Zurich, and P. Honegger, "Comparison of the Developmental Effects of Two Mercury Compounds on Glial Cells and Neurons in Aggregate Cultures of Rat Telencephalon," *Brain Res.*, Vol. 741, 1996, pp. 52–59.

[17] Monnet-Tschudi, F., M.-G. Zurich, and P. Honegger, "Neurotoxicant-Induced Inflammatory Response in Three-Dimensional Brain Cell Cultures," *Human Exp. Toxicol.*, Vol. 26, 2007, pp. 339–346.

[18] van Vliet, E., et al., "A Novel In Vitro Metabolomics Approach for Neurotoxicity Testing, Proof of Principle for Methyl Mercury Chloride and Caffeine," *Neurotoxicology*, Vol. 29, 2008, pp. 1–12.

CHAPTER

4

Approaches Towards a Multiscale Model of Systemic Inflammation in Humans

P. T. Foteinou,[1] J. D. Scheff,[1] S. E. Calvano,[2] S. F. Lowry,[2] and I. P. Androulakis,[1,3]

[1]Biomedical Engineering, Rutgers University, Piscataway, 599 Taylor Road Piscataway, NJ 08854
[2]Department of Surgery, UMDNJ-Robert Wood Johnson Medical School, New Brunswick, NJ 08901
[3]Chemical and Biochemical Engineering Department, Rutgers University, NJ 08854
Corresponding author: I. P. Androulakis, e-mail: yannis@rci.rutgers.edu, phone: 732-445-4500 ext. 6212, fax: 732-445-3753

Abstract

The intricacies in translating basic research to clinical practice are recognized as challenges that need to be overcome in order to successfully transfer the information from the preclinical stage to the clinical stage. Integrated initiatives are identified as valuable in uncovering the mechanisms underpinning the progression of human diseases. Significant opportunities emerge in the context of systems biology, which aims at the deconvolution of complex phenomena to their constituent elements and the quantification of the dynamic interactions between these components through the development of appropriate computational and mathematical models. On the other hand, a central problem in computational systems biology is to identify those critical components and the rules that define their interactions and give rise to the emergent behavior of a host response. In order to address these pressing issues, we have undertaken an integrated approach related to the identification of networks of functional modules describing disease progression across multiple scales. In this chapter, we will discuss the potential role that a systems-based approach can have in the quest to better understand and create a module for the inflammatory response. We focus on inflammation as a key physiological response of clinical and translational importance.

Key terms heart rate variability
human
infection
inflammation
mathematical modeling

4.1 Introduction

More than 40 million major surgical operations are performed annually in the United States, of which as many as 2 million are complicated by surgical site infections [1]. Surgical adverse effects contribute significantly to postoperative morbidity by substantially transforming the immune system into a severely suppressed state that promotes sepsis [2]. Surgical patients account for approximately 30% of all sepsis patients [3] and the present incidence of acquired surgical site infections is likely to continue to increase among nontrauma surgical patients [4]. Sepsis is a syndrome resulting from massive, acute activation of the systemic inflammatory response (SIRS) in the setting of severe infection and remains one of the leading causes of death in United States [3, 5]. The manifestation of SIRS criteria is the common clinical phenotype of stressed surgical patients and reflects the presence of consequential systemic inflammation [4]. However, systemic inflammation is not inherently detrimental. Inflammatory processes are required for immune surveillance and regeneration after injury, during which multiple cell types are deployed to locate pathogens, recruit cells, and eventually eliminate the offenders and restore homeostasis [6]. It is the dysregulation of the resolution of inflammation that has undesirable implications.

Despite our growing understanding of the cellular and molecular mechanisms of SIRS and the success of preclinical studies, most generated hypotheses related to the management and treatment of severe human inflammation have failed clinical testing [7]. Even the improved capacity to acquire quantitative data in a clinical setting has generally failed to improve outcomes in acutely ill patients. A key reason for this conundrum is the difficulty in predicting potential unintended consequences of therapeutic interventions against human diseases. Such failures were often attributed to invoking the single variable assumption in a clinical scenario. It has been argued that prediction of the behavior of complex diseases derived from local insights may be impossible [8]. Thus, the intricacies in translating basic research to clinical practice are recognized as a challenge impeding the successful transfer of information from the preclinical stage to the clinical stage [9, 10]. Significant opportunities emerge in the context of systems biology, which aims at the deconvolution of complex phenomena into their constituent elements and the quantification of the dynamic interactions between these components through the development of appropriate computational and mathematical models [11, 12].

A systems-oriented mathematical modeling approach, as a means of dynamic knowledge representation, offers the promising possibility of improving the interpretation of quantitative, patient-specific information and help better target therapy [13–16]. The potential for studying complex phenomena in a model-based manner opens the possibility of generating and exploring simultaneously multiple hypotheses for deciphering complex modes of action and the possibility of proposing combination therapies. However, such models are typically complex and nonlinear, which precludes the identification of unique parameters and states of the model that best represent available data [17]. It has been argued, however, that the awkwardly posed inverse problem in quantitative physiology is not merely a technical obstacle, but rather reflects clinical reality and, when addressed adequately in the solution process, provides a novel link between mathematically described physiological knowledge and the clinical concept of differential diagnoses. As a result, there is a growing research effort towards the development of systems-based, quantitative models of the inflammatory response at various degrees of complexity driven by the premise that such models can potentially enable the translation of knowledge from bench to bedside [18].

Inflammation can be studied in the absence of complex pathophysiology and comorbidities of human sepsis by using surrogate models [19]. The human endotoxin challenge is one well-accepted surrogate model for studying the acute inflammatory response as it captures many of the clinically observed features of the systemic inflammatory phenotype [19–23]. Endotoxin is a major component of the outer membrane of gram-negative bacteria, and the inflammation caused by the activation of the innate immune system by this moiety can be a complicating factor in a variety of situations including trauma, burns, invasive surgery, and organ-specific illnesses. The prototypical examples of endotoxin are lipopolysaccharides (LPS). The response following endotoxin administration in human subjects includes core temperature, cardiac, vasomotor, hematologic, metabolic, hormonal, acute phase reactant, and cytokine components that have been well described [19, 24–26]. In particular, innate immune cell activation leads to the production and release of pro-inflammatory cytokines, which are proximal mediators of the systemic inflammatory response.

Mechanisms aiming at the regulation of the inflammatory response involve not only the local release of anti-inflammatory cytokines but also hormonal influences [4]. Some studies indicate that the central nervous system (CNS) is a pivotal regulator of the immune response and controls inflammation at various levels [27–31]. The primary stress response pathway by which the CNS regulates the immune system is the hypothalamic-pituitary-adrenal axis (HPA), through the production of glucocorticoids. Furthermore, activation of the sympathetic division (SNS) of the autonomic nervous system regulates the immune function primarily via the release of adrenergic neurotransmitters [32]. Although most research has focused on the sympathetic immunomodulatory output, it recently became clear that the other arm of CNS, the parasympathetic division (PNS), is also involved in the reflex regulation of inflammation [33]. All these functions are centrally integrated through an extensive and coherent circuitry involving all areas of the medulla, the hypothalamic regions, and the autonomic system [34].

Disruptions may occur at any level of autonomic functions and thereby have a deleterious effect on the host. Heart rate variability (HRV), assessed by evaluating the standard deviation of normal-to-normal interbeat intervals, is a noninvasive means of quantifying the cardiac autonomic input and is a predictor of trauma patient outcome [35, 36]. Decreases in HRV (i.e., increases in regularity) have also been extensively studied and characterized as generalized responses to human endotoxemia [37, 38]. The loss of a high-level signal variability reflects a detectable systemic-level loss of adaptability and fitness. Thus, the development of surrogate multiscale, nonlinear, dynamic, complex models gains ever-increasing acceptance as a means of deciphering the intricacies of critical illness, emphasizing the emerging loss ("decomplexification") of rich variability, observed in normal physiology, in critically ill patients [39].

Dissecting the role of the CNS in controlling inflammatory processes requires an understanding of the complex bidirectional relationship between autonomic activity and the immune response. A vital enabler in that respect is the development of a systems-based approach that integrates processes across multiple scales and subsequently models the emerging host response as the outcome of orchestrated interactions of critical modules. To meet this challenge, a systems-based model of human inflammation is proposed that bridges the initiating signal (LPS) and phenotypic expressions (HRV) through semimechanistic host response models that include transcriptional (cellular) dynamics, signaling cascades, and physiological (hormonal) components. Thus, this modeling effort intends to establish direct communication links from the cellular to the systemic host response level integrating bidirectional influences between central systems and physiological systems that should not be viewed as distinct functional domains. Such a modeling approach could potentially provide invaluable insights into how disruption within these compartments contributes to morbidity and mortality in severely stressed patients, thus making it a critical enabler for the generation of alternative testable hypotheses that could clarify how cellular events and inflammatory processes mediate the links between patterns of autonomic control (HRV) and clinical outcomes.

4.2 Materials

4.2.1 Human endotoxin model and data collection

Gene expression data used in this study were generated as part of the *Inflammation and Host Response to Injury Large Scale Collaborative Project* funded by the USPHS, U54 GM621119 [40]. Human subjects were injected intravenously with endotoxin (CC-RE, lot 2) at a dose of 2-ng/kg body weight (endotoxin-treated subjects) or 0.9% sodium chloride (placebo-treated subjects). Following the lysis of erythrocytes and isolation of total RNA from leukocyte pellets [20], biotin-labeled cRNA was hybridized to the Hu133A and Hu133B arrays containing a total of 44,924 probes for measuring the expression level of genes that can be either activated or repressed in response to endotoxin. A set of 5,093 probe sets were characterized by significant variation (corresponding to 0.1% false discovery rate) across the time course of the experiment using the SAM software [41]. The data are publicly available through the GEO Omnibus Database (http://www.ncbi.nlm.nih.gov/geo/) under the accession number GSE3284.

In addition to transcriptional profiling analysis, blood samples are also collected and analyzed to determine the plasma concentration of counter-regulatory hormones, such as epinephrine, after endotoxin administration at the following four time points: 0, 2, 4, and 6 hours [42]. The purpose of this study was broader and attempted to assess the effect of acute hypercortisolemia on hormonal and cytokine responses to human endotoxemia. Furthermore, human volunteers were injected with the same amount of LPS, while vital signs, including heart rate variability (HRV) indices, were recorded [43]. In particular, human subjects received an initial recording of heart rate and electrocardiogram (ECG) to screen for any arrhythmic patterns or irregular heartbeats. During the analysis of heart rate variability, parameters and interbeat intervals were collected using ECG data at a rate of 256 samples/s. Therefore, in a continuous ECG record, each QRS complex (which corresponds to the depolarization of the ventricles) was detected and the normal-to-normal (NN) intervals were tabulated. The overall heart rate variability was assessed by calculating the standard deviation of normal interbeat intervals (SDNN). Therefore, at the systemic level HRV measurements are gathered from [43], while plasma cortisol concentrations are also recorded under the systemic inflammatory manifestations of human endotoxemia. Therefore, at the level of circulating hormones, the plasma concentration of stress hormones including cortisol and epinephrine are employed [42, 43]. The data have been appropriately de-identified, and appropriate IRB approval and informed, written consent were obtained by the glue grant investigators [20].

4.3 Methods

4.3.1 Transcriptional dynamics and intrinsic responses

Intravenous administration of endotoxin elicits dynamic and reproducible changes in the circulating leukocyte population as well as significant changes in blood leukocyte gene expression patterns, with a resolution of all clinical manifestations within 24 hours [20]. The response following endotoxin administration in human subjects includes changes in core temperature, cardiac, vasomotor, hematologic, metabolic, hormonal, acute phase reactant, and cytokine components that have been well described [19, 24–26]. Innate immune cell activation leads to production and release of proinflammatory cytokines, which are proximal mediators of the systemic inflammatory response. Although the bulk of this pro-inflammatory mediator release likely originates in cells of the reticuloendothelial system [44], the leukocytes present in peripheral blood are also activated and, more importantly, are available for sampling with minimal invasiveness.

In order to study the underlying complexity of the dynamics of inflammation and to establish quantifiable relationships among the various components of the inflammatory response, model-based approaches have been proposed [45–47]. A number of excellent prior studies [46, 48–52] have placed significant emphasis on simulating inflammation based on the kinetics of well-defined features of the overall response. One of the key characteristics of these models is the a priori postulation of certain components that are consistent with biological knowledge and are known to play a major role in triggering the inflammatory response [46, 47]. Thus, there is emphasis on reducing the complexity of the models of inflammatory response by identifying a limited number of time-dependent interactions of key elements that are highly sensitive to specific modes of initiation and modulation of the response.

However, one of the big challenges is the systematic identification of such representative biological features, based on experimental data that can adequately represent the complex dynamics of a host undergoing an inflammatory response. This requires the decomposition of the nonlinear dynamics of the response into an elementary set that can serve as a surrogate for predicting the collective behavior of the system. A possible answer to this problem can be identified through the analysis of gene expression data aimed at monitoring the dynamics of the host response to an inflammatory agent, exploring the idea that cellular responses correspond to dynamically converging high-dimensional transcriptional trajectories [53]. Decomposing the intrinsic dynamics of the entire system into a reduced set of responses enables us to both project and understand the complex dynamics of the system by studying the properties of its essential dynamic parts. Given the transcriptional profiling analysis of human blood leukocytes, we hypothesized that the genes that are most responsive to LPS are governed by a definite mechanism and have concerted changes in their expression profile.

We recently proposed a systematic computational framework that decomposes high-dimensional microarray data into an elementary set of temporal responses [54–56]. We are interested in unraveling a critical set of informative temporal responses that are characterized as the blueprints of the orchestrated dynamics of the perturbed biological system. In doing so, we hypothesize that there is a definite underlying mechanism that describes the emerging dynamic inflammatory response and capturing the *essential inflammatory responses* might serve as surrogates for the dynamic evolution of the host response due to endotoxin stimulus. Based on our prior work, we first apply a microclustering approach, which is based on a symbolic transformation of time series data that assigns a unique integer identifier (*hash value*) to each expression motif [54]. Having assigned the temporal expression profiles to distinct motifs, the next task is to select expression motifs that appear to be highly nonrandom based on their gene population size. Having identified the statistically significant expression motifs from the initial large set of microclusters, we need to identify a discriminating set of critical temporal shapes that best characterizes the intrinsic dynamic response of the system. Due to global nature of the transcriptional measurements and the fact that we do not a priori select a limited set of responsive genes, the entirety of the transcriptional response is expected to exhibit a rather Gaussian type of response with no clear defining responses [57]. We define the TS [54] of the system as the overall distribution of expression values at a specific time point; then, the deviation of the TS can be quantified at each time point with respect to a baseline distribution (t = 0 hours) by applying the Kolmogorov-Smirnov test [58]. Thus, through the use of the concept of TS it is possible to tease out the essential components of the cellular response in response to an external disturbance. Given the aforementioned metric, we are interested in identifying the minimum number of expression motifs that characterize the maximum deviation of the transcriptional state of the system. This selection problem defines a combinatorial optimization problem for which we apply a stochastic optimization algorithm, based on simulated annealing [59].

We identify three critical expression motifs enriched in critical and relevant biological pathways:

1. Early upregulation response (proinflammatory component, P): Genes in this major temporal class are important in cytokine-cytokine receptor interaction as well as in a toll-like receptor signaling pathway crucial in activating transcription factors that act synergistically with proinflammatory transcription factors such as members of NFkB/RelA family.
2. Late upregulation response (anti-inflammatory component, A): Genes in this functional class participate in the JAK-STAT cascade which is essential to regulate the expression of target genes that counterreact the inflammatory response. In addition to this, it is emphasized [60] that a STAT pathway from a receptor signaling system is a major determinant of key regulatory systems, including feedback loops such as SOCS induction, which subsequently suppresses the early induced cytokine signaling and essential activators for IL10 signaling [61]. Moreover, we identified the late increased expression of IL10RB, which is assumed to be indicative of the IL10 signaling cascade.
3. Downregulation response (Energetic component, E): The downregulated essential response is characterized by a set of genes, which are mainly involved in the cellular bioenergetic processes. In addition to this, a large set of genes, which are essential to ribosome biogenesis and assembly (RPL/RPS family), are repressed coupled with those genes, which participate in protein synthesis machinery, oxidative phosphorylation, and pyruvate metabolism. Endotoxin-induced inflammation

causes the dysregulation of leukocyte bioenergetics and a persistent decrease in mitochondrial activity leads to reduced cellular metabolism and a subsequent decline in organ function [62]. A restoration of organ function should be associated with an increase in bioenergetics and metabolic activity [63] and we are assuming that a persistent shutdown of these genes might lead to multiple organ dysfunction.

These transcriptional responses effectively decompose the overall dynamic and present the constitutive elements of the overall response, Figure 4.1. They correspond to the cellular signatures in response to LPS administration and manifest the integrated systemic response. In order to reproduce the experimental data, we select the transcriptional signature of specific genes representative of each essential response. IL1B is selected to serve as the representative biomarker of the proinflammatory response. The gene transcript of IL10RB is considered to be indicative of the immune-regulatory signal of the anti-inflammatory response. Finally, a subunit of NADH ubiquinone dehydrogenase complex (mitochondrial component) NDUFC2 is considered as the proxy for the energetic component. These essential transcriptional signatures are normalized by taking the ratio of the measured mRNA level at each time point with respect to the control time point ($t = 0$ hours). Selecting any other gene that belongs to the aforementioned essential inflammatory responses can very well be used as a surrogate for a representative of the response and will not alter the qualitative characteristic of our semimechanistic mathematical model to be described in the following section.

4.3.2 Modeling inflammation at the cellular level

Considering the leukocytes as a well-defined system, the purpose of human studies like the ones described in the previous section is to characterize qualitatively the cellular

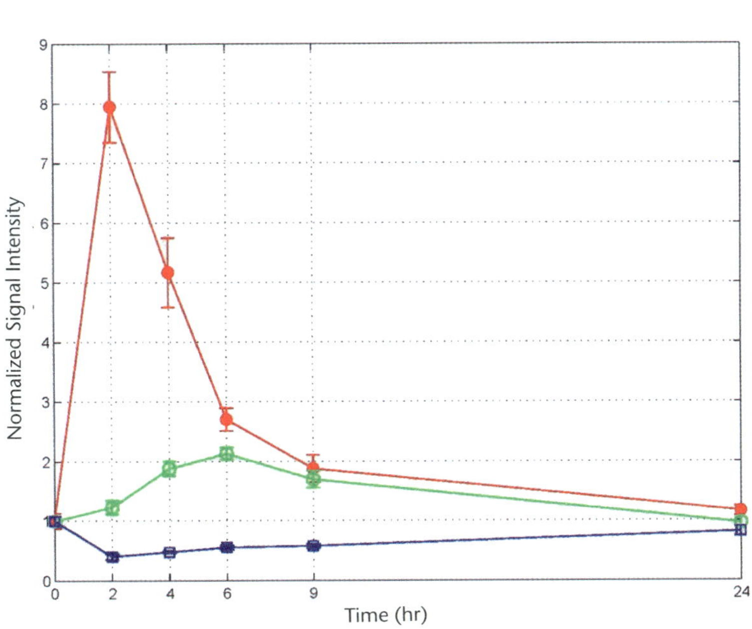

Figure 4.1 Essential leukocyte transcriptional elements: (solid markers expressed as mean ± SEM) proinflammatory response (P), (open circles) anti-inflammatory response (A), and (square markers) energetic response (E).

dynamics. The purpose of a systems biology approach, on the other hand, is to reverse-engineer quantifiable representations of the intracellular dynamics [64] by identifying: (1) appropriate constitutive elements; (2) the topology of the interactions among these elements; and (3) the quantitative relations among these elements. In the aforementioned section we addressed the first issue, whereas now we will discuss how to construct the topology and dynamics of the underlying network describing the dynamics at a single scale (i.e., that of the leukocytes). Vodovotz et al. recently discussed the state of the art in mechanistic simulations of inflammation [18]. Such models are a critical enabler in advancing the translational potential of clinical research, a subject recently reviewed [15]. We have recently advocated a cellular, mechanistic-based, modeling approach that explores three unique aspects. First, through the analysis of the leukocyte gene expression data, we identify the essential responses characterizing the cellular (transcriptional) dynamics. Second, we explore the concepts of physicochemical modeling [65] to express the intertwined relations and dynamics that connect extracellular signals and intracellular signaling cascades eventually leading to the emergent transcriptional dynamics (identified in the previous step). Finally, we explore the pharmacodynamic concept of indirect response (IDR) [66] in order to establish implicit interaction among signaling molecules and emerging transcriptional responses. In this section we will summarize the basic concepts and models.

In our injury model, the inflammatory response is activated when endotoxin is recognized by pathogen recognition receptors [67]. The LPS-induced stimulation initiates a complex signaling cascade that ultimately targets the transcription initiation of proinflammatory cytokines [68]. During the recognition process LPS binds to the LPS-binding protein in plasma and is delivered to the surface receptor CD14. Subsequently, LPS is transferred to the signaling receptor toll-like receptor 4 (TLR4) with the recruitment of its essential accessory protein MD2 [69–71]. Downstream of the receptor-ligand complex there are intermediate secondary messages that involve both the amplification of the signaling and its diversification in the cytoplasmic region. Ultimately, such a signal transduction cascade will activate critical signaling modules that will subsequently lead to the translocation (activation) of proinflammatory transcription factors (e.g., NF-κB) responsible for the transcription initiation of inflammatory genes. Therefore, LPS interacts with its signaling receptor (TLR4) in order to induce the signal transduction cascade that triggers essential signaling modules for the activation of proinflammatory transcription factors. This cascade of events is eventually manifested through the coordinated transcriptional changes measured via high-throughput microarray analyses [72]. Our inability to precisely model the complex signaling events that characterize system's adaptation process to environmental changes makes IDR modeling appealing [73, 74]. Thus, we propose to model such a transcriptional event using the basic principles of indirect response modeling (IDR) widely used in developing pharmacodynamic and pharmacogenomic models [75, 76]. The underlying assumption is that external signals indirectly affect the synthesis and/or degradation term of the response of interest, in our case the transcriptional dynamics. Therefore, in order to establish quantifiable relationships among the previously identified "essential" components of human inflammation, an indirect response model was proposed in [77, 78]. The proposed model couples receptor-mediated phenomena with transcriptional effects based on ligand-receptor kinetics, while each transcriptional motif is assumed to be the manifestation of a process involving a zeroth-order synthesis rate and a first-order degradation term.

The underlying assumption of IDR models is that external signals affect indirectly the synthesis and/or degradation term of the response of interest, in our case the transcriptional dynamics. As a result, the existence of such signals can either stimulate or inhibit the production and degradation rate of the response. The underlying assumption of this modeling effort is that intracellular signaling cascades activate inflammation specific transcriptional responses [79]. Although a large family of transcription factors is known to be involved in inflammation, we focus on a particular family, NFκB, for two reasons. First, the nuclear factor κB family is known to be a major player in the inflammatory response [80] and as such it has been widely studied as a major contributor. Second, the fact the NFκB plays an important role has led to the development of numerous independent modeling approaches in order to quantify the expected response of its signaling cascade [81]. We therefore hypothesize that NFκB serves as a proxy for the inflammation specific transcription factor that initiates the expression of proinflammatory genes as a result of the formation of an activating signal associated with the binding of LPS to its signaling receptor. The binding interaction between endotoxin (LPS) and its receptor (R) is assumed to be a standard ligand-receptor interaction [82].

The proinflammatory response can be characterized as the first-line transcriptional response that is triggered upon the recognition of the extracellular ligand (LPS) by the pattern recognition receptors (PRR, e.g., TLR4) [79]. Proinflammation will serve as the signal that will further stimulate the down-regulation of genes that are associated with the cellular energetic processes [83]. We hypothesize that the proinflammatory response acts as the stimulatory factor for the energetic response, while a dysregulation in the cellular bioenergetics can serve as a positive feedback danger signal to the proinflammatory response. The anti-inflammatory response serves as the essential immunoregulatory signal that aims at restoring homeostasis in the host defense system. Thus, it will be stimulated by the activation of the inflammatory components, which are the proinflammatory and energetic responses. It will serve as the inhibitory signal on the production rate of the proinflammation and the energetic response.

4.3.2.1 A physicochemical cellular host response model

The dynamics of the inflammatory stimulus (LPS) are described in (4.1a) as a convolution of two terms: a first-order elimination with a rate $k_{lps,2}$ and a logistic-type function with growth rate $k_{lps,1}$ [84]. Effectively, despite the presence of various mediators that are activated in response to LPS (e.g., LPS binds to LBP plasma protein during its recognition from the host), we model a single compartment pharmacodynamics model for LPS assuming a homogeneous circulating blood compartment. The logistic function is usually used to model a variety of physical situations in which a quantity's growth is self-limited, which means that the initial growth is approximately exponential and as saturation begins the growth stops [84]. Therefore, depending on the relative magnitude of the two rate parameters of the clearance of LPS, we can simulate situations where the bacterial concentration is not fully eliminated. In human subjects the endotoxin is cleared within the first 1–2 hours after LPS administration with an approximate average half-time of $\tau_{1/2}$ ~ 8–15 minutes [85]. The two parameters $k_{lps,1}$ and $k_{lps,2}$ have been independently estimated so that the LPS profile decays within 2 hours in the absence of any complications [85].

The dynamics of the TLR4 receptor (R), (4.1b), depend on the association/dissociation parameters of the ligand-receptor interaction [82] whose corresponding parameters k_1 and k_2 are based on literature values [86] and the translation of its $mRNA_{,R}$ to surface protein (k_{syn}). The rate of translation k_{syn} of the $mRNA_{,R}$ to the corresponding surface protein describes the dynamic evolution of synthesis of new receptors; hence, this parameter is estimated so that the dynamic profile of the surface free receptor is downregulated based on the premise that under the inflammatory stimulus, the surface free receptors are occupied. The dynamics of the gene transcript of the receptor, (4.1c), is characterized by a production rate ($K_{in,mRNA,R}$) and a degradation rate ($K_{out,mRNA,R}$), but is assumed to be indirectly stimulated by the proinflammatory signaling. The measured $mRNA_{,R}$ is characterized by an upregulation for the first 4 hours post-LPS administration and a return to baseline. As a result, the two parameters are estimated so that we can best fit the available $mRNA_{,R}$. The equilibrium (LPSR) complex, (4.1d), is characterized by the binding parameters k_1 and k_2 as well as by the k_3 parameter that shows the rate of formation of the activated signal (IKK) that eventually converges to the upregulation of NF-κB transcription factor. Therefore, we introduce the NF-κB signal transduction cascade as the prototypical module for initiating and controlling the expression of proinflammatory genes.

Numerous signaling molecules and reactions participate in the NF-κB signaling pathway [81]. However, sensitivity analysis [87] demonstrated that the activity of NF-κB is maximally modulated by a reduced set of basis signaling molecules (IKK, IKBa and NF-κB). As such, [88] proposed a minimal model of NF-κB that accounts for the propensity of oscillations in the dynamic behavior of NF-κB activity. However, instead of simulating the kinase activity as a constant parameter and incorporating saturation degradation rates as discussed in [88], we propose to model IKK as a transient signal. Thus, the cellular surface complex (LPSR) induces the activation of kinase activity (IKK) with a rate k_3, while being eliminated with a rate k_4, (4.1e). The nonlinear function of Hill-type is an essential functional form in order to achieve a bistability response in the dynamics of the probed system [89–92]. Such a bistability is an essential characteristic of the nonlinear dynamics of inflammation, as suggested by various animal studies [89–93]. In chronic inflammatory diseases, several cytokines might be responsible for perpetuating and amplifying the inflammatory reaction through the critical node (IKK) [94]. Therefore, we simulate such an interaction by the presence of a positive feedback loop in the kinetics of kinase (IKK) activity.

Assuming that $NF\kappa B_n$, (4.1f) is the fraction of total NFκB that is in the nucleus, the term ($1\text{-}NF\kappa B_n$) denotes the available free cytoplasmic concentration of NFκB and herein the nuclear concentration ($NF\kappa B_n$) and nuclear activity are used interchangeably. The import rate of cytoplasmic NFκB into the nucleus depends on the availability of the free cytoplasmic concentration ($1\text{-}NF\kappa B_n$) stimulated by the kinase activity (IKK). However, its degradation rate depends on the presence of its primary inhibitor (IkBa), as it retrieves nuclear NFκB by forming an inactive complex in the cytoplasmic region [95]. The dynamics of the gene transcript of IKBa ($mRNA_{,IKBa}$), (4.1g), are characterized by a zero order production rate ($K_{in,IkBa}$) and a first-order degradation rate ($K_{out,IkBa}$), which is stimulated by NFκB [94]. The protein inhibitor IkBa, (4.1h), is the product of translation of its gene transcript ($mRNA_{,IkBa}$) and it degrades at a rate $k_{I,2}$, which is stimulated by the kinase activity (IKK). Based on the premise that IkB_a forms a complex with the available cytoplasmic NFκB, mathematically we expressed this as the product (1-NFkBn) IkBa. From the modeling point of view, in order to achieve a zero steady state for the protein inhibitor IkBa we need the additional negative term $-k_{I,1}$.

At the transcriptional response level, the nuclear activity of NFκB ($NF\kappa B_n$) serves as the active signal that indirectly stimulates the production rate of the essential proinflammatory response (P), which quantitatively is expressed by the linear function ($H_{P,\ NFkBn}$), (4.1i). We are also assuming that the energetic response variable will be responsible for more inflammation ($H_{P,E}$) [83]. The essential anti-inflammatory signaling component, (4.1j), is assumed to inhibit the production rate of the proinflammatory transcriptional signature. The essential anti-inflammatory signal (A) is stimulated by the activated proinflammatory response ($H_{A,P}$) as well as by the energetic response ($H_{A,E}$), and it decays with rate $K_{out,A}$. The energetic variable (E), (4.1k), is stimulated by the proinflammatory response (P) and we are also assuming that the crucial anti-inflammatory component (A) counter-regulates both the proinflammatory and energetic responses of the system.

Thus, the human inflammation model consists of 11 state variables: (1) the inflammatory instigator (LPS), (2) the endotoxin signaling free protein receptor (R, TLR4), (3) the mRNA of TLR4 (mRNA,R), (4) the formed complex (LPSR), (5) the active kinase activity (IKK), (6) the nuclear concentration of NFκB (NFkBn), (7) the mRNA of IKBa ($mRNA_{,IKBa}$), (8) the translated protein inhibitor of N-FκB (IKBa), and finally the transcript abundance of the essential transcriptional responses: (9) proinflammation (P), (10) anti-inflammation (A), and (11) the energetic response (E). Given that anti-inflammatory drugs, such as corticosteroids, play a pivotal role in modulating the progression of inflammation, their contribution is further assessed in this modeling effort. As such, the contribution of corticosteroids, to be used as a template for assessing anti-inflammatory intervention strategies, is assessed by incorporating appropriate PK/PD models. Significant prior research efforts have attempted to elucidate the mechanisms driving corticosteroid activity [96–103]. Such studies simulate the pharmacogenomic effect of glucocorticoids at the transcriptional level, taking their mechanistic (signaling) action into account [104, 105]. In an attempt to demonstrate the capability of our model to generate a behavior via interacting modules, we opt to integrate the regulatory signaling information with the anti-inflammatory mechanism of corticosteroids, as putative controllers of inflammation. The corticosteroid intervention envelope consists of a set of elementary interactions that involve: (1) the binding of the corticosteroid drug (F) to its cytosolic receptor (R_F), (2) the subsequent formation of the drug-receptor complex (FR), (3) the translocation of the cytosolic complex to the nucleus (FR(N)) that alters the transcriptional machinery activating or repressing numerous genes, and finally (4) the autoregulation of the gene transcript of the glucocorticoid receptor (R_m).

The drug disposition is modeled either via a monoexponential kinetic model that quantifies the plasma concentration of the steroid (F) under conditions of intravenous injection or via the time invariant parameter (R_{in}) that reflects the constant infusion rate while being eliminated with a rate k_{el}. The dynamics of the gene transcript of the corticosteroid drug (R_m), (4.1m), are characterized by a zero order production rate ($k_{syn\ Rm}$) and a first order degradation rate (k_{deg}). The active drug-receptor complex (FR(N)) exerts an inhibitory effect towards the mRNA of the glucocorticoid receptor. The parameter

IC_{50_Rm} denotes the concentration of the nuclear drug-receptor complex FR(N) at which the synthesis rate of the receptor drops at 50% of its baseline value. The dynamics of the free cytosolic receptor density, R_F, is modeled in (4.1n) where k_{syn_R} is the synthesis rate of receptor that stems from its transcription, r_f is the fraction of the drug that is recycled, k_{re} is the parameter that shows the recycling of drug from the nucleus to the cytosol, and k_{on} is a parameter associated with the drug-receptor binding. In addition to this, k_{dgr_R} is the degradation rate of the receptor (R_F). The formed cytosolic complex (FR), (4.1o), depends upon the binding interaction k_{on} of the ligand (F) with its receptor (R_F) and on its translocation rate k_T to the nucleus. Therefore, the translocation of the drug-receptor complex to the nucleus accounts for the nuclear receptor complex FR(N), which is the active complex that mediates the transcriptional induction of various genes. By incorporating an appropriate PK/PD model, we can evaluate the possibility of a corticosteroid-dependent regulation of either the inhibitor of NFkB, (4.1p), or the anti-inflammatory component of the response, (q). We recently demonstrated [78, 106] that this indirect model properly captures the onset and resolution of inflammation, but it also predicts a number of responses outside the range of parameter estimation. This justifies the fundamental assumptions that established the functional relationships between the individual components that define the architecture and interactions of the constitutive disease model. The mathematical representation of this model is succinctly presented in (4.1a–q) as follows.

$$\frac{dLPS}{dt} = k_{lps,1} \cdot LPS \cdot (1 - LPS) - k_{lps,2} \cdot LPS \tag{4.1a}$$

$$\frac{dR}{dt} = k_{syn} \cdot mRNA,R + k_2 \cdot (LPSR) - k_1 \cdot LPS \cdot R - k_{syn} \cdot R \tag{4.1b}$$

$$\frac{d(LPSR)}{dt} = k_1 \cdot LPS \cdot R - k_3 \cdot (LPSR) - k_2 \cdot (LPSR) \tag{4.1c}$$

$$\frac{d(mRNA,R)}{dt} = k_{in,mRNA,R} \cdot (1 + H_{P,R}) - K_{out,mRNA,R} \cdot mRNA,R \tag{4.1d}$$

$$\frac{dIKK}{dt} = k_3 \cdot (LPSR) \Big/ (1 + IkBa) - k_4 \cdot IKK + P\left(\frac{IKK^2}{1 + IKK^2}\right) \tag{4.1e}$$

$$\frac{dNFkBn}{dt} = \frac{k_{NFkB,1} \cdot IKK \cdot (1 - NFkBn)}{(1 + IkBa)} - k_{NFkB,2} \cdot NFkBn \cdot IkBa \tag{4.1f}$$

$$\frac{dmRNA_{IkBa}}{dt} = K_{in,iKBa} \cdot (1 + k_{IkBa,1} \cdot NFkBn) - K_{out,IkBa} \cdot mRNA_{IkBa} \tag{4.1g}$$

$$\frac{dIkBa}{dt} = k_{I,1} \cdot mRNA_{IkBa} - k_{I,2} \cdot (1 + IKK) \cdot (1 - NFkBn) \cdot IkBa - k_{I,1} \tag{4.1h}$$

$$\frac{dP}{dt} = K_{in,P} \cdot (1 + H_{P,NFkBn}) \cdot (1 + H_{P,E}) / A - K_{out,P} \cdot P \tag{4.1i}$$

$$\frac{dA}{dt} = K_{in,A} \cdot (1 + H_{A,P}) \cdot (1 + H_{A,E}) - K_{out,A} \cdot A \tag{4.1j}$$

$$\frac{dE}{dt} = K_{in,E} \cdot (1 + H_{E,P}) / A - K_{out,E} \cdot E \tag{4.1k}$$

$$\frac{dF}{dt} = \begin{cases} R_{in,F} - k_{el} \cdot F, \text{infusion} \\ -k_{el} \cdot F, \text{injection} \end{cases} \tag{4.1l}$$

$$\frac{dR_m}{dt} = k_{syn\_Rm} \cdot \left(1 - \frac{FR(N)}{IC_{50\_Rm} + FR(N)}\right) - k_{\deg} \cdot R_M \tag{4.1m}$$

$$\frac{dR_F}{dt} = k_{syn\_R} \cdot R_m + r_f \cdot k_{re} \cdot FR(N) - k_{on} \cdot F \cdot R_F - k_{dgr\_R} \cdot R_F \tag{4.1n}$$

$$\frac{dFR}{dt} = k_{on} \cdot F \cdot R_F - k_T \cdot FR, \frac{dFR(N)}{dt} = k_T \cdot FR - k_{re} \cdot FR(N) \tag{4.1o}$$

$$\begin{cases} \frac{dmRNA_{IkBa}}{dt} = K_{in,IkBa} \cdot (1 + k_{IkBa,1} \cdot NFkBn) \cdot (1 + FR(N)) - K_{out,IkBa} \cdot mRNA_{IkBa} \\ \frac{dA}{dt} = K_{in,A} \cdot (1 + H_{A,P}) \cdot (1 + H_{A,E}) \cdot (1 + FR(N)) - K_{out,A} \cdot A \end{cases} \tag{4.1p}$$

$$H_{i,j} = k_{i,j} \cdot J \tag{4.1q}$$

4.3.3 Modeling inflammation at the systemic level

The human response to endotoxin evokes not only peripheral blood leukocyte transcriptional alterations but also a neuroendocrine response characteristic of acute injury and sepsis [107]. Accordingly, a rise in circulating endocrine hormones is manifested 2–4 hours following endotoxin administration [19]. Moreover, diminished heart rate variability, as a component of autonomic dysfunction, is also induced by the intravenous administration of low-dose endotoxin to human subjects [7]. The physicochemical host response model described in (4.1) demonstrated how we can link the initiating signal (LPS) with the cellular (leukocyte) transcriptional dynamics. However, one of the simplifying hypotheses made in this model is that the immunomodulatory role of autonomic (hormonal) influences, including endogenous cortisol, is not accounted for. Therefore, the additional scale we wish to explore is associated with critical aspects of the neuroendocrine-immune system axis. We opt therefore to extend our prior modeling effort by incorporating critical aspects of the neuroendocrine-immune crosstalk connecting the cellular response level with neural-based pathways whose disturbance is quantified through phenotypic expressions such as HRV.

4.3.3.1 Modeling neuroendocrine-immune system interactions

The primary stress response pathway by which the CNS regulates the immune system is the hypothalamic-pituitary-adrenal axis (HPA), through the production of glucocorticoids. The HPA axis, as one of the peripheral limbs of the stress system, responds to proinflammatory cytokines by releasing cortisol, which subsequently inhibits a proinflammatory cytokine expression [108]. In addition to regulation of immune function by glucocorticoids, other counter-regulatory hormones associated with distress, including catecholamines (epinephrine, EPI), can modulate a range of immune functions [109]. Such hormones are secreted by the sympathetic nervous system pathway (SNS) and act via adrenergic receptors on immune cells. There is considerable evidence indicating that the proinflammatory response (P) acts as the immune signal that stimulates central components of the stress system through the afferent vagus nerve [32]. Thus, it is reasonable to assume that the proinflammatory response (P) acts as the peripheral immune signal that stimulates both the secretion of cortisol and epinephrine.

In order to mathematically describe the secretion of stress hormones (cortisol, epinephrine), the principles of IDR models are employed. In particular, the plasma concentration of cortisol (F) is assumed to be the manifestation of a process that involves a zero-order production rate ($K_{in,Fen}$) and a first-order degradation term ($K_{out,,F}$), (4.2a). Upon inflammatory stimulation, the proinflammatory response (P) serves as the intercellular messenger that indirectly stimulates the HPA axis through the linear function ($H_{Fen,P}$). Furthermore, the contribution of exogenous cortisol upon total cortisol concentrations is assessed through the stimulatory parameter ($R_{in,F}$), where (w_{Fex}) is a binary variable that is active under conditions of a hydrocortisone infusion. Thus, the total cortisol concentration (F) is defined as the joint effect between endogenous and exogenous steroid as outlined in [110].

The dynamics of epinephrine are characterized by a zero-order production rate ($K_{in,EPI}$) stimulated by the proinflammatory response (P) through the linear stimulatory function ($H_{EPI,P}$) and a first-order degradation rate ($K_{out,EPI}$), (4.2b). Furthermore, in the case of prior stress hormone infusion, the stimulatory effect of exogenous epinephrine is simulated in the same manner as in case of exogenous cortisol via the parameter ($R_{in,EPI}$). Though the production of proinflammatory cytokines stimulates the production of epinephrine, the latter attenuates the proinflammatory manifestations of human endotoxemia as supported by reduced TNF levels [42, 111]. The anti-inflammatory influence of epinephrine is shown to be mediated by β adrenergic stimulation resulting in an increase in cAMP intracellular levels followed by potentiation in the production rate of IL-10 signaling (A) [111, 112]. In order to describe such postreceptor effect, a precursor-dependent indirect response model [113] is proposed in which the precursor reflects the signaling receptor of epinephrine (R_{EPI}). Accordingly, the dynamic changes of epinephrine receptor (R_{EPI}) depend on an apparent zero-order production rate $k_{0,REPI}$; $k_{1,REPI}$, and $k_{2,REPI}$ represent first-order rate constants for the loss of the receptor (4.2c). Since the response is triggered as a result of the formation of an activating complex associated with the binding of EPI to its receptor, EPIR represents the formed signaling complex which decays with a first-order rate $k_{3,EPI}$, (4.2d). Furthermore, the stimulatory postadrenergic effect of sympathetic activity in favoring the production of cAMP signaling is described by the principles of a signal transduction model as outlined in [114, 115]. In particular, the production and loss of the cAMP signaling depends on first-order rate constants, which are equivalent to the reciprocal of the transit times (τ) consistent with the transit compartment model while n is a shaping (scaling) factor, (4.2e) [113]. Such a scaling factor is used to amplify the signal transduction cascade associated with the postadrenergic effect of epinephrine on the host.

Given a quantification of the cortisol and epinephrine dynamics, the influence of neuroendocrine hormones on the host response to endotoxin can be further incorporated. Regarding the immunosuppressive effects of cortisol, we have previously demonstrated that hypercortisolemia increases plasma IL-10 concentrations during human endotoxemia [116]. As such, we will assume that cortisol modulates the host response to endotoxin primarily via potentiation of IL-10 signaling (A). Such steroid-dependent immunomodulatory effect is quantified via the stimulatory function $H_{A,FRN}$, (4.2f). Furthermore, epinephrine also potentiates IL-10 signaling (A) via a cAMP-dependent mechanism, which is quantified through the linear stimulatory function $H_{A,cAMP}$, (4.2f).

In addition to the neuroendocrine response evoked by endotoxin is the evolving concept of autonomic dysfunction as assessed by HRV indices. Recent studies imply that disordered neuroendocrine functions are associated with diminished HRV in stressed patients [7]. In an effort to quantify disturbances in autonomic activities, clinical measurements, at the systemic level, of HRV will be incorporated.

4.3.3.2 A model for the assessment of reduced heart rate variability in human endotoxemia

Clinical data associated with HRV measurements [19] establish that the host response to endotoxin causes depressions in cardiac-vagal tone and in overall HRV; these results are consistent with prior studies [37, 38] indicating diminished physiologic variability as a generalized response to human endotoxemia. Heart rate variability analysis has been extensively used to evaluate autonomic modulation of sinus node in normal subjects as well as in critically ill patients to identifying patients at increased risk with respect to mortality. Thus, the prognostic significance of HRV has made it a critical enabler to detect either physiologic deterioration or response to therapy [117].

In order to quantify the effect of acute endotoxin injury on heart rate variability, a critical question arises that involves the relationship between proinflammatory markers and autonomic dysfunction. There is considerable human evidence indicating that systemic low-grade proinflammatory activity is associated with reduced heart rate variability [118–120]. Assuming a linear relationship between proinflammation and HRV would imply that any modulations in the peripheral immune response will subsequently drive changes in hemodynamic parameters. However, it is important to realize that it cannot be taken for granted that factors modifying the magnitude of the immune response should affect all circulatory parameters equally [121].

Such nonlinearity in the sinus node transduction processes may arise from sensitivity of pacemaker discharge to the timing of pulsatile neural activity and from functional inhomogeneity within the sinus node tissue [122]. In addition to this, the concept of nonlinearity in cardiovascular variability has been stressed in the hemodynamic parameters of endotoxin-induced systemic inflammation under conditions of prior endocrine stress hormone infusion [43, 123]. One of the interesting observations was that the anti-inflammatory influence of endocrine hormones, including cortisol and epinephrine, does not extend to changes in heart rate variability induced by the inflammatory stimulus (LPS). Thus, reduction in endotoxin-induced proinflammation does not influence autonomic dysfunction (HRV). On the other hand, the potential role of hormone replacement therapy in severe inflammation has been emphasized in a large study of trauma patients exhibiting diminished HRV [124]. It was suggested that among injured patients with diminished HRV, there exists a subset of patients with established adrenal insufficiency who, upon treatment with exogenous steroid, recover diminished HRV activity and attain an improved clinical outcome.

In order to quantify such nonlinear interactions, the effect of peripheral proinflammation upon HRV response to endotoxin will be assumed that it can be mathematically approximated by employing appropriate sigmoid activation functions as outlined in [125]. Although the overall heart rate variability is assessed by evaluating the standard deviation of normal-to-normal interbeat intervals, the physiological background for such variation involves the activation of signal transduction mechanisms in the sinus node of the heart associated with the modulation of neuromediator concentrations [126]. The possible nonlinear modulatory effect of proinflammation (P) upon HRV is described by the dynamics of f_P, (4.2g), where the switch-like behavior is determined by the sigmoid function ($\tanh(P - w)$) and w is a parameter greater than the proinflammatory response (P) elicited upon endotoxin-induced inflammation. This nonlinear gain modulatory function should be active under conditions of an inflammatory response and inactive when the system lies in its homeostasis. We therefore model this event based on the function $\tanh(P^{\varphi} - 1)^{\varphi}$, which takes a value of zero when proinflammation (P) lies in its baseline (homeostasis) and one otherwise. The underlying rationale for this function is predicated upon a neurocomputational model [127] that aims at simulating the firing rate of neuronal activity. In our model, f_P could qualitatively reflect the activation of efferent nerve activity on the heart eventually leading to the up-regulation of intracellular mediators, (S_f). The loss and production of such mediators (2 hours) are thereby described by the principles of time-dependent transduction systems [113] depending on two parameters, a first-order parameter (τ_S) and a scaling factor (n_S). The dynamics of HRV, (4.2i), are described by a zero-order production rate ($K_{in,HRV}$) and a first-order degradation rate ($K_{out,HRV}$) stimulated by the effector biological signal (S_f). Taken together, the integrated module that describes critical aspects of the neuroendocrine-immune system interactions is presented in (4.2a–j).

$$\frac{dF}{dt} = w_{F_{ex}} \cdot R_{in} + K_{in,\,F_{en}} \cdot \left(1 + H_{F_{en} \cdot P}\right) - K_{out,\,F} \cdot F \tag{4.2a}$$

$$\frac{dEPI}{dt} = w_{EPI_{ex}} \cdot R_{in,\,EPI} + K_{in,\,EPI} \cdot \left(1 + H_{EPI,P}\right) - K_{out,\,EPI} \cdot EPI \tag{4.2b}$$

$$\frac{dR_{EPI}}{dt} = k^{0}_{R_{EPI}} - \left[k_{1,R_{EPI}} \cdot \left(1 + H_{R_{EPI},\,EPI}\right) + k_{2,R_{EPI}}\right] \cdot R_{EPI} \tag{4.2c}$$

$$\frac{dEPIR}{dt} = k_{1,R_{EPI}} \cdot \left(1 + H_{R_{EPI},\,EPI}\right) \cdot R_{EPI} - k_{3,\,EPIR} \cdot EPIR \tag{4.2d}$$

$$\frac{dcAMP}{dt} = \frac{1}{\tau} \cdot \left(EPIR^{n} - cAMP\right) \tag{4.2e}$$

$$\frac{dA}{dt} = K_{in,A} \cdot \left(1 + H_{A,cAMP}\right) \cdot \left(1 + H_{A,\,E}\right) \cdot \left(1 + H_{A,\,FRN}\right) - K_{out,A} \cdot A \tag{4.2f}$$

$$\frac{df_P}{dt} = \left(1 + \tanh(P - w) - f_P\right) \cdot \tanh\left(P^{\phi} - 1\right)^{\phi} \tag{4.2g}$$

$$\frac{dS_f}{dt} = \frac{1}{\tau_S} \cdot \left(\tanh\left(P^{\phi} - 1\right)^{\phi} \cdot \left(f_P\right)^{n_S} - S_f\right) \tag{4.2h}$$

$$\frac{dHRV}{dt} = K_{in,HRV} - K_{out,HRV} \cdot \left(1 + k_{HRV,S} \cdot S_f\right) \cdot HRV \tag{4.2i}$$

$$H_{i,j} = k_{i,j} \cdot J, \quad w_{i_{ex}} = \begin{cases} 1, \text{infusion} \\ 0, \text{elsewhere} \end{cases}, \quad i = \begin{cases} F \\ EPI \end{cases} \tag{4.2j}$$

4.4 Results

4.4.1 Elements of the multiscale host response model of human inflammation

We have previously demonstrated that the transcriptional dynamics of human leukocytes exposed to bacterial endotoxin can be decomposed into to three elementary comprehensive responses [77]. Unlike previous approaches [8, 46–51] that concentrate on specific biomarkers, these elementary responses capture the functional dynamics and were shown to be related to proinflammatory (P), anti-inflammatory (A), and energetic (E) transcriptional events associated with the overall host response. The response is triggered by the activation of the NF-κB signaling module as a result of the formation of an activating signal associated with the binding of LPS to appropriate receptors (R). In order to introduce higher-level biological information, we opt to take essential aspects of the crosstalk between the peripheral immune response and the central nervous system into account. A schematic illustration of the network architecture that constitutes the multilevel host response model is presented in Figure 4.2. At the cellular level, interacting components are associated with elementary signaling pathways that propagate extracellular signals to the emergent transcriptional response level. At the autonomic level, essential modules associated with the release of endocrine stress hormones are incorporated. Such hormones are integral parts of the bidirectional communication pathway between peripheral inflammation (cellular level) and the neuroendocrine axis (HPA, SNS). As such, the essential proinflammatory response (P) acts as the peripheral intercellular messenger that stimulates the hardwired brain network, particularly at the neuroendocrine level. Activation of the neuroendocrine-immune axis releases stress hormones including cortisol and epinephrine that interact with appropriate receptors in immune cells, thereby potentiating the production rate of humoral anti-inflammatory cytokines (A). Finally, clinical measurements, at the systemic level, of HRV are also incorporated to assess disruptions in autonomic activities, which reflect the manifestation of systemic decomplexification and correlate with the severity of the host.

4.4.2 Estimation of relevant model parameters

Given the available human experimental data, standard parameter estimation techniques are applied in order to evaluate appropriate model parameters [128]. In particular, we estimate those parameters that are involved in the dynamics of proinflammation (P), anti-inflammation (A), energetics (E), NF-kB inhibitor ($mRNA_{IkBa}$), TLR4 ($mRNA_R$), cAMP mediator (cAMP), epinephrine (EPI), cortisol (F), and heart rate response (HRV). Furthermore, parameter estimation is performed to estimate the parameter ($R_{in,F}$) under conditions of hydrocortisone infusion (exogenous cortisol), reproducing human plasma

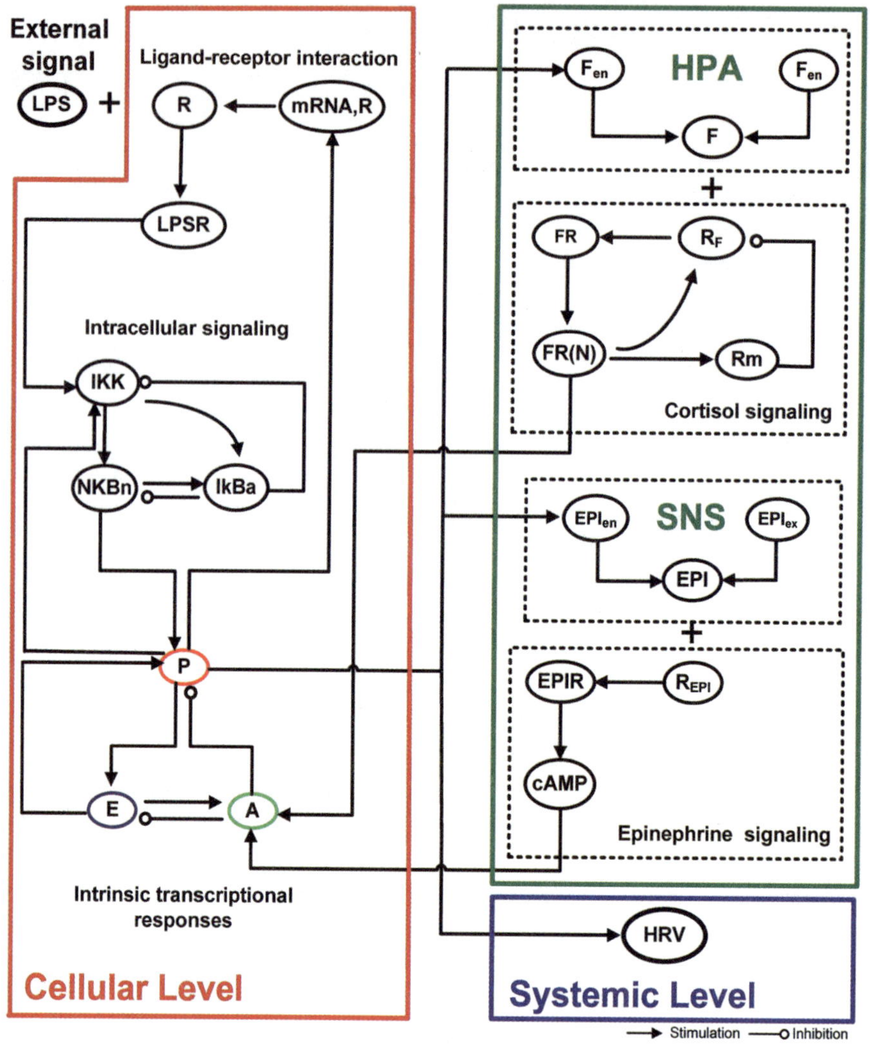

Figure 4.2 Topological interactions composing the multiscale model of inflammation. At the cellular level, interacting components involve the propagation of LPS signaling on the transcriptional response level (P, A, E) through the activation of endotoxin signaling receptor (R) and elementary signaling pathways (NFκB signaling module). At the level of circulating hormones, essential modules are associated with the release of endocrine stress hormones from neuroendocrine axis (HPA, SNS) coupled with their anti-inflammatory influence on the host. The dynamics of cortisol and epinephrine signaling involve components interacting at the cellular level. At the systemic level, physiologic deterioration of the host is quantified by HRV.

cortisol levels in subjects pre-exposed 6 hours prior to LPS and continued for another 6 hours after the endotoxin injection. All the parameters associated with the propagation of LPS signaling on the transcriptional and neuroendocrine levels are shown in Table 4.1, while the performance of the multilevel human inflammation model is shown in Figure 4.3. In addition to this, the reconstruction of plasma cortisol levels under conditions of prior cortisol infusion is shown in Figure 4.4.

In our computational model the host restores homeostasis without any external perturbation. A self-limited inflammatory response to the endotoxin stimulus corresponds to resolved dynamic profiles for all the elements that constitute our model. In essence, a

Table 4.1 Estimated Values of the Parameters Composing the Multilevel Human Inflammation Model

Parameter	Value	Parameter	Value	Parameter	Value
$k_{lps,1}$	4.500	$K_{in,A}$	0.461	$K_{EPI,P}$	0.231
$k_{lps,2}$	6.790	$K_{out,A}$	0.809	$k_{1,REPI}$	2.657
k_{syn}	0.020	$k_{A,cAMP}$	0.145	$k_{R,EPI}$	0.649
k_1	3.000	$k_{A,E}$	0.534	$k_{2,REPI}$	2.213
k_2	0.040	$k_{A,FRN}$	0.401	$k_{3,REPI}$	2.500
k_3	5.000	$K_{in,E}$	0.080	τ	0.053
k_4	2.240	$K_{out,E}$	0.280	n	5.509
$k_{in,mRNA,R}$	0.090	$k_{P,R}$	1.740	$K_{out,EPI}$	7.286
$k_{out,mRNA,R}$	0.250	$k_{P,1}$	29.740	k^0_{REPI}	6.594
$k_{NFkB,1}$	16.290	$k_{P,2}$	9.050	w	10
$k_{NFkB,2}$	1.180	$k_{E,P}$	2.210	τ_S	0.723
$K_{in,IKBa}$	0.460	$R_{in,F}(w_{Fex}=0)$	0	n_S	1.185
$k_{IkBa,1}$	13.270	$K_{in,Fen}$	0.842	$K_{in,HRV}$	0.038
$k_{I,1}$	1.400	$k_{Fen,P}$	0.256	$K_{out,HRV}$	0.038
$k_{I,2}$	0.870	$K_{out,F}$	1.058	$k_{HRV,S}$	35.254
$K_{in,P}$	0.030	$R_{in,F}(w_{Fex}=1)$	2.922		
$K_{out,P}$	0.330	$K_{in,EPI}$	5.921		

Table 4.2 Summarizing the Input and Output Data Relevant to Model Predictions

Figure Legend	Input Data	Output Data (Responses)
Figure 4.3	LPS(t = 0h) = 1	LPS, LPSR, IKK, $mRNA_R$, R, $NFkB_n$, $mRNA_{IkBa}$, Cyt_{IkBa}, P, A, E, EPI, R_{EPI}, EPIR, cAMP, F, FR(N), f_p, S_f, HRV
Figure 4.4	LPS(t = 0 hours) = 1, w_{Fex} = 1	F, FR(N)
Figure 4.5	LPS(t = 0 hours) = 4	P, A, E, EPI, F, HRV
Figure 4.6	LPS(t = 0 hours) =1, $k_{lps,2}$=3.395	LPS, $NFkB_n$, P, A, F, HRV
Figure 4.7	LPS(t = 0 hours) = 1, $k_{IkBa,1}$ = 0	$mRNA_{IkBa}$, IKK, $NFkB_n$, P, A, E
Figure 4.8	LPS(t = 0 hours) = 1, $K_{in,P}$ = 0.06	P, $mRNA_{IkBa}$, $NFkB_n$, A
Figure 4.9	Dashed line: LPS(t = 0h) = 0 and LPS(t = 4 hours) = 1 Solid line: LPS(t = 0 hours) = 0.2 and LPS(t = 4 hours) = 1	$NFkB_n$, P, A, E
Figure 4.10	Dashed line: LPS(t = 0 hours) = 0 and LPS(t = 0.2 hour) = 2 Solid line: LPS(t = 0 hours) = 1 and LPS(t = 0.2 hour) = 2	$NFkB_n$, P, A, E
Figure 4.11	Dashed line: LPS(t = 0 hours) = 4, $R_{in,EPI}$ = 15 (w_{EPIex} = 1) Solid line: LPS(t = 0 hours) = 4, w_{EPIex} = 0	EPI, cAMP, P, HRV
Figure 4.12	Dashed line: LPS(t = 0 hours) = 4 $R_{in,F}$ = 2.922 (w_{Fex} = 1) Solid line: LPS(t = 0 hours) = 4 w_{Fex} = 0	F, FR(N), P, HRV
Figure 4.13	Dashed line: LPS(t = 0 hours) = 1 $R_{in,F}$ = 2.922 (w_{Fex} = 1) Solid line: LPS(t = 0 hours) = 1 w_{Fex} = 0	F, FR(N), P, A, EPI, HRV
Figure 4.14	Dashed line: LPS(t = 0 hours) = 1, $R_{in,EPI}$ = 20 (w_{EPIex} = 1) Solid line: LPS(t = 0 hours) = 1, w_{EPIex} = 0	EPI, EPIR, cAMP, A, P, HRV

self-limited inflammatory response involves the successful elimination of the inflammatory stimulus within the first 2 hours postendotoxin administration while followed by a subsequent resolution within 24 hours. Though the kinetic parameters associated with the epinephrine-receptor interactions are not calibrated, the dynamic profile of

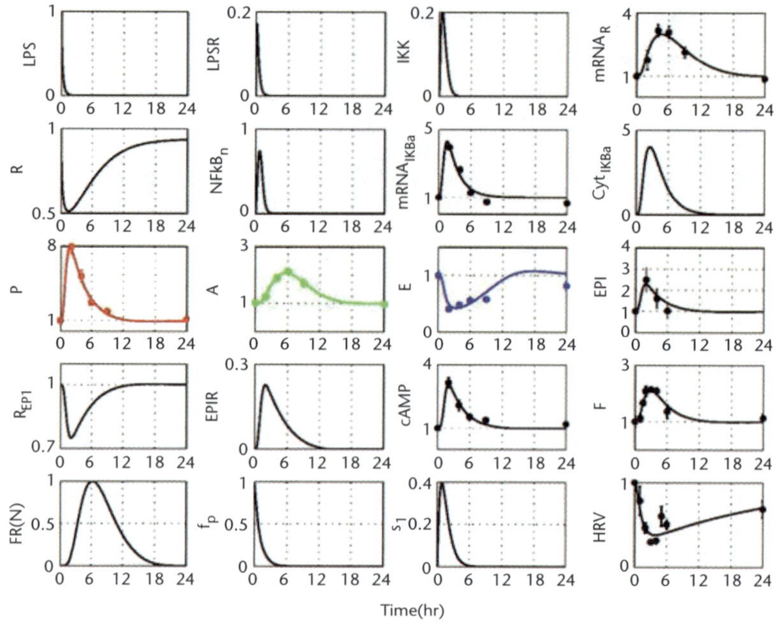

Figure 4.3 Estimation of relevant model parameters intending to reproduce experimental measurements. Solid lines (—) correspond to model predictions under LPS administration, while the symbols () refer to experimental data expressed as mean ±SEM. The initial condition of endotoxin (LPS(t = 0 hours) = 1) refers to LPS concentration relative to 2-ng/kg body weight.

β-adrenergic receptor (R_{EPI}) lies in qualitative agreement with the basis of receptor occupancy theory [114] in that the concentration of free adrenergic receptors decreases in the presence of the ligand (EPI). Regarding the gain modulatory effect of peripheral proinflammation in the heart (f_P), such an exponential decrease would biologically reflect the decay rate of cardiac neuronal activity [129], which is eventually "translated" to the upregulation of neuromediator concentrations (S_f) in the heart. While comparing Figure 4.4(a, b) and Figure 4.4(c, d), both plasma cortisol levels (F_{Fex}) and the steroid active signal ($FR(N)_{Fex}$) are expected to be greater under conditions of hypercortisolemia [Figure 4.4(c, d)] relative to the baseline cortisol profiles [Figure 4.4(a, b)].

4.4.3 Qualitative assessment of the model

Building a mathematical model that can predict relevant biological implications to the host response to endotoxin allows us to identify ways of both controlling and modulating such a complex phenomenon.

In the following sections, we demonstrate the ability of our model to enable such "predictions" and provide further evidence of the appropriateness of the assumptions invoked in the development of the model. First, we explore the implications of increasing levels of initial insult (LPS) since this would probably constitute the most obvious irreversible disturbance. Second, we explore possible mechanistic dysregulation, which may reflect secondary effects that lead to potential malfunction of the response leading to sustained inflammation. Then we explore the emergence of "memory" effects and evaluate the implication of priming the host with controlled level of endotoxin stimulus as a prequel to the main inflammatory stimulus. Finally, we evaluate a possible modula-

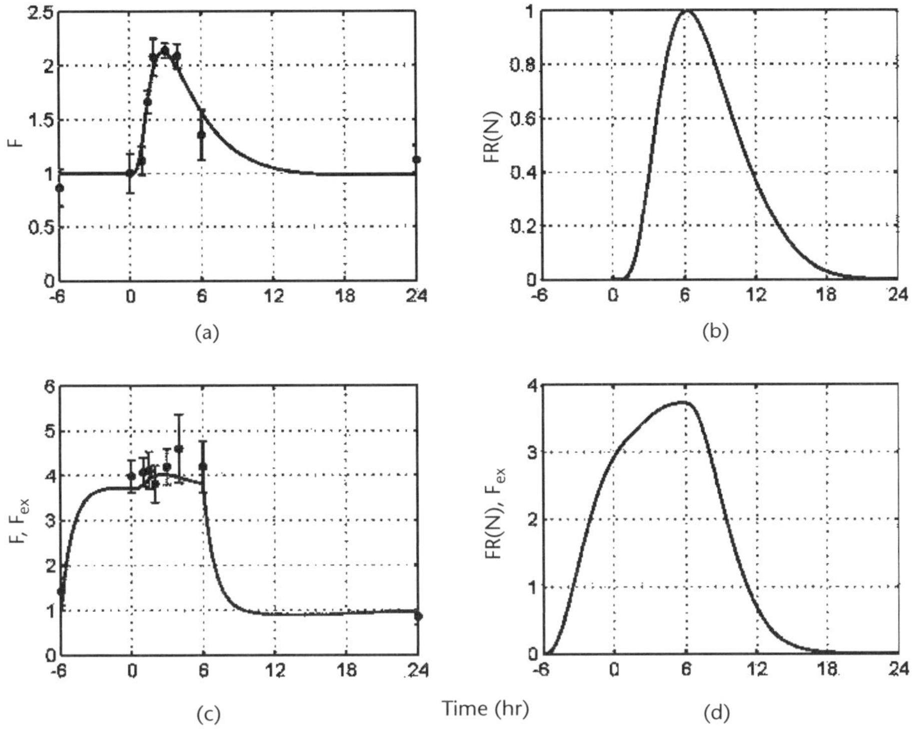

Figure 4.4 (a) Plasma cortisol levels (F) and (b) steroid active signal (FR(N)) under conditions of acute endotoxin injury (LPS(t = 0 hours) = 1); and (c) simulated plasma cortisol levels (F_{Fex}) and (d) steroid active signals $FR(N)_{Fex}$ under conditions of low-dose steroid infusion (w_{Fex} = 1), which is initiated at t = –6 hours before LPS and continued for 6 hours after LPS. Solid lines correspond to model predictions, while solid markers represent experimental data expressed as mean ± SEM.

tion in the dynamics of the host in response to an acute hormone stress infusion (glucocorticoid, catecholamine excess) under conditions of either high or low infectious challenge.

4.4.3.1 Implications of increased insult

An increase in the dose of the inflammatory stimulus can be responsible for an overwhelming inflammatory response. Such a situation in which the initial levels of endotoxin are increased is simulated in Figure 4.5. We observe that when the concentration of LPS is high enough, the response does not abate. Such computational result implicates the role of the host rather than the inflammatory stimulus itself, which is eventually cleared from the system, accounting for the progression of an unconstrained inflammatory response. The uncontrolled (overwhelming) production of pro-inflammatory signaling in turn contributes to derangements in autonomic activity characterized by endocrine dysfunction and a loss of adrenergic responsiveness that drives an overall diminished heart rate variability [120].

4.4.3.2 Modes of dysregulation of the inflammatory response

The dynamics of the host response to an inflammatory stimulus are highly complex and nonlinear, suggesting that a dysregulation in the dynamics of the inflammatory

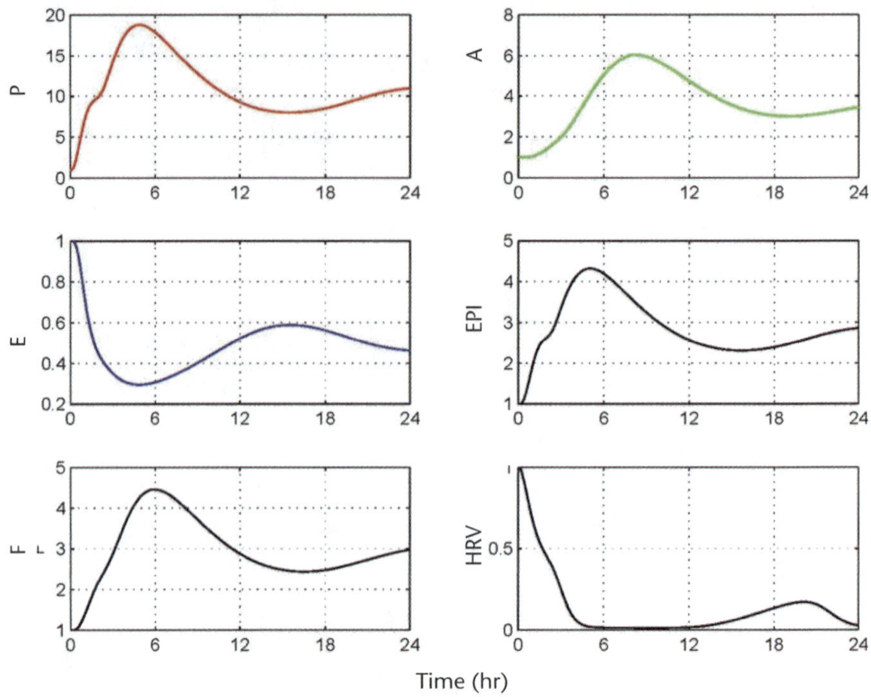

Figure 4.5 Simulation of an unresolved inflammatory response due to a high endotoxin concentration [(LPS(t = 0 hours) = 4)]. Such a high concentration of LPS dysregulates the NFκB signaling module, giving rise to an unconstrained immune response followed by abnormal hormonal responses that macroscopically are translated into diminished physiologic variability.

response in restoring homeostasis can have multiple causes. One such possibility is associated with a malfunction in the clearance rate of endotoxin (which corresponds to a higher exposure of the host response to the stimulus followed by persistence in the concentration of LPS). A dysregulation in the intracellular dynamics can occur and cause an aberrant response, given that improper regulations of NFκB signaling might also be involved in the inflammatory component of a disease like sepsis [130]; such a mode of dysregulation separates the host response dynamics from the insulting agent. The model was probed by appropriately manipulating parameters that include: (1) a reduction in the first-order degradation rate of LPS, (2) a knockout reduction in the transcriptional rate of the primary inhibitor of NF-κB (IkBa), and (3) a potentiation in the production rate of proinflammatory mediators.

A persistent inflammatory response which corresponds to an increased exposure of the host response to the inflammatory stimulus (LPS) is simulated in Figure 4.6. As such, the inferred NF-κB activity can be characterized as a "two-wave" response; initially it increases due to the inflammatory stimulus while trying to adapt its regulatory activity at 2–3-hour postendotoxin administration. However, at $t > 3$ hours the activity of NF-κB cannot be regulated successfully and it settles to a sustained elevated state that drives downstream the overexcitation of both pro- and anti-inflammatory mediators, leading to an unconstrained inflammatory response. Interestingly, in [131] Klinke et al. explored experimentally the possibility of modulating the temporal control of NF-κB activation. Macrophages are exposed to a persistent inflammatory stimulus (LPS) and

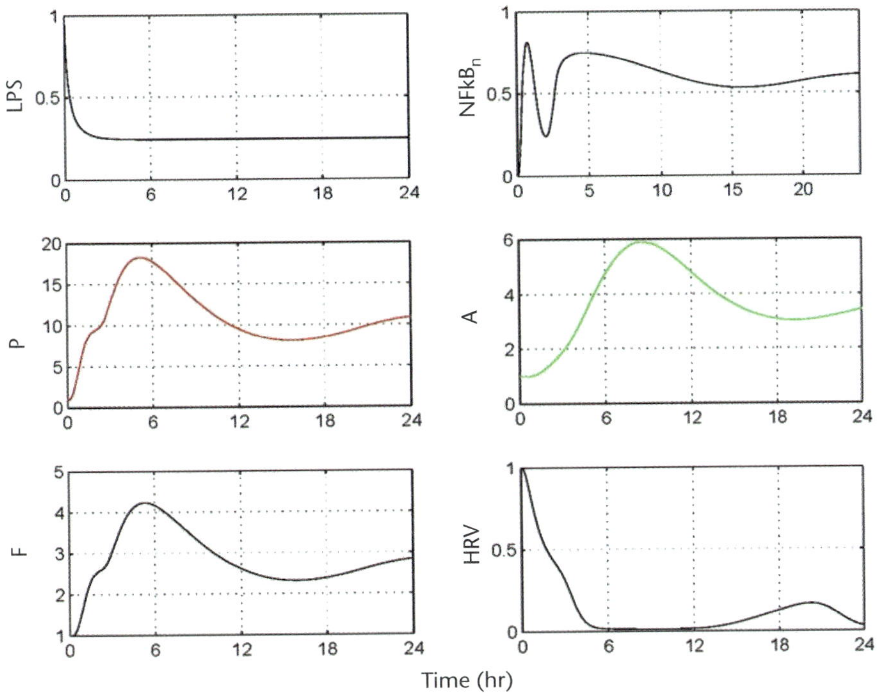

Figure 4.6 Temporal responses of inflammatory components in persistent infectious inflammatory response where the inflammatory stimulus cannot be eliminated responsible for the observed persistence in the dynamic profiles of the inflammatory constituents. Reducing the degradation rate of LPS to half of its initial value the inflammatory stimulus cannot be cleared.

the available experimental data show the presence of a "damped" oscillatory behavior in NF-κB activity.

The protein inhibitor of NF-κB (IkBa) aims at retrieving nuclear concentration of NF-κB with the formation of an inactive complex in the cytoplasm regulating the expression of various inflammatory genes. The transcription factor NF-κB upregulates the gene transcript of IkBa ($mRNA_{,IkBa}$) so that the translated protein IkBa serves as the major component for regulating its transcriptional activity. Thus, the case of no transcriptional activity of NF-κB in the promoter region of IkBa is simulated in Figure 4.7. In the absence of NF-κB inhibitor ($IkBa^{-/-}$), there is an aberrant regulatory activity of NF-κB that leads to its persistent nuclear activity driving an inflammatory response that fails to restore homeostasis. Such an in silico result has been experimentally tested annotating the impact of such a knockout in inducing a chronic inflammatory response [81]. Additionally, a pre-existence of proinflammatory cytokines due to the presence of a prior "insult" may deregulate the intracellular dynamics responsible for an amplification of the inflammatory response, as shown in Figure 4.8. In our model such a scenario can be simulated due to the positive feedback interaction between the intracellular critical node (IKK activity) and the proinflammatory response that disturbs the bistable behavior of the system. Thus, such mode of perturbation deregulates the NFκB signaling module leading to a persistent NFκB activity. Such persistence implies that the nuclear concentration of NFκB cannot be further constrained by its primary inhibitor, IkBa and eventually settle to a steady state far away from their equilibrium (homeostasis). We simulate such a scenario by manipulating the zero order production rate of the

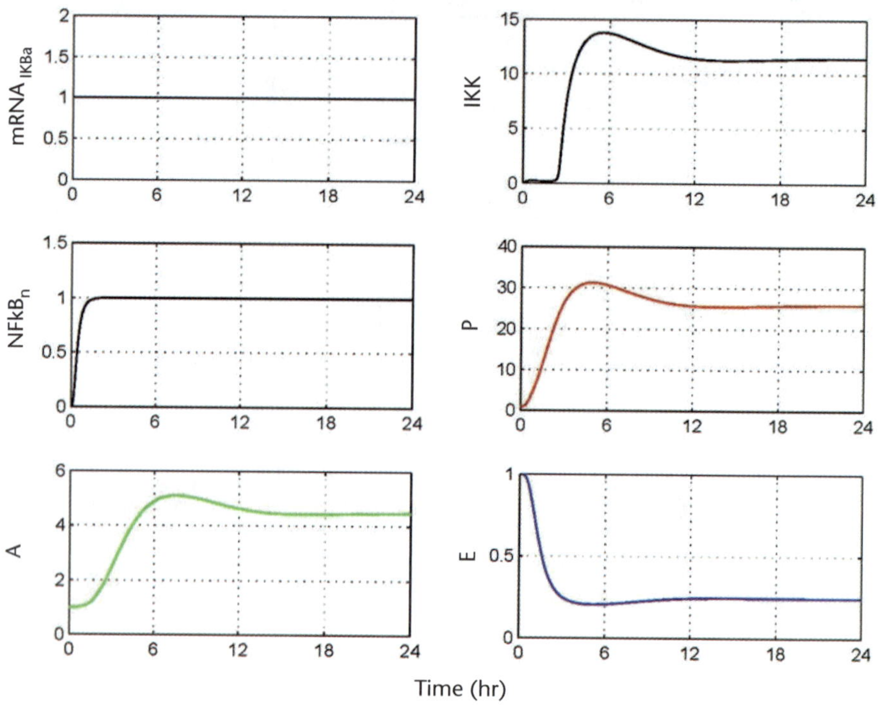

Figure 4.7 Simulation of a knockout in silico experiment (IkBa$^{-/-}$). Manipulating the model so that there is no de novo transcriptional synthesis of NF-κB inhibitor (IkBa), which is responsible for the absence of NF-κB autoregulatory feedback loop. Such a scenario accounts for maladapted activity of NFkBn that triggers an uncompensated inflammatory response.

proinflammatory response ($K_{in,P}$) and particularly increasing it twice its initial value. Clinically, such an increased rate in the production of proinflammatory mediators might be the outcome of a surgical trauma followed by bacterial infection, a so-called two hit scenario [132].

4.4.3.3 The emergence of memory effects

Repeated doses of endotoxin stimulus in many instances are characterized by a less vigorous immune system, which is known as endotoxin tolerance [133]. Even though the phenomenon of endotoxin tolerance involves the administration of low, repeated doses of endotoxin over periods of time ranging from 1 day to 1 week [134], herein we opt to investigate the response of the system being pre-exposed to low dose of LPS for less than a day. This is because in our proposed model all the interacting components do resolve within the first 24 hours while the system has not acquired a reprogramming dynamic state. However, published studies [135, 136] have reported that "rapid" endotoxin tolerance can be induced when the system is pre-exposed to a low endotoxin challenge for between 3–6 hours. Thus, we simulate such a scenario in Figure 4.9. When the system is pre-exposed to a lower inflammatory stimulus for about 4 hours before the main endotoxin challenge, our model predicts a much less vigorous inflammatory response. In particular, such an event, which can be characterized either as a short-time attenuation effect or else a rapid tolerance, is experimentally observed by the decreased concentrations of various proinflammatory mediators (e.g., TNF-α), IL1B in response to a sec-

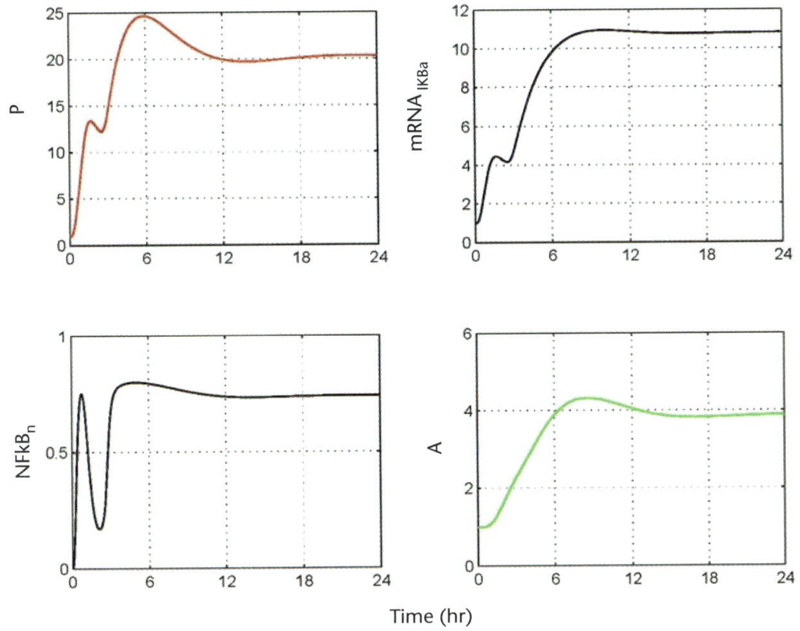

Figure 4.8 Preexistence of proinflammatory mediators may enhance abnormally the intracellular signaling through IKK. Such a response leads to an unconstrained activity of NFkBn that drives downstream a persistent proinflammatory response which cannot be counter-regulated by the anti-inflammatory arm of the host defense system. Such a mode of dysregulation is simulated by manipulating the zero production rate of proinflammation (Kin,P) so that Kin,P(unhealthy response) ~2* Kin,P (healthy response).

ondary ex vivo whole blood stimulation with LPS [135]. In addition to this, in the experimental study [136], concentrations of the particular proinflammatory mediator (TNF-α) were decreased profoundly ex vivo at 3–6 hours after in vivo endotoxin administration. However, by 24 hours the endotoxin tolerance had completely resolved. Such preconditioning results in an attenuation of the inflammatory response characterized by a less vigorous intracellular signaling coupled with the decreased peak level of the proinflammatory response.

The magnitude and the timing of repeated doses of endotoxin are key determinants for discriminating between endotoxin tolerance and potentiation. The successive administration of low doses of LPS may perturb the system's homeostasis towards the progression of an unresolved inflammatory response. Thus, if the repeated doses are characterized by a very short time interval, it is possible for the dynamics of the system to result in an overwhelming inflammatory response. Therefore, the successive administration of two inflammatory insults that individually account for constrained ("self-limited") inflammatory responses might be detrimental to the outcome of sepsis (unresolved inflammatory response). Such an event can occur because of the absence of a "protective" memory in the system so that the system has not elicited its regulatory mechanism to compensate for the cumulative result of two successive doses. Such an abrupt insult might dysregulate the dynamics of the host response to infection, having a detrimental effect in the physiological state of the system as seen in Figure 4.10.

These results indicate that the cellular response is critically affected by the mode of exposure, thus demonstrating the need for an appropriate, quantifiable model to

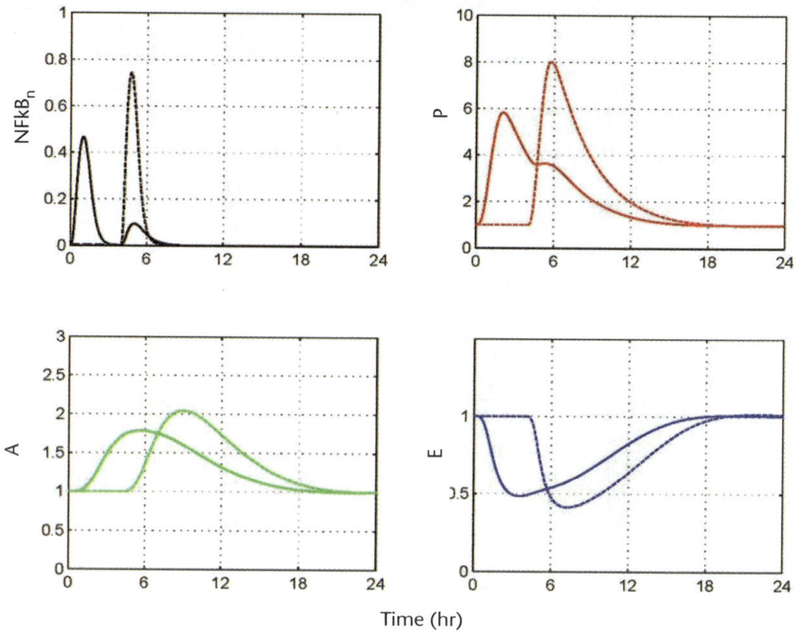

Figure 4.9 "Rapid" endotoxin tolerance. Pre-exposing the system into a smaller inflammatory insult results in a reduction in the cell capacity to respond to the main endotoxin challenge, which is characterized as a short-time attenuation scenario. Solid line: LPS(t = 0 hours) = 0.2 and LPS(t = 4 hours)=1 Dashed line: LPS(t = 0 hours)=0 and LPS(t = 4 hours) = 1.

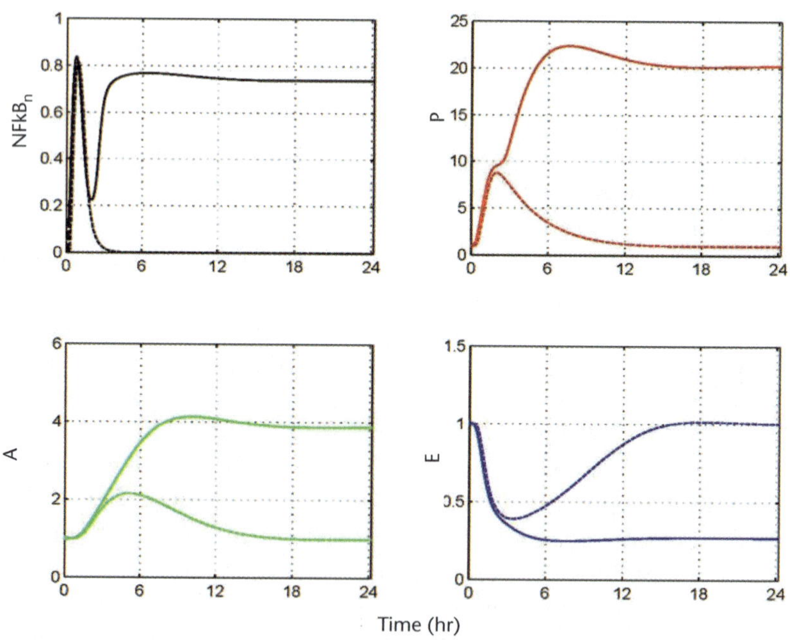

Figure 4.10 Lethal potentiation. Successive administration of small doses of endotoxin can lead to an unresolved inflammatory response (due loss of "regulatory" memory). Solid line: LPS(t = 0 hours) = 1 and LPS(t = 0.2 hour) = 2; dashed line: LPS(t = 0 hours) = 0 and LPS(t = 0.2 hour) = 2.

account for and integrate the various components constituting the response. Furthermore, our results clearly indicate that the dynamics of the response are definitely affected by the parameters defining the exposure to the inflammatory agent.

4.4.3.4 Evaluation of stress hormone infusion in modulating the inflammatory response

We have demonstrated the ability of our model to simulate the trajectory of an unconstrained inflammatory response, and the potential of the proposed model is also demonstrated through its capability to respond to systematic perturbations that intend to modulate the dynamics in favor of a balanced immune response coupled with a restoration in autonomic balance. Considerable attention has been given to the effectiveness of pharmacological agents such as ligands of adrenergic receptors in influencing the production rate of both pro- and anti-inflammatory cytokines [137, 138]. In particular, significant modulations in the cytokine network were observed in human subjects exposed to epinephrine infusion [111], underscoring the role of neuroendocrine activity in dampening excessive proinflammatory effects. In particular, we opt to simulate the mode of an intervention strategy that mimics the activity of the sympathetic nervous system pathway. Such an intervention strategy results in the potentiation of the total plasma concentration of epinephrine (EPI), which further increases the intracellular cAMP signaling (dashed lines) in Figure 4.11. Based on the anti-inflammatory effect of acute EPI infusion via the cAMP-dependent mechanism, it is expected that an increase in intracellular cAMP levels will attenuate the proinflammatory response (P) and ultimately restore autonomic activity (HRV), which serves as a proxy indicator of improved survival [139]. Such improvement in autonomic activity underscores the role of epinephrine in improving cardiac index under severe conditions (i.e., low-output septic shock) as supported by Court et al [140].

Furthermore, the autonomic nervous system controls inflammation at various levels that involves not only the release of sympathetic neurotransmitters but also the central secretion of glucocorticoids. During the progression of a systemic inflammatory response syndrome, there is an important problem of adrenal insufficiency, which is common in the ICU [141]. Recently, a possible association between reduced HRV and adrenal insufficiency in trauma patients was explored in [124], where an increase in HRV was observed in patients responding to steroid therapy. We opt to simulate such a scenario in Figure 4.12 assuming that the trajectory of an unconstrained response (high LPS concentration) would qualitatively reflect critically ill patients that are diagnosed with relative adrenal insufficiency. The initiation of such intervention strategy increases the total amount of endogenous cortisol (F) which subsequently potentiates the active steroid signal (FR(N)). This increase in total cortisol levels potentiates the anti-inflammatory arm of the system (A) immediately after the administration of LPS, and thereby attenuates the proinflammatory response (P). Thus, the indirect attenuation of the proinflammatory signaling (P) via potentiation of the humoral anti-inflammatory signaling (A) suffices to reverse the inflammatory dynamics and eventually restore autonomic balance.

Although the immunosuppressive effects of corticosteroids upon the systemic inflammatory manifestations of human endotoxemia have been well described, the influence of this anti-inflammatory intervention on overall autonomic dysfunction is

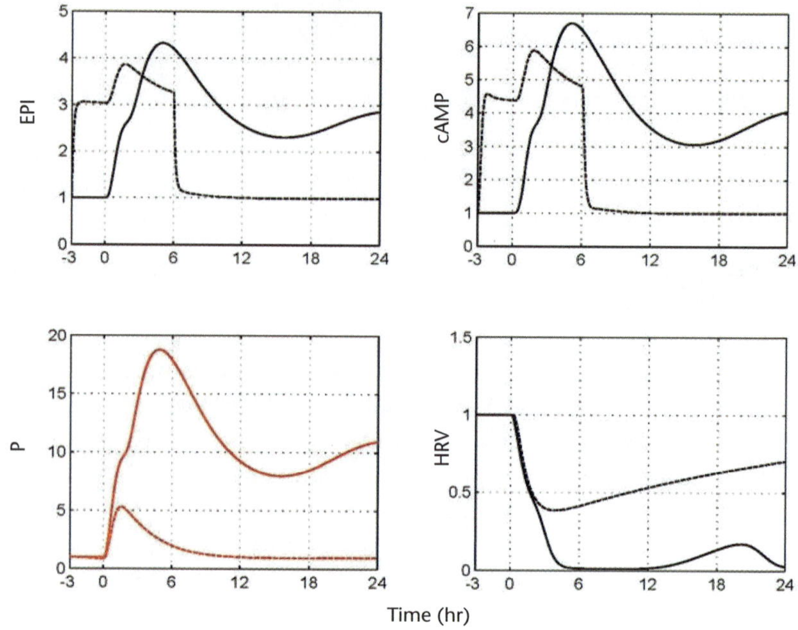

Figure 4.11 Simulating the effect of acute epinephrine infusion ($w_{EPI,ex}$=1, $R_{in,EPI}$=15) that is initiated 3 hours prior to the main endotoxin challenge (LPS(t = 0 hours) = 4) and continued for 6 hours after LPS. Dashed and solid lines represent the progression of a balanced (due to the system's pre-exposure into epinephrine infusion) and unconstrained inflammatory response (due to high infectious challenge), (LPS(t = 0 hours) = 4), respectively.

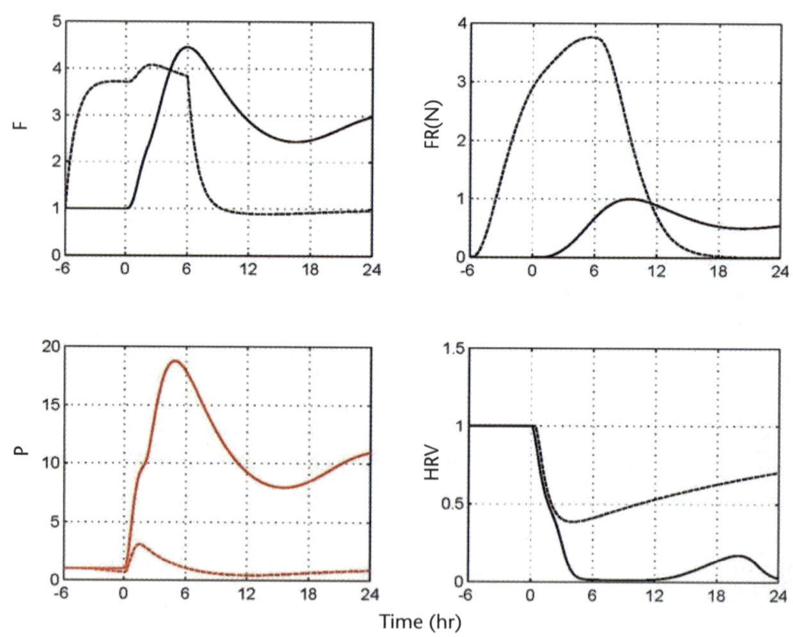

Figure 4.12 Explore the effect of low-dose steroid administration 6 hours prior to the endotoxin challenge (dashed lines) while continued for another 6 hours after LPS (w_{Fex}=1). Solid lines simulate the progression of a systemic inflammatory response syndrome (high LPS concentration, LPS(t = 0 hours) = 4) while dashed lines reflect the protective effect that can be exerted by such hormonal(steroid) replacement therapy that augments the production of humoral anti-inflammatory mediators (A)

not well understood. Predicated upon this, the influence of steroid administration on a self-limited endotoxin-induced inflammatory response is simulated in Figure 4.13. Accordingly, a potentiation in IL-10 signaling (A) is observed under conditions of hypercortisolemia, followed by attenuation in the proinflammatory response (P). Such attenuation further induces hormonal changes, including a reduction in plasma epinephrine concentration. Experimentally, a decrease in endogenous epinephrine secretion under acute hypercortisolemia has been demonstrated in [42], thus validating the assumptions invoked in the development of the proposed integrated model. In addition to cytokine and hormonal measurements under hypercortisolemia, phenotypic responses including HRV parameters are further obtained in [43]. Remarkably, although acute hypercortisolemia significantly attenuated proinflammatory cytokines, such attenuation does not contribute to any alterations in HRV indices.

In addition to the influence of low-dose steroid on endotoxin-induced inflammation, recent data document that prior EPI exposure may attenuate the proinflammatory response, but such anti-inflammatory influence does not extend to changes in the system's overall adaptability (HRV) [123]. Since increased catecholamine secretion accompanies modest infection and the effect of EPI in inhibiting LPS-induced proinflammatory response has been documented in [142], we sought to simulate whether antecedent EPI infusion would alter the cytokine responses to endotoxin, as shown in Figure 4.14. In particular, increasing plasma EPI levels modulates the innate immune system activation and particularly attenuates the proinflammatory response

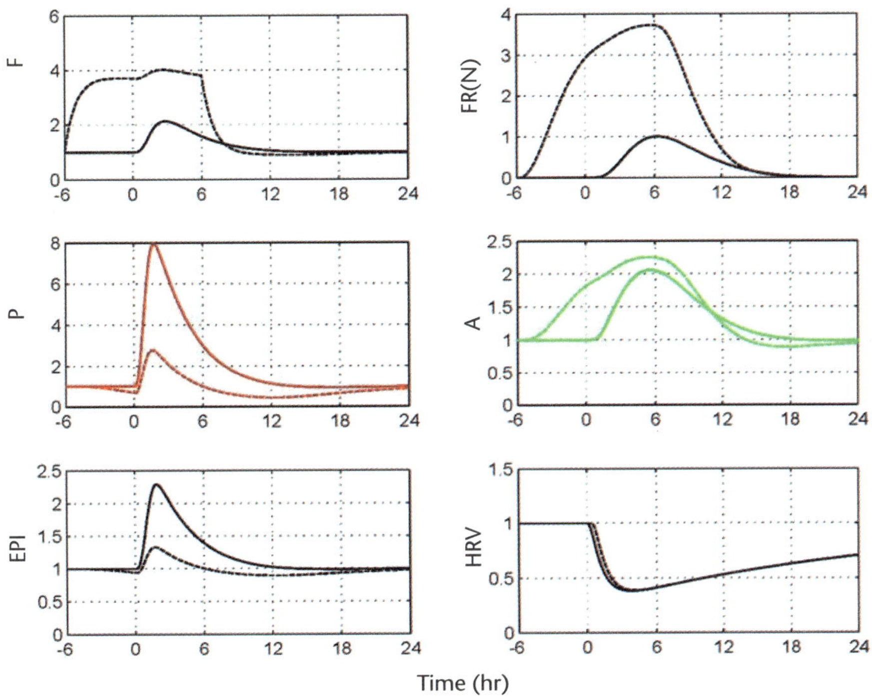

Figure 4.13 Explore the effect of hypercortisolemia on autonomic dysfunction under the systemic inflammatory manifestations of human endotoxemia. Solid lines simulate a self-limited inflammatory response (low-dose endotoxin, LPS(t = 0 hours) = 1) while dashed lines reflect the potentiation of IL-10 signaling due to continuous steroid infusion initiated 6 hours prior to LPS administration and continued for 6 hours after endotoxin (w_{Fex} = 1).

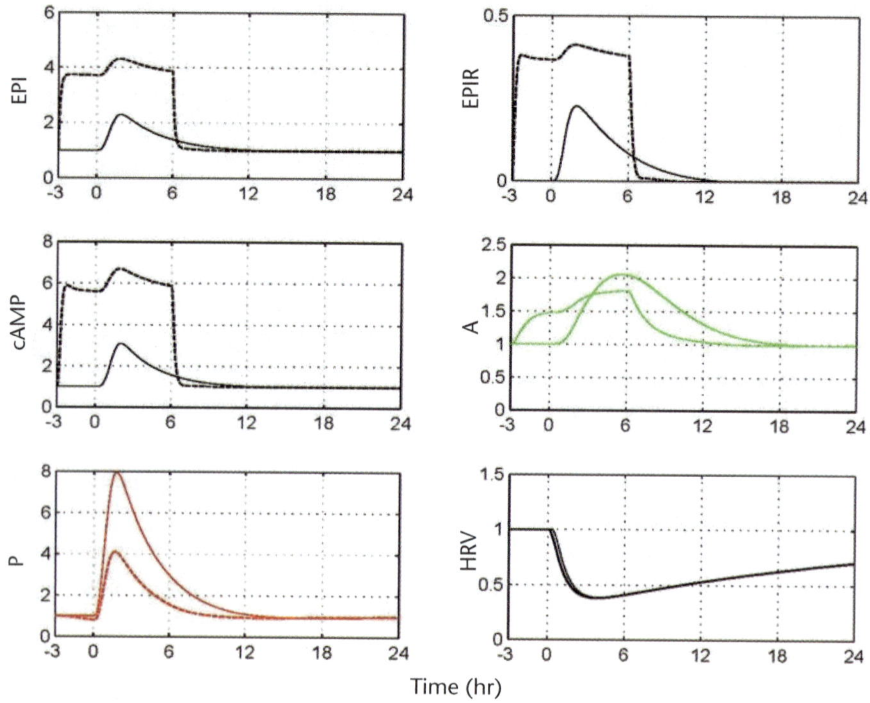

Figure 4.14 Modulation in the progression of the inflammatory reaction due to short-term epinephrine infusion (initiated 3 hours before LPS and continued for 6 hours after LPS). Such intervention potentiates the secretion of epinephrine from SNS that through cAMP anti-inflammatory signaling can protect in part the host response attenuating the proinflammatory response (P). Solid lines simulate a self-limited endotoxin-induced inflammatory reaction while dashed lines reflect the scenario of prior epinephrine infusion ($w_{EPIex} = 1$, $R_{in,EPI} = 20$).

through potentiation of the anti-inflammatory effect of cAMP signaling. However, such attenuation in the progression of the inflammatory response does not contribute to any changes in HRV response, which is consistent with the aforementioned results with steroid administration before LPS. From a modeling standpoint, such responses are captured due to the possible nonlinear crosstalk between peripheral proinflammation and heart rate response, thus making our modeling effort a critical enabler for evaluating the influence of neuroendocrine activity on the cellular and systemic levels.

In summary, a multiscale model of human endotoxemia, as a prototype model for systemic inflammation in humans, is proposed that couples critical aspects of the complex bidirectional relationship between the neuroendocrine axis and the immune system. This modeling effort aims at exploring the emergence of interaction networks at the (low) level of intracellular signaling, regulation and the (high) level of interacting hormonal and physiological components that give rise to an overall systemic response. At the (low) cellular level interacting components are associated with elementary signaling pathways that propagate extracellular signals to the transcriptional response level. Furthermore, essential modules associated with the immunomodulatory role of endocrine stress hormones (cortisol, epinephrine) are considered. Finally, at the systemic (high) level, phenotypic expressions such as HRV are further incorporated to assess autonomic dysfunction indicative of the severity of illness. Since both corticosteroids and catecholamines are used clinically in the context of systemic inflammation, the proposed model has the potential for direct clinical relevance. Thus, such a modeling effort lays the foundation for a translational systems-based model of inflammation that could clarify the clinical contexts in which autonomic dysfunction contributes to morbidity and mortality in severely stressed patients.

It is the goal of this study to demonstrate the feasibility of the proposed approach as a notional template for multiscale modeling in human physiology. By incorporating biological information in the form of critical signaling cascades and kinetic rules we would probably be able to explore the possibility of the generalization of this framework in a wide range of disease progression models. In this study we evaluated human data of immune competent populations with reversible inflammatory challenges; however, it is in our future plans to further assess our methods on the Trauma Related Database (TRDB) currently being assembled as part of the Inflammation and Host Response to Injury Glue grant (http://www.gluegrant.org/).

4.5 Conclusions

In this chapter we discussed the potential of systems-based approaches in developing appropriate network models. We demonstrated how data analysis can yield significant insights and enable the generation of testable hypotheses. The overarching goals of this research are to demonstrate the feasibility of a clinically relevant, mechanistic-based, predictive, multiscale human inflammation model coupling essential aspects of the complex bidirectional relationship between the CNS and the immune response and to propose a template for multiscale modeling extendable to a variety of clinically important conditions. It is important to realize that in silico models will never replace either biological or clinical research. They could, however, rationalize the decision-making process by establishing the range of validity and predictability of intervention strategies, thus paving the way for improving the working feedback loop between "dry" and "wet" experiments. As our understanding of the intricacies of biological systems increases, in silico modeling will become an increasingly attractive tool for applying that knowledge to decrease our reliance on in vivo testing. Finally, we would like to argue that possibly a very significant, and often overlooked, success of systems-based research is that through the universal language of mathematics and the opportunity of formalizing and quantifying abstract concepts of complex physiological phenomena, albeit with significant simplifications, oftentimes it has managed to establish communication bridges between scientists from a variety of fields with a common goal: to develop a better understanding of a physiological condition. This could be one of the most significant impacts of systems-based translational research.

Acknowledgments

We wish to acknowledge the invaluable input of our collaborators R. R. Almon and W. J. Jusko (Biological Sciences, SUNY Buffalo). We also acknowledge financial support from the NIH under grant GM082974, NSF, under grant 0519563, the EPA, under grant GAD R 832721-010, and a Busch Biomedical Research Award. SEC and SFL are supported, in part, from USPHS Grant GM34695. The investigators acknowledge the contribution of the Inflammation and the Host Response to Injury Large-Scale Collaborative Project Award # 2-U54-GM062119 from the National Institute of General Medical Sciences. The Inflammation and the Host Response to Injury "Glue Grant" program is supported by the National Institute of General Medical Sciences. This manuscript was prepared using a dataset obtained from the Glue Grant program and does not necessarily reflect the opinions or views of the Inflammation and the Host Response to Injury Investigators or the NIGMS.

References

[1] Vogel, T. R., V. Y. Dombrovskiy, and S. F. Lowry, "Trends in Postoperative Sepsis: Are We Improving Outcomes?" *Surg. Infect. (Larchmt)*, Vol. 10, No. 1, 2009, pp. 71–78.

[2] Bruce, J., et al., "The Measurement and Monitoring of Surgical Adverse Events," *Health Technol. Assess.*, Vol. 5, No. 22, 2001, pp. 1–194.

[3] Angus, D. C., et al., "Epidemiology of Severe Sepsis in the United States: Analysis of Incidence, Outcome, and Associated Costs of Care," *Crit. Care Med.*, Vol. 29, No. 7, 2001, pp. 1303–1310.

[4] Lowry, S. F., "The Stressed Host Response to Infection: The Disruptive Signals and Rhythms of Systemic Inflammation," *Surg. Clin. North Am.*, Vol. 89, No. 2, 2009, pp. 311–326.

[5] Decker, T., "Sepsis: Avoiding Its Deadly Toll," *J. Clin. Invest.*, Vol. 113, No. 10, 2004, pp. 1387-9.

[6] Cavaillon, J. M., and D. Annane, "Compartmentalization of the Inflammatory Response in Sepsis and SIRS," *J. Endotoxin Res.*, Vol. 12, No. 3, 2006, pp. 151–170.

[7] Lowry, S. F., and S. E. Calvano, "Challenges for Modeling and Interpreting the Complex Biology of Severe Injury and Inflammation," *J. Leukoc. Biol.*, Vol. 83, No. 3, 2008, pp. 553–557.

[8] Clermont, G., et al., "In Silico Design of Clinical Trials: A Method Coming of Age," *Crit. Care Med.*, Vol. 32, No. 10, 2004, pp. 2061–2070.

[9] Marshall, J. C., et al., "Preclinical Models of Shock and Sepsis: What Can They Tell Us?" *Shock*, Vol. 24, Suppl. 1, 2005, pp. 1–6.

[10] Marshall, J. C., "Modeling MODS: What Can Be Learned from Animal Models of the Multiple-Organ Dysfunction Syndrome?" *Intensive Care Med.*, Vol. 31, No. 5, 2005, pp. 605–608.

[11] Kitano, H., "Systems Biology: A Brief Overview," *Science*, Vol. 295, No. 5560, 2002, pp. 1662–1664.

[12] Rajasethupathy, P., S. J. Vayttaden, and U. S. Bhalla, "Systems Modeling: A Pathway to Drug Discovery," *Curr. Opin. Chem. Biol.*, Vol. 9, No. 4, 2005, pp. 400–406.

[13] An, G., "Introduction of an Agent-Based Multi-Scale Modular Architecture for Dynamic Knowledge Representation of Acute Inflammation," *Theor. Biol. Med. Model*, Vol. 5, 2008, p. 11.

[14] An, G., J. Faeder, and Y. Vodovotz, "Translational Systems Biology: Introduction of an Engineering Approach to the Pathophysiology of the Burn Patient," *J. Burn Care Res.*, Vol. 29, No. 2, 2008, pp. 277–285.

[15] Foteinou, P. T., et al., "Translational Potential of Systems-Based Models of Inflammation," *Clinical and Translational Science*, Vol. 2, No. 1, 2009, pp. 85–89.

[16] Vodovotz, Y., et al., "Translational Systems Biology of Inflammation," *PLoS Comput. Biol.*, Vol. 4, No. 4, 2008, p. e1000014.

[17] Zenker, S., J. Rubin, and G. Clermont, "From Inverse Problems in Mathematical Physiology to Quantitative Differential Diagnoses," *PLoS Comput. Biol.*, Vol. 3, No. 11, 2007, p. e204.

[18] Vodovotz, Y., et al., "Mechanistic Simulations of Inflammation: Current State and Future Prospects," *Math Biosci.*, Vol. 217, No. 1, 2009, pp. 1–10.

[19] Lowry, S. F., "Human Endotoxemia: A Model for Mechanistic Insight and Therapeutic Targeting," *Shock*, Vol. 24, Suppl. 1, 2005, pp. 94–100.

[20] Calvano, S. E., et al., "A Network-Based Analysis of Systemic Inflammation in Humans," *Nature*, Vol. 437, No. 7061, 2005, pp. 1032–1037.

[21] Fannin, R. D., et al., "Differential Gene Expression Profiling in Whole Blood During Acute Systemic Inflammation in Lipopolysaccharide-Treated Rats," *Physiol. Genomics*, Vol. 21, No. 1, 2005, pp. 92–104.

[22] Talwar, S., et al., "Gene Expression Profiles of Peripheral Blood Leukocytes After Endotoxin Challenge in Humans," *Physiol. Genomics*, Vol. 25, No. 2, 2006, pp. 203–215.

[23] Wittebole, X., et al., "Nicotine Exposure Alters In Vivo Human Responses to Endotoxin," *Clin. Exp. Immunol.*, Vol. 147, No. 1, 2007, pp. 28–34.

[24] Copeland, S., et al., "Acute Inflammatory Response to Endotoxin in Mice and Humans," *Clin. Diagn. Lab. Immunol.*, Vol. 12, No. 1, 2005, pp. 60–67.

[25] Van Zee, K. J., et al., "Influence of IL-1 Receptor Blockade on the Human Response to Endotoxemia," *J. Immunol.*, Vol. 154, No. 3, 1995, pp. 1499–1507.

[26] van Deventer, S. J., et al., "Experimental Endotoxemia in Humans: Analysis of Cytokine Release and Coagulation, Fibrinolytic, and Complement Pathways," *Blood*, Vol. 76, No. 12, 1990, pp. 2520–2526.

[27] Blalock, J. E., "Harnessing a Neural-Immune Circuit to Control Inflammation and Shock," *J. Exp. Med.*, Vol. 195, No. 6, 2002, pp. F25–F28.

[28] Jara, L. J., et al., "Immune-Neuroendocrine Interactions and Autoimmune Diseases," *Clin. Dev. Immunol.*, Vol. 13, No. 2-4, 2006, pp. 109–123.

[29] Taub, D. D., "Neuroendocrine Interactions in the Immune System," *Cell Immunol.*, Vol. 252, No. 1-2, 2008, pp. 1–6.

[30] Elenkov, I. J., "Neurohormonal-Cytokine Interactions: Implications for Inflammation, Common Human Diseases and Well-Being," *Neurochem. Int.*, Vol. 52, No. 1-2, 2008, pp. 40–51.

[31] Elenkov, I. J., et al., "Cytokine Dysregulation, Inflammation and Well-Being," *Neuroimmunomodulation*, Vol. 12, No. 5, 2005, pp. 255–269.

[32] Elenkov, I. J., et al., "The Sympathetic Nerve—An Integrative Interface Between Two Supersystems: The Brain and the Immune System," *Pharmacol. Rev.*, Vol. 52, No. 4, 2000, pp. 595–638.

[33] Pavlov, V. A., and K. J. Tracey, "Neural Regulators of Innate Immune Responses and Inflammation," *Cell Mol. Life Sci.*, Vol. 61, No. 18, 2004, pp. 2322–2331.

[34] Sharshar, T., et al., "Science Review: The Brain in Sepsis—Culprit and Victim," *Crit. Care*, Vol. 9, No. 1, 2005, pp. 37–44.

[35] Lombardi, F., "Clinical Implications of Present Physiological Understanding of HRV Components," *Card. Electrophysiol. Rev.*, Vol. 6, No. 3, 2002, pp. 245–249.

[36] Norris, P. R., et al., "Heart Rate Variability Predicts Trauma Patient Outcome as Early as 12 H: Implications for Military and Civilian Triage," *J. Surg. Res.*, Vol. 129, No. 1, 2005, pp. 122–128.

[37] Rassias, A. J., et al., "Decreased Physiologic Variability as a Generalized Response to Human Endotoxemia," *Crit. Care Med.*, Vol. 33, No. 3, 2005, pp. 512–519.

[38] Godin, P. J., et al., "Experimental Human Endotoxemia Increases Cardiac Regularity: Results from a Prospective, Randomized, Crossover Trial," *Crit. Care Med.*, Vol. 24, No. 7, 1996, pp. 1117–1124.

[39] Buchman, T. G., "Nonlinear Dynamics, Complex Systems, and the Pathobiology of Critical Illness," *Curr. Opin. Crit. Care*, Vol. 10, No. 5, 2004, pp. 378–382.

[40] Cobb, J. P., et al., "Application of Genome-Wide Expression Analysis to Human Health and Disease," *Proc. Natl. Acad. Sci. USA*, Vol. 102, No. 13, 2005, pp. 4801–4806.

[41] Storey, J. D., et al., "Significance Analysis of Time Course Microarray Experiments," *Proc. Natl. Acad. Sci. USA*, Vol. 102, No. 36, 2005, pp. 12837–12842.

[42] Barber, A. E., et al., "Glucocorticoid Therapy Alters Hormonal and Cytokine Responses to Endotoxin in Man," *J. Immunol.*, Vol. 150, No. 5, 1993, pp. 1999–2006.

[43] Alvarez, S. M., et al., "Low-Dose Steroid Alters In Vivo Endotoxin-Induced Systemic Inflammation But Does Not Influence Autonomic Dysfunction," *J. Endotoxin Res.*, Vol. 13, No. 6, 2007, pp. 358–368.

[44] Fong, Y. M., et al., "The Acute Splanchnic and Peripheral Tissue Metabolic Response to Endotoxin in Humans," *J. Clin. Invest.*, Vol. 85, No. 6, 1990, pp. 1896–1904.

[45] Cross, A. S., and S. M. Opal, "A New Paradigm for the Treatment of Sepsis: Is It Time to Consider Combination Therapy?" *Ann. Intern. Med.*, Vol. 138, No. 6, 2003, pp. 502–505.

[46] Chow, C. C., et al., "The Acute Inflammatory Response in Diverse Shock States," *Shock*, Vol. 24, No. 1, 2005, pp. 74–84.

[47] Lagoa, C. E., et al., "The Role of Initial Trauma in the Host's Response to Injury and Hemorrhage: Insights from a Correlation of Mathematical Simulations and Hepatic Transcriptomic Analysis," *Shock*, Vol. 26, No. 6, 2006, pp. 592–600.

[48] Day, J., et al., "A Reduced Mathematical Model of the Acute Inflammatory Response II. Capturing Scenarios of Repeated Endotoxin Administration," *J. Theor. Biol.*, Vol. 242, No. 1, 2006, pp. 237–256.

[49] Kumar, R., et al., "The Dynamics of Acute Inflammation," *J. Theor. Biol.*, Vol. 230, No. 2, 2004, pp. 145–155.

[50] Prince, J. M., et al., "In Silico and In Vivo Approach to Elucidate the Inflammatory Complexity of CD14-Deficient Mice," *Mol. Med.*, Vol. 12, No. 4-6, 2006, pp. 88–96.

[51] Reynolds, A., et al., "A Reduced Mathematical Model of the Acute Inflammatory Response: I. Derivation of Model and Analysis of Anti-Inflammation," *J. Theor. Biol.*, Vol. 242, No. 1, 2006, pp. 220–236.

[52] Vodovotz, Y., et al., "In Silico Models of Acute Inflammation in Animals," *Shock*, Vol. 26, No. 3, 2006, pp. 235–244.

[53] Huang, S., et al., "Cell Fates as High-Dimensional Attractor States of a Complex Gene Regulatory Network," *Phys. Rev. Lett.*, Vol. 94, No. 12, 2005, p. 128701.

[54] Yang, E., et al., "Bioinformatics Analysis of the Early Inflammatory Response in a Rat Thermal Injury Model," *BMC Bioinformatics*, Vol. 8, No. 1, 2007, p. 10.

[55] Yang, E., et al., "Extracting Global System Dynamics of Corticosteroid Genomic Effects in Rat Liver," *J. Pharmacol. Exp. Ther.*, Vol. 324, No. 3, 2008, pp. 1243–1254.

[56] Yang, E., et al., "Identification of Global Transcriptional Dynamics," *PLoS ONE*, 2009. Accepted for publication.

[57] Vemula, M., et al., "Expression Profiling Analysis of the Metabolic and Inflammatory Changes Following Burn Injury in Rats," *Physiol. Genomics*, Vol. 18, No. 1, 2004, pp. 87–98.

[58] Lampariello, F., "On the use of the Kolmogorov-Smirnov Statistical Test for Immunofluorescence Histogram Comparison," *Cytometry*, Vol. 39, No. 3, 2000, pp. 179–188.

[59] Kirkpatrick, S., C. D. Gelatt, Jr., and M. P. Vecchi, "Optimization by Simulated Annealing," *Science*, Vol. 220, No. 4598, 1983, pp. 671–680.

[60] Murray, P. J., "The JAK-STAT Signaling Pathway: Input and Output Integration," *J. Immunol.*, Vol. 178, No. 5, 2007, pp. 2623–2629.

[61] Brightbill, H. D., et al., "A Prominent Role for Sp1 During Lipopolysaccharide-Mediated Induction of the IL-10 Promoter in Macrophages," *J. Immunol.*, Vol. 164, No. 4, 2000, pp. 1940–1951.

[62] Singer, M., et al., "Multiorgan Failure Is an Adaptive, Endocrine-Mediated, Metabolic Response to Overwhelming Systemic Inflammation," *Lancet*, Vol. 364, No. 9433, 2004, pp. 545–548.

[63] Brealey, D., et al., "Association Between Mitochondrial Dysfunction and Severity and Outcome of Septic Shock," *Lancet*, Vol. 360, No. 9328, 2002, pp. 219–223.

[64] Zamir, E., and P. I. Bastiaens, "Reverse Engineering Intracellular Biochemical Networks," *Nat. Chem. Biol.*, Vol. 4, No. 11, 2008, pp. 643–647.

[65] Aldridge, B. B., et al., "Physicochemical Modelling of Cell Signalling Pathways," *Nat. Cell Biol.*, Vol. 8, No. 11, 2006, pp. 1195–1203.

[66] Jusko, W. J., and H. C. Ko, "Physiologic Indirect Response Models Characterize Diverse Types of Pharmacodynamic Effects," *Clin. Pharmacol. Ther.*, Vol. 56, No. 4, 1994, pp. 406–419.

[67] Wells, C. A., T. Ravasi, and D. A. Hume, "Inflammation Suppressor Genes: Please Switch Out All the Lights," *J. Leukoc. Biol.*, Vol. 78, No. 1, 2005, pp. 9–13.

[68] Kishore, R., et al., "Lipopolysaccharide-Mediated Signal Transduction: Stabilization of TNF-Alpha Mrna Contributes to Increased Lipopolysaccharide-Stimulated TNF-Alpha Production by Kupffer Cells After Chronic Ethanol Feeding," *Comp. Hepatol.*, Vol. 3, Suppl. 1, 2004, p. S31.

[69] Van Amersfoort, E. S., T. J. Van Berkel, and J. Kuiper, "Receptors, Mediators, and Mechanisms Involved in Bacterial Sepsis and Septic Shock," *Clin. Microbiol. Rev.*, Vol. 16, No. 3, 2003, pp. 379–414.

[70] Guha, M., and N. Mackman, "LPS Induction of Gene Expression in Human Monocytes," *Cell Signal*, Vol. 13, No. 2, 2001, pp. 85–94.

[71] Du, X., et al., "Analysis of Tlr4-Mediated LPS Signal Transduction in Macrophages by Mutational Modification of the Receptor," *Blood Cells Mol. Dis.*, Vol. 25, No. 5-6, 1999, pp. 328–338.

[72] Calvano, S. E., et al., "A Network-Based Analysis of Systemic Inflammation in Humans," *Nature*, 2005.

[73] Sharma, A., and W. J. Jusko, "Characteristics of Indirect Pharmacodynamic Models and Applications to Clinical Drug Responses," *Br. J. Clin. Pharmacol.*, Vol. 45, No. 3, 1998, pp. 229–239.

[74] Dayneka, N. L., V. Garg, and W. J. Jusko, "Comparison of Four Basic Models of Indirect Pharmacodynamic Responses," *J. Pharmacokinet. Biopharm.*, Vol. 21, No. 4, 1993, pp. 457–478.

[75] Krzyzanski, W., and W. J. Jusko, "Mathematical Formalism for the Properties of Four Basic Models of Indirect Pharmacodynamic Responses," *J. Pharmacokinet. Biopharm.*, Vol. 25, No. 1, 1997, pp. 107–123.

[76] Jin, J. Y., et al., "Modeling of Corticosteroid Pharmacogenomics in Rat Liver Using Gene Microarrays," *J. Pharmacol. Exp. Ther.*, Vol. 307, No. 1, 2003, pp. 93–109.

[77] Foteinou, P. T., et al., "An Indirect Response Model of Endotoxin-Induced Systemic Inflammation," *Journal of Critical Care*, Vol. 22, No. 4, 2007, pp. 337–338.

[78] Foteinou, P. T., et al., "Modeling Endotoxin-Induced Systemic Inflammation Using an Indirect Response Approach," *Math. Biosci.*, Vol. 217, No. 1, 2009, pp. 27–42.

[79] Aderem, A., and K. D. Smith, "A Systems Approach to Dissecting Immunity and Inflammation," *Semin. Immunol.*, Vol. 16, No. 1, 2004, pp. 55–67.

[80] Saklatvala, J., J. Dean, and A. Clark, "Control of the Expression of Inflammatory Response Genes," *Biochem. Soc. Symp.*, Vol. 70, 2003, pp. 95–106.

[81] Hoffmann, A., et al., "The IkappaB-NF-kappaB Signaling Module: Temporal Control and Selective Gene Activation," *Science*, Vol. 298, No. 5596, 2002, pp. 1241–1245.

[82] Lauffenburger, D. A., and J. J. Linderman, "Receptors. Models for Binding, Trafficking, and Signalling," *The International Journal of Biochemistry and Cell Biology*, Vol. 28, 1996, pp. 1418–1418.

[83] Protti, A., and M. Singer, "Strategies to Modulate Cellular Energetic Metabolism During Sepsis," *Novartis Found. Symp.*, Vol. 280, 2007, pp. 7–16; discussion pp. 16–20, 160–164.

[84] Zwietering, M. H., et al., "Modeling of the Bacterial Growth Curve," *Appl. Environ. Microbiol.*, Vol. 56, No. 6, 1990, pp. 1875–1881.

[85] Greisman, S. E., et al., "The Role of Endotoxin During Typhoid Fever and Tularemia in Man. IV. The Integrity of the Endotoxin Tolerance Mechanisms During Infection," *J. Clin. Invest.*, Vol. 48, No. 4, 1969, pp. 613–629.

[86] Shin, H. J., et al., "Kinetics of Binding of LPS to Recombinant CD14, TLR4, and MD-2 Proteins," *Mol. Cells*, Vol. 24, No. 1, 2007, pp. 119–124.

[87] Ihekwaba, A. E., et al., "Sensitivity Analysis of Parameters Controlling Oscillatory Signalling in the NF-kappaB Pathway: The Roles of IKK and IkappaBalpha," *Syst. Biol. (Stevenage)*, Vol. 1, No. 1, 2004, pp. 93–103.

[88] Krishna, S., M. H. Jensen, and K. Sneppen, "Minimal Model of Spiky Oscillations in NF-kappaB Signaling," *Proc. Natl. Acad. Sci. USA*, Vol. 103, No. 29, 2006, pp. 10840–10845.

[89] Kerschen, E. J., et al., "Endotoxemia and Sepsis Mortality Reduction by Non-Anticoagulant Activated Protein C," *J. Exp. Med.*, Vol. 204, No. 10, 2007, pp. 2439–2448.

[90] Lehmann, V., M. A. Freudenberg, and C. Galanos, "Lethal Toxicity of Lipopolysaccharide and Tumor Necrosis Factor in Normal and D-Galactosamine-Treated Mice," *J. Exp. Med.*, Vol. 165, No. 3, 1987, pp. 657–663.

[91] Rifkind, D., "Prevention by Polymyxin B of Endotoxin Lethality in Mice," *J. Bacteriol.*, Vol. 93, No. 4, 1967, pp. 1463–1464.

[92] Tschaikowsky, K., J. Schmidt, and M. Meisner, "Modulation of Mouse Endotoxin Shock by Inhibition of Phosphatidylcholine-Specific Phospholipase C," *J. Pharmacol. Exp. Ther.*, Vol. 285, No. 2, 1998, pp. 800–804.

[93] Wang, H., et al., "HMG-1 as a Late Mediator of Endotoxin Lethality in Mice," *Science*, Vol. 285, No. 5425, 1999, pp. 248–251.

[94] Barnes, P. J., and M. Karin, "Nuclear Factor-kappaB: A Pivotal Transcription Factor in Chronic Inflammatory Diseases," *N. Engl. J. Med.*, Vol. 336, No. 15, 1997, pp. 1066–1071.

[95] Carmody, R. J., and Y. H. Chen, "Nuclear Factor-kappaB: Activation and Regulation During Toll-Like Receptor Signaling," *Cell Mol. Immunol.*, Vol. 4, No. 1, 2007, pp. 31–41.

[96] Almon, R. R., et al., "Pharmacodynamics and Pharmacogenomics of Diverse Receptor-Mediated Effects of Methylprednisolone in Rats Using Microarray Analysis," *J. Pharmacokinet. Pharmacodyn.*, Vol. 29, No. 2, 2002, pp. 103–129.

[97] Almon, R. R., et al., "Pharmacogenomic Responses of Rat Liver to Methylprednisolone: An Approach to Mining a Rich Microarray Time Series," *AAPS J.*, Vol. 7, No. 1, 2005, pp. E156–E194.

[98] Almon, R. R., D. C. DuBois, and W. J. Jusko, "A Microarray Analysis of the Temporal Response of Liver to Methylprednisolone: A Comparative Analysis of Two Dosing Regimens," *Endocrinology*, Vol. 148, No. 5, 2007, pp. 2209–2225.

[99] Almon, R. R., et al., "Corticosteroid-Regulated Genes in Rat Kidney: Mining Time Series Array Data," *Am. J. Physiol. Endocrinol. Metab.*, Vol. 289, No. 5, 2005, pp. E870–E882.

[100] DuBois, D. C., et al., "Differential Dynamics of Receptor Down-Regulation and Tyrosine Aminotransferase Induction Following Glucocorticoid Treatment," *J. Steroid Biochem. Mol. Biol.*, Vol. 54, No. 5-6, 1995, pp. 237–243.

[101] Jusko, W. J., "Receptor-Mediated Pharmacodynamics of Corticosteroids," *Prog. Clin. Biol. Res.*, Vol. 387, 1994, pp. 261–270.

[102] Sun, Y. N., et al., "Fourth-Generation Model for Corticosteroid Pharmacodynamics: A Model for Methylprednisolone Effects on Receptor/Gene-Mediated Glucocorticoid Receptor Down-Regulation and Tyrosine Aminotransferase Induction in Rat Liver," *J. Pharmacokinet. Biopharm.*, Vol. 26, No. 3, 1998, pp. 289–317.

[103] Xu, Z. X., et al., "Third-Generation Model for Corticosteroid Pharmacodynamics: Roles of Glucocorticoid Receptor mRNA and Tyrosine Aminotransferase mRNA in Rat Liver," *J. Pharmacokinet. Biopharm.*, Vol. 23, No. 2, 1995, pp. 163–181.

[104] Jusko, W. J., D. DuBois, and R. Almon, "Sixth-Generation Model for Corticosteroid Pharmacodynamics: Multi-Hormonal Regulation of Tyrosine Aminotransferase in Rat Liver," *J. Pharmacokin. Pharmacodyn.*, 2005.

[105] Ramakrishnan, R., et al., "Fifth-Generation Model for Corticosteroid Pharmacodynamics: Application to Steady-State Receptor Down-Regulation and Enzyme Induction Patterns During Seven-Day Continuous Infusion of Methylprednisolone in Rats," *J. Pharmacokinet. Pharmacodyn.*, Vol. 29, No. 1, 2002, pp. 1–24.

[106] Foteinou, P. T., et al., "In Silico Simulation of Corticosteroids Effect on an NFkB-Dependent Physicochemical Model of Systemic Inflammation," *PLoS One*, Vol. 4, No. 3, 2009, p. e4706.

[107] Lin, E., and S. F. Lowry, "The Human Response to Endotoxin," *Sepsis*, Vol. 2, 1998, pp. 255–262.

[108] Webster, J. I., L. Tonelli, and E. M. Sternberg, "Neuroendocrine Regulation of Immunity," *Annu. Rev. Immunol.*, Vol. 20, 2002, pp. 125–163.

[109] Padgett, D. A., and R. Glaser, "How Stress Influences the Immune Response," *Trends Immunol.*, Vol. 24, No. 8, 2003, pp. 444–448.

[110] Mager, D. E., et al., "Dose Equivalency Evaluation of Major Corticosteroids: Pharmacokinetics and Cell Trafficking and Cortisol Dynamics," *J. Clin. Pharmacol.*, Vol. 43, No. 11, 2003, pp. 1216–1227.

[111] van der Poll, T., et al., "Epinephrine Inhibits Tumor Necrosis Factor-Alpha and Potentiates Interleukin 10 Production During Human Endotoxemia," *J. Clin. Invest.*, Vol. 97, No. 3, 1996, pp. 713–719.

[112] van der Poll, T., "Effects of Catecholamines on the Inflammatory Response," *Sepsis*, Vol. 4, 2000, pp. 159–167.

[113] Mager, D. E., E. Wyska, and W. J. Jusko, "Diversity of Mechanism-Based Pharmacodynamic Models," *Drug Metab. Dispos.*, Vol. 31, No. 5, 2003, pp. 510–518.

[114] Mager, D. E., and W. J. Jusko, "Pharmacodynamic Modeling of Time-Dependent Transduction Systems," *Clin. Pharmacol. Ther.*, Vol. 70, No. 3, 2001, pp. 210–216.

[115] Sun, Y. N., and W. J. Jusko, "Transit Compartments Versus Gamma Distribution Function to Model Signal Transduction Processes in Pharmacodynamics," *J. Pharm. Sci.*, Vol. 87, No. 6, 1998, pp. 732–737.

[116] van der Poll, T., et al., "Hypercortisolemia Increases Plasma Interleukin-10 Concentrations During Human Endotoxemia—A Clinical Research Center Study," *J. Clin. Endocrinol. Metab.*, Vol. 81, No. 10, 1996, pp. 3604–3606.

[117] Winchell, R. J., and D. B. Hoyt, "Spectral Analysis of Heart Rate Variability in the ICU: A Measure of Autonomic Function," *J. Surg. Res.*, Vol. 63, No. 1, 1996, pp. 11–16.

[118] Marsland, A. L., et al., "Stimulated Production of Proinflammatory Cytokines Covaries Inversely with Heart Rate Variability," *Psychosom. Med.*, Vol. 69, No. 8, 2007, pp. 709–716.

[119] Aronson, D., M. A. Mittleman, and A. J. Burger, "Interleukin-6 Levels Are Inversely Correlated with Heart Rate Variability in Patients with Decompensated Heart Failure," *J. Cardiovasc. Electrophysiol.*, Vol. 12, No. 3, 2001, pp. 294–300.

[120] Malave, H. A., et al., "Circulating Levels of Tumor Necrosis Factor Correlate with Indexes of Depressed Heart Rate Variability: A Study in Patients with Mild-to-Moderate Heart Failure," *Chest*, Vol. 123, No. 3, 2003, pp. 716–724.

[121] Bendixen, H. H., et al., "Dose-Dependent Differences in Catecholamine Action on Heart and Periphery," *J. Pharmacol. Exp. Ther.*, Vol. 145, 1964, pp. 299–306.

[122] Brown, G. L., and J. C. Eccles, "The Action of a Single Vagal Volley on the Rhythm of the Heart Beat," *J. Physiol.*, Vol. 82, No. 2, 1934, pp. 211–241.

[123] Jan, B. U., et al., "Influence of Acute Epinephrine Infusion on Endotoxin-Induced Parameters of Heart Rate Variability: A Randomized Controlled Trial," *Ann. Surg.*, Vol. 249, No. 5, 2009, pp. 750–756.

[124] Morris, J. A., Jr., et al., "Adrenal Insufficiency, Heart Rate Variability, and Complex Biologic Systems: A Study of 1,871 Critically Ill Trauma Patients," *J. Am. Coll. Surg.*, Vol. 204, No. 5, 2007, pp. 885–892; discussion pp. 892–893.

[125] Berntson, G. G., J. T. Cacioppo, and K. S. Quigley, "Autonomic Determinism: The Modes of Autonomic Control, the Doctrine of Autonomic Space, and the Laws of Autonomic Constraint," *Psychol. Rev.*, Vol. 98, No. 4, 1991, pp. 459–487.

[126] Zaza, A., and F. Lombardi, "Autonomic Indexes Based on the Analysis of Heart Rate Variability: A View from the Sinus Node," *Cardiovasc. Res.*, Vol. 50, No. 3, 2001, pp. 434–442.

[127] Gutkin, B. S., S. Dehaene, and J. P. Changeux, "A Neurocomputational Hypothesis for Nicotine Addiction," *Proc. Natl. Acad. Sci. USA*, Vol. 103, No. 4, 2006, pp. 1106–1111.

[128] Contreras, M., and L. M. Ryan, "Fitting Nonlinear and Constrained Generalized Estimating Equations with Optimization Software," *Biometrics*, Vol. 56, No. 4, 2000, pp. 1268–1271.

[129] Lameris, T. W., et al., "Epinephrine in the Heart: Uptake and Release, But No Facilitation of Norepinephrine Release," *Circulation*, Vol. 106, No. 7, 2002, pp. 860–865.

[130] Arnalich, F., et al., "Predictive Value of Nuclear Factor kappaB Activity and Plasma Cytokine Levels in Patients with Sepsis," *Infect. Immun.*, Vol. 68, No. 4, 2000, pp. 1942–1945.

[131] Klinke, D. J., et al., "Modulating Temporal Control of NF-kappaB Activation: Implications for Therapeutic and Assay Selection," (unedited manuscript), *Biophys. J. BioFAST*, 2008.

[132] Romascin, A. D., D. M. Foster, and J. C. Marshall, "Let the Cells Speak: Neutrophils as Biologic Markers of the Inflammatory Response," *Sepsis*, Vol. 2, 1998, pp. 119–125.

[133] Fan, H. and J. A. Cook, "Molecular Mechanisms of Endotoxin Tolerance," *J. Endotoxin Res.*, Vol. 10, No. 2, 2004, pp. 71–84.

[134] Wysocka, M., et al., "IL-12 Suppression During Experimental Endotoxin Tolerance: Dendritic Cell Loss and Macrophage Hyporesponsiveness," *J. Immunol.*, Vol. 166, No. 12, 2001, pp. 7504–7513.

[135] McCall, C. E., et al., "Tolerance to Endotoxin-Induced Expression of the Interleukin-1 Beta Gene in Blood Neutrophils of Humans with the Sepsis Syndrome," *J. Clin. Invest.*, Vol. 91, No. 3, 1993, pp. 853–861.

[136] Poll, V. D., "Changes in Endotoxin-Induced Cytokine Production by Whole Blood After In Vivo Exposure of Normal Humans to Endotoxin," *J. Infect. Disease*, Vol. 174, 1996, p. 1356–1360.

[137] Hasko, G., et al., "Differential Effect of Selective Block of Alpha 2-Adrenoreceptors on Plasma Levels of Tumour Necrosis Factor-Alpha, Interleukin-6 and Corticosterone Induced by Bacterial Lipopolysaccharide in Mice," *J. Endocrinol.*, Vol. 144, No. 3, 1995, pp. 457–462.

[138] Ignatowski, T. A. and R. N. Spengler, "Regulation of Macrophage-Derived Tumor Necrosis Factor Production by Modification of Adrenergic Receptor Sensitivity," *J. Neuroimmunol.*, Vol. 61, No. 1, 1995, pp. 61–70.

[139] Stein, P. K., and R. E. Kleiger, "Insights from the Study of Heart Rate Variability," *Ann. Rev. Med.*, Vol. 50, 1999, pp. 249–261.

[140] Court, O., et al., "Clinical Review: Myocardial Depression in Sepsis and Septic Shock," *Crit. Care*, Vol. 6, No. 6, 2002, pp. 500–508.

[141] Keh, D., and C. L. Sprung, "Use of Corticosteroid Therapy in Patients with Sepsis and Septic Shock: An Evidence-Based Review," *Crit. Care Med.*, Vol. 32, No. 11, Suppl., 2004, pp. S527–S533.

[142] Van der Poll, T., and S. F. Lowry, "Epinephrine Inhibits Endotoxin-Induced IL-1 Beta Production: Roles of Tumor Necrosis Factor-Alpha and IL-10," *Am. J. Physiol.*, Vol. 273, No. 6, Pt. 2, 1997, pp. R1885–R1890.

CHAPTER

5

A Liposome Assay for Evaluating the Ocular Toxicity of Chemicals

Brett A. Howell, Yash Kapoor, and Anuj Chauhan

Department of Chemical Engineering, University of Florida, Gainesville, FL
Corresponding author: Anuj Chauhan, Department of Chemical Engineering, University of Florida, CHE P.O. Box 116005, Gainsville, FL, 32611-6005, e-mail: Chauhan@che.ufl.edu, phone: 352-392-2592, fax: 352-392-9513

Abstract

Ocular toxicity is primarily assessed with the Draize score, which is determined from a qualitative assessment of the ocular damage to animal models after potential ocular irritants are introduced on the ocular surface. The Draize scores obtained from the in vivo testing have been found to vary between laboratories and significant resources are needed to carry out such studies. A liposome-based in vitro assay can be used as an alternative to in vivo testing for ocular toxicity assessment. Key challenges and improvements on this assay are presented here for easy, fast, and reliable testing of possible eye irritants for which the in vivo data are not available. This in vitro assay is conducted with liposomes composed of lipids that mimic the composition of the lipid bilayers of the corneal epithelial cells. The liposomes are loaded with a hydrophilic dye and dye leakage from the liposomes upon introduction of possible irritants can be used as a response reflecting the possible level of irritancy when these irritants are applied to the ocular surface. Quantitative dye leakage can be correlated with Draize scores of previously studied chemicals and used as a calibration curve for assessing the ocular toxicity of new chemicals. When used appropriately and in conjunction with other in vitro alternatives, this method can serve as an initial step for determining the toxicity response of chemicals used for ocular applications.

Key terms cornea damage, Draize alternative, in vitro alternative, liposomes, membrane damage, ocular toxicity, surfactants

5.1 Introduction

The propensity for chemical compounds to cause eye irritation is a major concern for a variety of applications ranging from pharmaceutical devices applied directly to the eyes to fear of accidental exposure for nonocular compounds. Probing compounds for their potential ocular toxicity with cost-effective, ethical, and efficient approaches has continued to be a major challenge. The liposome-based assay described herein is one in vitro alternative for making such assessments. Liposomes are prepared so that their composition mimics the composition of the corneal epithelial cell membranes with the inherent assumption that ocular toxicity is caused by cornea cell damage and the subsequent release of cytotoxic substances [1–3]. The liposomes are prepared in the presence of a hydrophilic fluorescent dye, allowing for the quantification of the leakage of the liposome bilayer after a possible ocular irritant has been exposed to the liposomes. Liposome-based assays for ocular toxicity have been used in several studies over the past 25 years [1, 4–8]. Many of the studies reported only minor success when correlating in vivo data to liposome leakage results and gross outliers were often present, especially with nonionic surfactants. As a result, the use of the liposome assay for ocular toxicity has not been widespread. Recent advances in the experimental design have made the assay more closely resemble the true physiology encountered in vivo and have improved correlations between the in vitro and the in vivo data [9]. When carefully designed, the liposome assay provides an excellent way to measure the ocular toxicity of a variety of chemical substances.

A variety of ocular toxicity assays have been studied, and the most well-known in vivo assay is the Draize test [10]. The Draize test involves applying the compound in question to an eye of an animal model such as rabbits followed by qualitative assessments of the induced injury leading to ordinal scores. This method has many obvious drawbacks including a lack of reproducibility, observer-dependent variability, a need to use and possibly harm animals, and significant cost. The liposome assay avoids all of these drawbacks. Alternative in vitro methods developed to replace the Draize test include the EYTEX test (Ropak Labs, Irvine, California), which assesses protein denaturation. Correlations obtained from this test for surfactants were poor and the method was ineffective with cationic surfactants [11]. Several methods utilizing cultured cells have been developed that measure the effect of test substances on cell vitality, growth, or dye leakage or uptake [3, 12, 13]. All these methods have had a limited success in evaluating ocular toxicities of various compounds. The liposome assay has several distinct advantages over other methods, such as low cost, simplicity, and reproducibility. In general, it may be best to conduct a battery of in vitro methods to best assess ocular toxicity rather than rely on a single in vitro approach. In vitro methods of assessing ocular toxicity have been reviewed in detail elsewhere [14].

The liposome assay involves three major steps. First, the liposomes are prepared in the presence of a highly concentrated fluorescent dye so that the cores of the liposomes contain the dye solution. Next, the bulk dye unentrapped by liposomes is removed. Finally, test substances are exposed to the liposomes and dye leakage is assessed. The dye leakage can be correlated to Draize data for substances for which Draize data is already available, and the ocular toxicity of new compounds can then be estimated by determining the dye leakage and utilizing the correlation.

5.2 Experimental Design

The liposome assay correlates chemical-induced leakage of a fluorescent dye from the core of liposomes to the ocular toxicity of the chemical substance. It offers several advantages, including affordability, high throughput, and reproducibility. The procedures and lab equipment required are also relatively simple and commonly found in most labs. The method is based on the assumption that the permeation of a substance through lipid bilayers on the epithelial lining of the eye initiates ocular toxicity [1]. The area and volume of the initial injury caused by permeation are key factors in determining the ultimate ocular toxicity imposed [2]. Once the epithelial lining has been damaged, cells release lysosomal enzymes, histamine, and inflammatory mediators, and these compounds cause cell death and mediate an immune response, leading to local swelling and discomfort [3]. The liposomes used in this assay should be designed to mimic the lipid composition of the corneal epithelium in an effort to imitate the ocular injury [4]. To make use of the assay data, a calibration curve that relates leakage to toxicity must be generated. Calibration curves from literature may be used, but testing liposome leakage with well-studied chemicals and creating a new calibration curve is the best practice. Doing so ensures that the experimental protocol has been carried out correctly and removes any discrepancies arising from interlab differences such as raw materials variation, liposome size, or experience level. The simplest way to obtain a calibration curve of liposome leakage versus toxicity is to use Draize scores, which are assessments of toxicity based on experiments done with rabbits or other animals [9, 10]. Since Draize scores for a variety of chemicals are readily available in literature, it is relatively easy to create a curve of chemical-induced leakage versus Draize score [1, 3, 5, 6, 9, 11–13, 15, 16].

Additional considerations are required, however, before the liposome assay can be accurately correlated to Draize scores. The degree of chemical penetration on the ocular surface is proportional to the surface area available for the substance to diffuse into the epithelium. Similarly, when liposomes are used as an alternative, the amount of the chemical substance penetrating the lipid bilayer will be directly proportional to the surface area of the liposomes. The assay must therefore be designed with differences between liposomes and the corneal epithelium considered. The surface area to encapsulated volume ratio (A/V) for liposomes is much larger than that of epithelial cells due to the extremely small size of liposomes, and the larger A/V ratio will lead to a more rapid dye release compared to the release of compounds from the epithelial cells. If the dye release from the liposomes is very rapid, it may be difficult to acquire leakage data in the time regime in which the leakage fraction is relatively small, which is desirable to maximize sensitivity. This geometrical difference, which has not been accounted for in some instances [1, 4, 5, 15], can be circumvented by adjusting the concentration used in the liposome assay with a surface area to volume ratio correction factor. The reduction in concentration will slow down the leakage, thereby allowing measurements at times in which the leakage fraction is small. The following equation can be used to evaluate the effective concentration that should be tested in the liposome assay to correctly predict the irritancy of chemicals in vivo [9]:

$$C_{TEST} = \frac{A_{Cornea} V_{Lipsome}}{V_{Cornea} A_{Liposome}} C_{Ocular} \tag{5.1}$$

In (5.1), A_{Cornea} and $A_{Liposome}$ refer to the surface area of the cornea and the liposomes. Similarly, V_{Cornea} and $V_{Liposome}$ refer to the respective volumes. C_{TEST} is the test concentration of the chemical to be used in the liposomal assay and C_{Ocular} is the concentration of the tested chemical on the ocular surface for the Draize test. It is imperative to readjust for concentration tested in the liposomal assay when we consider special chemicals like surfactants which can self-assemble above a critical concentration referred to as the critical micelle concentration (CMC). When Draize scores are calculated for surfactants with a concentration on the ocular surface greater than CMC, C_{Ocular} should be replaced by the CMC value for the particular surfactant. This is explained more explicitly in later sections.

The average surface area to volume ratio for liposomes and the cornea can be estimated as 63,132, and 345, respectively. These ratios are based on a surface area of 62 $Å^2$ occupied by a single lipid [17–19], a computed liposome volume [20] based on a liposome radius of 55 nm [21], a bilayer thickness of 35–40Å [22], and cell data from literature [23, 24]. The resulting ratio of surface area to volume ratios for liposomes and the cornea is around 183. It is important to point out that the correction factor depends heavily on the mean diameter of the liposomes used for the study and new correction factors must be computed when working with liposomes of dramatically different sizes [21]. If liposome extrusion is carried out with 100-nm membranes, this correction factor of 183 is probably adequate. It is noted that a rough approximation of the correction factor is sufficient for designing the protocols.

Once the concentration correction factor has been computed using the ocular concentration (C_{Ocular}) and the diameter of the unilamellar liposomes is employed, the liposome leakage assay should provide a good indicator of the toxicity of the chemical tested.

5.3 Materials

Reagents and supplies

- Methanol, 99.9% (Sigma Aldrich, St. Louis, Missouri)
- Chloroform, 99.9% (Sigma Aldrich)
- Dulbecco's Phosphate Buffered Saline (PBS) without calcium chloride and magnesium chloride, liquid form (Sigma Aldrich)
- Sephadex G-50, fine (Sigma Aldrich)
- Cholesterol (CH) (Sigma Aldrich)
- Whatman GF/B glass microfiber filter, 21-mm diameter, 0.68-mm thickness, 1-μm particle retention (Fisher Scientific, Pittsburgh, Pennsylvania)
- 5-mL disposable syringe (Fisher Scientific)
- 15-mL disposable polypropylene centrifuge tube (Fisher Scientific)
- Calcein dye, also known as fluorexon (Fisher Scientific)
- Triton X-100 (Fisher Scientific)
- 1,2-Dioleoyl-*sn*-Glycero-3-Phosphoethanolamine (DOPE), dissolved in chloroform, must be stored at –20°C (Avanti Polar Lipids Inc., Alabaster, Alabama)

- 1,2-Dimyristoyl-*sn*-Glycero-3-[Phospho-*rac*-(1-glycerol)] (DMPG), in sodium salt powder form, must be stored at 4°C (Avanti Polar Lipids Inc.) (note: DPPG may be used as a substitute)
- 1,2-Dimyristoyl-*sn*-Glycero-3-Phosphocholine (DMPC), in powder form, must be stored at 4°C (Avanti Polar Lipids Inc.) (note: DPPC may be used as a substitute)

Equipment

- G112SP1 Special Ultrasonic Cleaner, 115 volt, 3.5 amp (Avanti Polar Lipids Inc.)
- Avanti Mini-Extruder with a heating block and a 100-nm membrane (Avanti Polar Lipids Inc.)
- Turner Quantech Digital Filter Fluorometer with excitation (490 nm) and emission (515 nm) filters
- Compressed nitrogen

5.4 Methods

5.4.1 Preparation of calcein-loaded liposomes

1. Liposomes designed to mimic the cornea are composed of a molar ratio of 8:6:1.5:1.5 of DMPC:DOPE:DMPG:CH. Weigh out the powder form lipids (DMPC = 11.8 mg, DMPG = 2.2 mg, cholesterol = 1.2 mg) into a clean, 20-mL glass vial. Add the appropriate amount of DOPE dissolved in chloroform so that 9.6 mg of DOPE is added. This will depend on the concentration of DOPE in chloroform (i.e., 0.385 mL if the concentration is 25 mg DOPE/mL solvent).
2. A total solvent volume of 1.25 mL will be added to the vial at a 9:1 ratio (by volume) of chloroform to methanol, including the volume of chloroform added in step 1 (1.125-mL chloroform, 0.125-mL methanol, 20-mg lipid/mL solvent). Add the additional chloroform needed and 0.125 mL of methanol (i.e., 0.74-mL additional chloroform and 0.125-mL methanol if 0.385 mL of chloroform was added in step 1).
3. Allow the solvent to evaporate under a vented hood by blowing a gentle stream of nitrogen into the vial for 45 minutes. This can be done using a glass pipette and compressed nitrogen. The disturbance of the liquid by the nitrogen stream should be slight and barely visible, with no splashing or liquid displacement.
4. Hydrate the dried lipid layer with 1 mL of a 100-mM calcein solution made with PBS for a final concentration of 25 mg lipid/mL. The protocols described next are based on 1 mL of liposome solution (25 mg lipid/mL), but these can be scaled to other volumes.
5. After capping and ensuring the vial top is securely in place, mix the solution for 1–2 minutes on medium speed using a vortex mixer.
6. Place the vial in the bath sonicator (Ultrasonic Cleaner). Adjust the water level in the sonicator so that it barely covers the liposome dispersion in the bottom of the vial. Sonicate for 20 minutes.
7. Cover the vial with foil. Using a small magnetic stirrer and hotplate, stir the dispersion overnight (18–24 hours) at approximately 300 rpm while heating to 30–35°C. Note: The dye solution and all liposomes encapsulating calcein should be

covered with aluminum foil whenever possible to prevent the weakening of the fluorescent signal due to light exposure.

8. Assemble the Avanti Mini-extruder [25]. Wet 4 filter supports with clean PBS and place two supports on each internal membrane support on the circular region surrounded by the o-rings. Place one internal membrane support inside the extruder outer casing with the o-ring facing up. Using tweezers, carefully lift a 100-nm polycarbonate membrane and place it inside the extruder outer casing on top of the internal membrane support. Place the second internal membrane support inside the casing with the o-ring facing down. Place the Teflon bearing inside the retainer nut and hand-tighten the extruder assembly. Notes: When handling the membranes, avoid touching any portion other than the outer edges. Ensure that the membrane is flush with the internal support in the casing by lightly tapping the casing on the bench top. Do not wrench when tightening the extruder assembly.
9. Retrieve 1 mL of clean PBS with a syringe from the extruder kit and insert the syringe into the extruder. Insert the other empty syringe into the extruder on the opposite side. Place the assembly in the heating block on a hot plate heated to 35–40°C. Slowly push the PBS through the membrane several times while checking for leaks. Discard the PBS and retrieve 1 mL of liposome dispersion. Return the extruder to the hot plate and slowly push the dispersion through the membrane 15 times (Figure 5.1). Note: The entire volume of liposome dispersion may not easily pass through the 100-nm membrane initially. Pass small amounts of the dispersion through the membrane and then back several times until all 1 mL of the dispersion passes through the membrane to the other syringe with medium pressure and then pass all 1 mL of the solution through the membrane 15 additional times.

5.4.2 Separation of bulk calcein from loaded liposomes with Sephadex

1. Weigh out 0.5g of fine Sephadex G-50 into a beaker of at least a 30-ml volume. Add 6.5 mL of PBS and allow the gel to soak for at least 3 hours at room temperature.

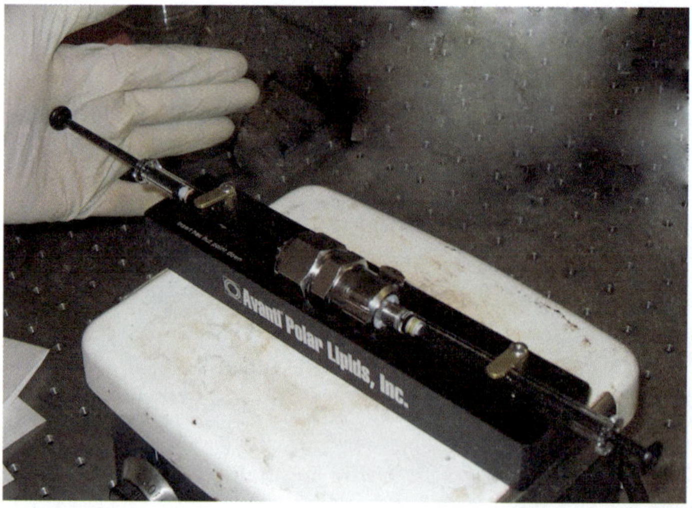

Figure 5.1 Pass the liposome dispersion from one syringe through the membrane holder to the other syringe approximately 15 times to form liposomes with a normally distributed size distribution.

Note: The gel may be soaked in advance but should be stored at 4°C after the initial soak and should not be stored for more than 24 hours.

2. Remove the plunger from a new 5-mL syringe. Fold a Whatman microfiber filter in half. Trim both sides of the folded membrane so that both sharp corners are removed (Figure 5.2). Place the folded membrane into the syringe and use a microspatula to push the membrane to the bottom of the syringe so that it covers the syringe outlet.
3. Place the syringe in a centrifuge tube. Pour the Sephadex gel into the syringe [26]. The gel may not flow easily and so some of the gel will need to be placed into the syringe manually with a spatula. Tap the syringe on the bench top lightly after all Sephadex has been put into the syringe.
4. Centrifuge the tube and syringe apparatus at 1,000g for 5 minutes [17]. Discard the liquid from the tube and ensure the centrifuge tube is replaced or cleaned. Place the syringe into the new/cleaned tube once again. Add the 1-mL liposome solution drop-wise to the Sephadex bed (Figure 5.3). Note: Some space will be present between the sides of the gel column and the syringe, and it is important to ensure that the liposomes are added to the bed slowly so that no liquid flows into the void space. The separation of calcein from the bulk may be improved by allowing the liposome-loaded column to sit for 5 minutes prior to centrifugation.
5. Centrifuge the tube and syringe with liposomes added for 7 minutes at approximately 500g and 3 minutes at 1,000g. Retrieve the dye-loaded liposomes from the centrifuge tube and place in a clean vial. Note: Visually inspect the liposome solution to ensure that no membrane debris passed into the tube. Concentrated liposome solution may be covered with foil and stored for 3–5 days at 4°C.
6. To ensure the presence of dye loaded liposomes, dilute the stock liposome solution by a factor of 1,000. The diluted solution should appear slightly red or orange. Add 0.1 mL of 20% (v/v) Triton X-100 solution to 2 mL of the diluted liposome solution. Stir gently with a pipette for 3–5 seconds. A fluorescent yellow color should be visible (Figure 5.4). Note: The solution must be diluted due to the linear detection regime of

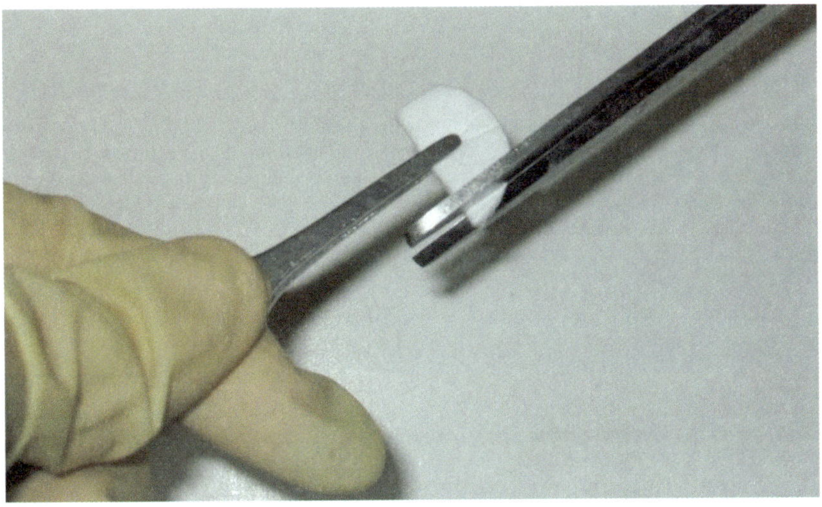

Figure 5.2 Fold the Whatman microfiber filter in half and trim the corners so that the folded filter can fit into the bottom of the syringe. The filter will act as a support for the gel column.

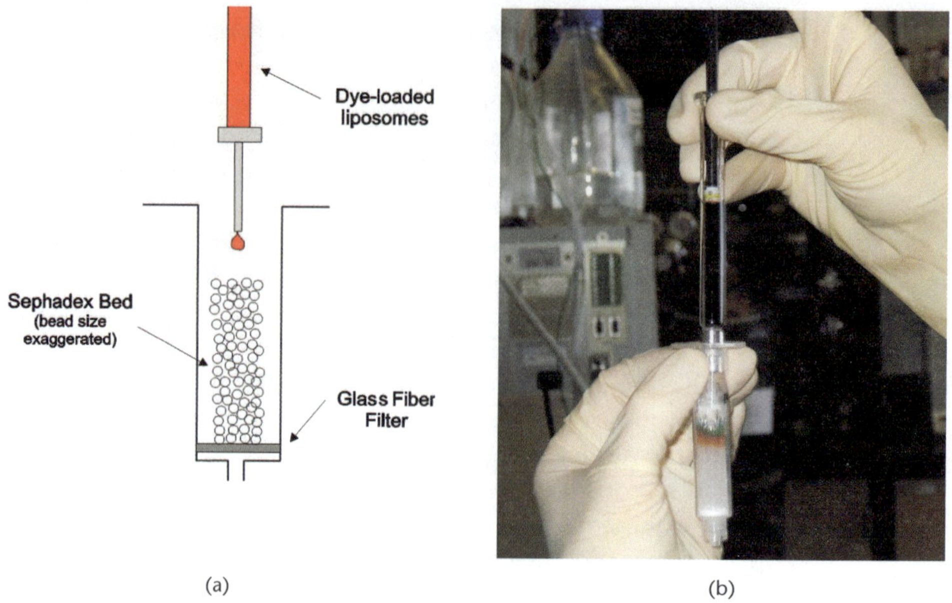

Figure 5.3 (a, b) Add the extruded liposomes drop-wise to the gel column. The gel beads are enlarged in (a).

Figure 5.4 Examples of the diluted liposome dispersion: (a) after dilution and (b) after 20% (v/v) Triton X-100 was added to the dispersion to break the liposomes. The dilution factor was reduced in the figure for the purposes of clarity. Actual color changes may be more difficult to visually discern when diluting the stock liposome solution coming from the gel column by a factor of 1,000.

calcein, which is roughly 0–10 μM, but the dilution factor may vary slightly depending upon experimental conditions.

5.4.3 Ocular toxicity experiments using dye-loaded liposomes

1. Prior to carrying out the liposome leakage experiments, a calibration curve for calcein fluorescence must be made. Several different linear detection regimes for calcein have been reported [17, 27, 28]. In general, calcein fluorescence increases

linearly with concentration until reaching some threshold concentration, beyond which the dye self-quenches. Above the critical concentration, the signal diminishes and finally disappears. In our lab, we found the linear detection regime to be 1–10 μM. If a fluorometer similar to the one listed herein is used, the proper excitation (490 nm) and emission filters (515 nm) must be chosen. The calibration curve for calcein fluorescence versus concentration should then be created and the linear detection regime should be determined.

2. Add 2 mL of the diluted liposome solution to a cuvette. Measure the fluorescence of the solution. If the separation of bulk calcein from the liposomes was successful, the baseline fluorescence prior to adding any chemical compounds or controls to the solution should be low (corresponding to $< 0.5\ \mu$M calcein). Break the liposomes by adding 0.1 mL of 20% v/v Triton X-100 to the liposome solution; mix gently for 2–3 seconds using a pipette, and measure the fluorescence of the solution. The final value of fluorescence should also be in the linear detection regime. If this is not the case, adjust the dilution factor. Repeat this step at least two additional times. The average value of the fluorescence after liposome breakage will serve as a control for the expected total calcein present in the solution at the conclusion of the leakage experiments. This step should be repeated for each new batch of diluted liposome solution made from the concentrated liposome stock. Note: This is done as a control due to observed fluctuations in calcein fluorescence with time which are caused due to excessive exposure to light or the detector. For more information, see the Troubleshooting Table.
3. Add 2 mL of the liposome solution to the cuvette. Measure the fluorescence prior to adding any control solutions or substances. Add 0.1 mL of the substance under investigation or a control solution, and then mix for 2–3 seconds by pipetting. Measure the fluorescence of the solution directly after mixing and then at predetermined time intervals. Cover the cuvette or other holding containers if the samples are transferred with foil at all times between measurements. Note: The length of time over which the dye signal remains stable will be determined by the degree of exposure to light and the substance added to the liposomes. This should be tested with controls. For more information, see the Troubleshooting Table.
4. At the final time point, add 0.1 mL of Triton X-100 to the solution and mix. Measure the final fluorescence. If the fluorescence is significantly different from the final value observed in step 2, the protocol must be reevaluated. Note: Ensure that all fluorescence values are corrected for dilutions prior to comparing.
5. The following formula is used to compute the percent release,

$$\%\,\text{Release} = \frac{F_t - F_o}{F_{total} - F_o} \times 100 \tag{5.2}$$

where F_t is the fluorescence measurement at the final time t, F_0 is the fluorescence at time zero, and F_{total} is the total calcein released, determined from steps 2 and 4. Corrections should be made for all dilutions.

6. These steps are repeated for all controls and compounds under investigation. To create a calibration curve of Draize scores versus leakage, carry out the above steps for several compounds for which Draize scores are available. Then repeat the method

using the substance in question and use the calibration curve to predict the ocular toxicity.

5.5 Data Acquisition, Anticipated Results, and Interpretation

The methods described above allow for the calculation of the amount of hydrophilic dye leaking from liposomes upon introduction of a chemical into the liposome solution. The leakage is expected to correlate with toxicity based on the hypothesis that the ocular toxicity is either initiated or manifested through the leakage of essential histamines and lysozomal enzymes from the corneal epithelium when in contact with foreign chemicals. There is no concrete evidence as to how the chemicals cause ocular toxicity, but it is generally accepted that the mechanism of ocular toxicity is closely related to the fluidity changes in the corneal epithelium and this phenomena is captured in the in vitro liposomal assay by making liposomes mimicking the corneal epithelium and challenging the liposome fluidity by introducing external chemicals.

A crucial step is to take into account the characteristic differences between the application of a chemical on the ocular surface and the subsequent testing in the in vitro liposomal assay. The absence of the tear pump and differences in surface area and volume in the in vitro and in vivo settings need to be understood before using this assay successfully for application purposes. Though liposomes can be synthesized closely mimicking the corneal surface, the significant increase in the surface area of the liposomes due to their smaller size should be taken into account when testing chemicals in this assay. This has been clearly stated in Section 5.2. The equation derived in (5.1) relating the concentration to be tested in the liposome assay to that tested on the ocular tissue to determine the Draize score becomes the most important consideration for this assay.

The amount of dye leaking from the liposomes should be measured at a suitable time so the dye does not lose activity due to prolonged exposure to light surrounding the experimental setup. This can be accomplished by measuring the amount of dye leakage after 10 minutes or less [4, 7, 9]. A percentage of dye leakage should then be correlated to the Draize scores for the chemicals to obtain a calibration/correlation curve. It should be understood that the correlation curves between the dye leakage in the liposomal assay and the Draize scores should be nonlinear as the Draize score saturates at a maximum value of 110. This has not been a common practice in some of the studies in the past. Spearman's rank correlation coefficient should be preferably used to determine the statistical significance of the correlations because of the ordinal nature of the Draize data. However, in the literature Pearson's correlation coefficient has also been utilized to correlate the in vivo and in vitro test data [1, 6, 9, 11–13, 15, 16, 29]. Once the correlation curve is established, liposome leakage can be used as a preliminary toxicity assessment tool for determining toxicity of new chemicals or for chemicals for which Draize scores/toxicity data is not available by predicting the Draize score from the correlation curve.

It is a common practice to divide the results for toxicity of various chemicals into broad categories that are scaled, in most cases, according to the researcher. Since there is no set protocol for a categorical division, we suggest a four-level gradation scheme for categorical division of the data for easier interpretation. These categories can be set depending on the Draize scores according to the following general scheme: mildly irritant (0–15), moderately irritant (15–25), irritant (25–50), and severely irritant (50–110). More varied schemes can be further found in other published works [11–13, 16, 30]. These schemes are used to qualitatively identify chemicals as possible irritants and unknown species can be assigned to a respective class based on the Draize score predicted from the calibration curve. Mildly irritant chemicals (Draize score < 15) are generally accepted as safe for ocular applications.

It is also important to understand that the ocular toxicity of a particular compound depends on the area under the curve (AUC) for the plot of tear film concentration versus time. Thus, a compound that is nontoxic at a given concentration could become toxic at a higher concentration or repeated dosing. To take this issue into account, it is required that the Draize scores that are utilized in the calibration curve are obtained at a fixed concentration for all the chemicals used in the correlation. The correlation chart can then be used to predict the ocular irritancy levels of new chemicals with the understanding that the Draize score obtained for the chemical from the correlation curve would correspond to the toxicity induced by the chemical when placed on the ocular surface at the concentration at which the Draize data was used to generate the curve. This makes the test slightly cumbersome as a separate calibration curve is required for a different desired concentration to be tested on the ocular surface. Hence, a battery of Draize scores need to be determined for each individual curve at fixed concentrations. The Draize score is a qualitative assessment of ocular tissues based on visual observations, which leads to an introduction of significant human errors. Hence, care should be taken in combining Draize data from different laboratories as it has been reported that significant differences can be found when Draize scores were reported for the same chemical and concentration at different laboratories [31]. Though it is important to consider the qualitative differences between the Draize scores from various laboratories, it has been found that the correlations from the overall fitting of the dye leakage percentage and the Draize scores lead to statistically similar calibration curves [9].

5.6 Discussion and Commentary

All chemicals can be broadly divided into three categories: positively charged, negatively charged, and nonionic. The mechanism of toxicity is dominated by and directly related to the class to which a particular chemical belongs. For a particular choice of potential ocular irritant, the affinity to the corneal epithelium should dictate the level of irritancy induced. The corneal epithelium has a net negative charge, which makes it more prone to injury when in contact with a positively charged substance. Other molecular factors that may play an important role in interactions between the test molecule and the epithelium include charge density, steric effects, and hydrophobic interactions. This interaction becomes more complicated when considering the interaction of special chemicals such as surfactants with the ocular surface (Figure 5.5). Surfactants self-assemble to form micelles above their critical micelle concentrations (CMC), and this self-assembly leads to special considerations before correlating results from the liposomal assay to the Draize test. It has been shown that as the concentration of surfactants is increased above their CMC in the test medium, the interaction between lipid

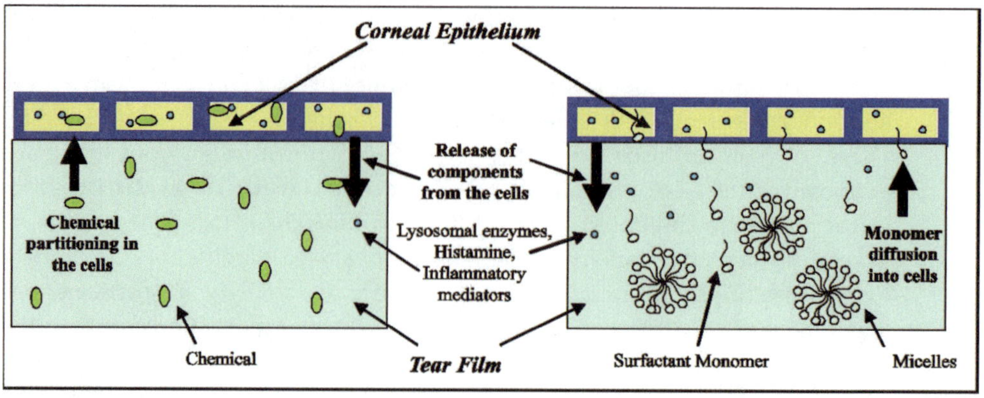

Figure 5.5 Schematic (not to scale) showing ocular toxicity of a chemical as compared to a surfactant.

bilayers and the surfactant saturates [1]. In Figure 5.6 the amount of surfactant adsorbed in the lipid bilayer as a function of concentration shows that the adsorption saturates above the CMC of the surfactant in the medium. This clearly indicates that the free surfactant or surfactant monomer is the dominant species interacting with the lipid bilayer and the surfactant micelles have little or no interaction [1, 32]. Thus, when surfactants are introduced on the eye, only the free surfactant present at the CMC interacts with the ocular surface. Also, unlike other chemicals, surfactant penetration inside the corneal epithelium induces micelle breakup on the ocular surface, and thus maintains a fixed concentration of surfactant monomers (at CMC) till all the surfactant micelles are drained from the eyes. Hence, when we utilize (5.1) to calculate the concentration to be used in the liposomal assay, C_{Ocular} should be taken as CMC for surfactants when they are introduced on the ocular surface above their respective CMCs. For some extensively studied surfactants, the user may find an array of CMC values reported in the literature and these values sometimes are different due to variation in sample purity and the method of determining the CMC [33, 34]. Therefore, it is instructive to carefully select and use the CMC values of surfactants from reliable sources. While the interactions

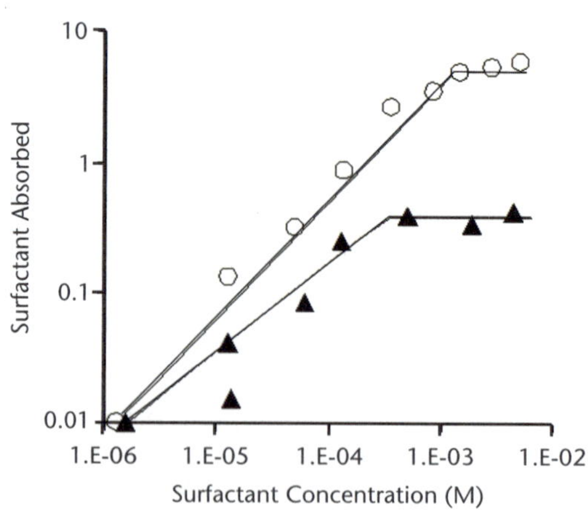

Figure 5.6 Saturation of lipid-bilayer uptake of the surfactants hexadecylpyridinium chloride (O) and Triton X-100 (▲). (Reproduced from Okahata and Ebato [1] with permission from the American Chemical Society.)

between the surfactants and the epithelia saturate beyond CMC, the toxicity continues to increase because of the increase in the residence time of surfactants and a consequent increase in the AUC with an increasing concentration.

One major drawback of utilizing correlations between liposomal leakage and Draize scores to estimate the ocular toxicity of chemicals whose Draize scores have not been evaluated is the assumption that the contribution of the conjunctiva and iris to the Draize score is negligible. This assumption can lead to significant underestimation of Draize scores, especially for chemicals, which are mild irritants and show an equivalent toxicity contribution from the cornea, conjunctiva, and iris. This problem can be circumvented if correlations are made between corneal Draize scores, which constitute 80 points out of the maximum allowable 110 points, to the liposome leakage. The major deterrent in this approach is the lack of information in most of the published studies in the past where only the combined Draize scores are reported. The importance of this issue can be clearly understood by considering the Draize scores (cornea, conjunctiva, iris) for polyoxyethylene (20) alkyl phenyl ether and alkyl ethoxysulphate from the work of Kato et al. [15]. The total Draize scores for these two chemicals are dominated by the conjunctival scores which may not be reflected in the liposomal assay since the liposomes are mimicking the corneal epithelium. Hence, careful scrutiny is required before using the liposomal assay for evaluating toxicity of potential irritants.

Preparation of liposomes is suggested to be done in PBS and though PBS is extensively used to replace the tear fluid, essential components of the tear fluid are not present in the liposome solution. These components can interact with certain chemicals and interfere during toxicity assessment in vivo leading to potential errors in the liposome assay. The interaction of tear components and a possible irritant have not been discussed in detail in the literature and it is thus very difficult to comment on the possible variations due to this difference between the in vitro and the in vivo assays.

5.7 Application Notes

The liposome assay discussed in this chapter represents an in vitro technique to be utilized for the estimation of ocular toxicity of a new chemical, which has not been tested on the ocular surface. The liposome assay, along with some other in vitro techniques, can combine to minimize and, in some cases, eliminate the harsh use of animals for determining the toxicity levels of eye irritants. The most important consideration while utilizing this assay is the choice of concentration to be used in vitro which can be closely related to the in vivo application. This idea has been discussed in a recently published work by our group [9]. It has been shown that the successful use of this assay requires a thorough consideration of the mechanistic differences between the liposomes and the ocular surface. Furthermore, it has been shown that this assay can be used successfully to determine the ocular toxicity of widely used surfactants for which ocular toxicity data is scarce. It is also noted that this assay would require the researcher to generate multiple correlation curves for complete ocular toxicity estimation. On the other hand, the simplicity and ease of application override this problem, and for application purposes, this assay can be a good starting point for estimating ocular toxicity.

Troubleshooting Table

Problem	Explanation	Potential Solutions
Lipids clump together and become difficult to weigh.	Unsaturated lipids such as DOPE in powder form absorb moisture.	Purchase unsaturated lipids dissolved in chloroform or use a dry box.
Polycarbonate extrusion membrane ruptures during extrusion. The rupture could be detected by noting that it will lead to a sudden decrease in the force needed to extrude the liposomes.	Lipid concentration is high, making extrusion difficult.	Incrementally pass increasing amounts of liposome dispersion through the membrane and back until all of the liposome dispersion passes through the membrane.
Baseline fluorescence of diluted liposome solution is too high after bulk calcein removal.	The separation of bulk calcein and liposomes was unsuccessful or dilution was insufficient.	Ensure that the liposomes are not added into the void volume around column; allow gel column with liposomes added to sit for 5 minutes before centrifuging; increase stock liposome dilution factor.
Calcein concentration in liposome solution after adding Triton X-100 to break liposomes is less than expected, but green fluorescence is optically visible.	The dilution factor is too low and the final concentration of calcein is outside the linear detection regime; confirm by diluting and checking for fluorescence increase.	Increase the factor by which the stock liposome solution from the gel column is diluted.
The fluorescence of liposome samples or even control samples with no investigative compounds added decreases over time.	The fluorescent signal from the dye decreases upon multiple exposures to the detector, certain test substances, and light.	Cover samples when not in use, limit the number of measurements per sample, use controls to distinguish compound effects and dye property changes, use dye solutions made within 1 week.

5.8 Summary Points

1. The liposomal assay can be utilized to determine the ocular toxicity of new chemicals if used appropriately with mechanistic considerations. A correction factor taking into account the surface area to volume ratio should be introduced in the liposomal assay for an effective implementation of this method.
2. Fluidity of the epithelium is challenged by the presence of possible ocular irritants, which is the underlying cause of toxicity on the ocular surface.
3. Using the liposomal assay for special chemicals like surfactants requires specific considerations as surfactants are present in two phases when their concentration is above the CMC. Surfactant monomers interact with the ocular surface, whereas the surfactant micelles have no interaction. As the surfactant monomers diffuse into the ocular surface, micelles break to populate the lost surfactant and maintain the surfactant concentration at the CMC.
4. Correlation curves between dye leakage and Draize scores can be used to determine the Draize scores for new substances.
5. A better correlation curve can result if the Draize scores used for the correlation are specific to the corneal epithelium alone.
6. The Draize score for a particular compound depends on the area under the curve (AUC) for the tear concentration versus time. The AUC varies with changes in concentrations of chemicals tested, which leads to varying Draize scores reported in literature.
7. Correlations generated for the liposomal study should be at a fixed concentration of the substances when used to determine the Draize scores as this is necessary for accurate prediction of the ocular irritancy for a new chemical. Conversely, the predicted Draize scores correspond to the irritancy level when that new substance is

put on the ocular substance at the concentration at which the correlation/calibration curve was generated. Hence, an array of correlation curves needs to be generated to determine ocular toxicity at varying concentrations.

References

[1] Okahata, Y., and H. Ebato, "Absorption Behaviors of Surfactant Molecules on a Lipid Coated Quartz-Crystal Microbalance. An Alternative to Eye-Irritant Tests," *Analytical Chemistry*, Vol. 63, No. 3, 1991, pp. 203–207.

[2] Jester, J. V., et al., "Area and Depth of Surfactant-Induced Corneal Injury Correlates with Cell Death," *Investigative Ophthalmology and Visual Science*, Vol. 39, No. 6, 1998, pp. 922–936.

[3] Tachon, P., et al., "Assessment of Surfactant Cytotoxicity: Comparison with the Draize Eye Test," *International Journal of Cosmetic Science*, Vol. 11, No. 5, 1989, pp. 233–243.

[4] Saarinen-Savolainen, P., et al., "Amphiphilic Properties of Pilocarpine Prodrugs," *International Journal of Pharmaceutics*, Vol. 133, 1996, pp. 171–178.

[5] Sunamoto, J., et al., "Possible Eye-Irritant Test Using Polysaccharide-Coated Liposomes as a Corneal Epithelium Model," *Chemical and Pharmaceutical Bulletin*, Vol. 35, No. 7, 1987, pp. 2958–2965.

[6] Saettone, M. F., et al., "Evaluation of Ocular Permeation Enhancers: In Vitro Effects on Corneal Transport of Four β-Blockers, and In Vitro/In Vivo Toxic Activity," *International Journal of Pharmaceutics*, Vol. 142, 1996, pp. 103–113.

[7] Saarinen-Savolainen, P., et al., "β-Cyclodextrin Derivatives (2-HP-β-CD, SBE4-β-CD) Decrease the Amphiphilicity and Membrane Perturbing Effects of Pilocarpine Prodrugs," *European Journal of Pharmaceutical Sciences*, Vol. 5, 1997, pp. 89–96.

[8] Bean, C. L., S. M. Galloway, and M. O. Bradley, "Evaluation of Three *In Vitro* Assays for Assessment of Membrane Damage by Surfactants," *In Vitro Toxicology*, Vol. 4, No. 2, 1991, pp. 133–144.

[9] Kapoor, Y., B. A. Howell, and A. Chauhan, "Liposome Assay for Evaluating the Ocular Toxicity of Surfactants," *Investigative Ophthalmology and Visual Science*, Vol. 50, No. 6, 2009, pp. 2727–2735.

[10] Wilhelmus, K. R., "The Draize Eye Test," *Survey of Ophthalmology*, Vol. 45, No. 6, 2001, pp. 493–515.

[11] Matsukawa, K., et al., "Interlaboratory Validation of the In Vitro Eye Irritation Tests for Cosmetic Ingredients. (11) EYTEX™," *Toxicology in Vitro*, Vol. 13, 1999, pp. 209–217.

[12] Cottin, M., and A. Zanvit, "Fluorescein Leakage Test: A Useful Tool in Ocular Safety Assessment," *Toxicology in Vitro*, Vol. 11, 1997, pp. 399–405.

[13] Kennah II, H. E., et al., "An Objective Procedure for Quantitating Eye Irritation Based upon Changes of Corneal Thickness," *Fundamental and Applied Toxicology*, Vol. 12, 1989, pp. 258–268.

[14] Vinardell, M. P., and M. Mitjans, "Alternative Methods for Eye and Skin Irritation Tests: An Overview," *Journal of Pharmaceutical Sciences*, Vol. 97, No. 1, 2008, pp. 46–59.

[15] Kato, S., et al., "Liposomes as an In Vitro Model for Predicting the Eye Irritancy of Chemicals," *Toxicology in Vitro*, Vol. 2, No. 2, 1988, pp. 125–130.

[16] Vian, L., et al., "Comparison of Three In Vitro Cytotoxicity Assays for Estimating Surfactant Ocular Irritation," *Toxicology in Vitro*, Vol. 9, No. 2, pp. 185–190.

[17] Zuidam, N. J., R. Vrueh, and D. J. A. Crommelin, "Characterization of Liposomes," in *Liposomes*, 2nd ed., V. P. Torchilin and V. Weissig, (eds.), New York: Oxford University Press, 2003, pp. 31–78.

[18] Kim, J. H., and M. W. Kim, "Temperature Effect on the Transport Dynamics of a Small Molecule Through a Liposome Bilayer," *The European Physical Journal E*, Vol. 23, 2007, pp. 313–317.

[19] Lis, L. J., et al., "Interactions Between Neutral Phospholipid Bilayer Membranes," *Biophysical Journal*, Vol. 37, No. 3, 1982, pp. 657–666.

[20] New, R. R. C., *Liposomes*, New York: Oxford University Press, 1990, p. 137.

[21] Howell, B., and A. Chauhan, "Amitriptyline Overdose Treatment by Pegylated Anionic Liposomes," *Journal of Colloid and Interface Science*, Vol. 324, 2008, pp. 61-70.

[22] Balgavy, P., et al., "Bilayer Thickness and Lipid Interface Area in Unilamellar Extruded 1,2-Diacylphosphatidylcholine Liposomes: A Small-Angle Neutron Scattering Study," *Biochimica et Biophysica Acta*, Vol. 1512, No. 1, 2001, pp. 40–52.

[23] Maric, K., et al., "Cell Volume Kinetics of Adherent Epithelial Cells Measured by Laser Scanning Reflection Microscopy: Determination of Water Permeability Changes of Renal Principal Cells," *Biophysical Journal*, Vol. 80, No. 4, 2001, pp. 1783–1790.

[24] Farinas, J., and A. S. Verkman, "Cell Volume and Plasma Membrane Osmotic Water Permeability in Epithelial Cell Layers Measured by Interferometry," *Biophysical Journal*, Vol. 71, No. 6, 1996, pp. 3511–3522.

[25] Avanti Polar Lipids, Inc., http://www.avantilipids.com/.

[26] Fry, D. W., J. C. White, and I. D. Goldman, "Rapid Separation of Low Molecular Weight Solutes from Liposomes Without Dilution," *Analytical Biochemistry*, Vol. 90, 1978, pp. 809–815.

[27] Hamann, S., et al., "Measurement of Cell Volume Changes by Fluorescence Self-Quenching," *Journal of Fluorescence,* Vol. 12, No. 2, 2002, pp. 139–145.

[28] Memoli, A., et al., "Effects of Surfactants on the Spectral Behaviour of Calcein (II): A Method of Evaluation," *Journal of Pharmaceutical and Biomedical Analysis,* Vol. 19, 1999, pp. 627–632.

[29] Okamoto, Y., et al., "Interlaboratory Validation of the In Vitro Eye Irritation Tests for Cosmetic Ingredients. (3) Evaluation of the Haemolysis Test," *Toxicology In Vitro*, Vol. 13, 1999, pp. 115–124.

[30] DeSousa, D. J., A. A. Rouse, and W. J. Smolon, "Statistical Consequences of Reducing the Number of Rabbits Utilized in Eye Irritation Testing: Data on 67 Petrochemicals," *Toxicology and Applied Pharmacology,* Vol. 76, 1984, pp. 234–242.

[31] Weil, C. S., and R. A. Scala, "Study of Intra- and Interlaboratory Variability in the Results of Rabbit Eye and Skin Irritation Tests," *Toxicology and Applied Pharmacology,* Vol. 19, 1971, pp. 276–360.

[32] Hall-Manning, T. J., et al., "Skin Irritation Potential of Mixed Surfactant Systems," *Food and Chemical Toxicology*, Vol. 36, No. 3, 1998, pp. 233–238.

[33] Klammt, C., et al., "Evaluation of Detergents for the Soluble Expression of Alpha-Helical and Betabarrel-Type Integral Membrane Proteins by a Preparative Scale Individual Cell-Free Expression System," *FEBS J.*, Vol. 272, 2005, pp. 6024–6038.

[34] Hait, S. K., and S. P. Moulik, "Determination of Critical Micelle Concentration (CMC) of Nonionic Surfactants by Donor-Acceptor Interaction with Iodine and Correlation of CMC with Hydrophile-Lipophile Balance and Other Parameters of the Surfactants," *J. Surfactants Deterg.*, Vol. 4, 2001, pp. 303–309.

CHAPTER

6

Prediction of Potential Drug Myelotoxicity by In Vitro Assays on Hematopoietic Progenitors

Augusto Pessina and Arianna Bonomi

Department of Public Health, Microbiology, Virology, University of Milan, Italy
Corresponding author: Augusto Pessina, Department of Public Health, Microbiology, Virology, University of Milan, Via Festa del Perdono 7, 20122, Milano, Italy, e-mail: augusto.pessina@unimi.it, phone: 0039 2 250315072, fax: 0039 2 50315068

Abstract

Clonogenic assays have been developed and used to investigate the proliferation and the differentiation in vitro both of pluripotent hemopoietic stem cells and of the different progenitors of blood cell lineages. Their application to the toxicology provides an essential tool to better understand the in vivo observation (experimental and clinical) by also helping the prediction of the degree of a possible in vivo myelotoxicity in drug-treated patients. A standard procedure developed and validated to determine the in vitro myelotoxic effect of drugs and chemicals both towards murine and human progenitors is described here. An algorithm is also suggested to be applied by using the in vitro CFU-GM assay for predicting the maximal tolerated dose (MTD) to enter a phase I clinical study. A miniaturized procedure in a 96-well plate may be also used for the high-throughput screening of compounds or if very small amounts are available.

Key terms
cord blood stem cells
drug myelotoxicity
GM-CFU assay
in vitro toxicity
MTD

6.1 Introduction

In vitro clonogenic assays have been used in preclinical safety studies on candidate drugs and are helpful in investigating the relative sensitivities of various animal species to hematotoxic effects [1, 2]. The type of hematotoxicity most frequently investigated is related to the acute effect of toxicants on bone marrow progenitors, such as granulocyte-macrophage (CFU-GM) and erythroid (CFU-E), and the effect is quantified from the number of surviving progenitors as a function of exposure level under maximally stimulatory cytokine concentrations. In vitro hematotoxicity assays are thought to have the potential to significantly reduce and refine the use of animals for hematotoxicity testing and also play a key role in bridging the gap between preclinical toxicology studies in animal models and clinical investigations [3]. Since the original technique suggested by Metcalf [4], many protocols have been developed and proposed for in vitro hematotoxicity testing [5–7]. The CFU-GM assay has been validated for testing drug hematotoxicity in both mouse bone marrow cells and human cord blood cells [8, 9], and a micromethod has been developed for testing a great number of molecules [10, 11]. More recently, an in vitro assay to check the drug toxicity on megakaryocytes (CFU-Mk) has been prevalidated [12], and the GM-CFU assay has been optimized to use cryopreserved cells from rat bone marrow [13] to be applied to test chemicals in rats, as this is very important in early stage toxicology assessment.

In this chapter we describe in detail the procedure for the GM-CFU assay applied to human cord blood and mouse bone marrow progenitors and the management of data according to an algorithm for predicting the in vivo human maximal tolerated dose (MTD) by adjusting in vivo data on mouse toxicity. The standard operating procedure (SOP) presented here has been also approved by the Scientific Advisory Committee of the European Centre for the Validation of Alternative Methods (ECVAM) [24].

6.2 Experimental Design

The experimental design suggested here is optimized to study three drugs/experiment.

At least three experiments (on different days) are performed: one of screening and two of inhibitory concentration (IC) determination. The screening experiment is to determine the range of the toxic activity of the compounds. The two IC determination experiments are performed at a more restricted range of drug concentrations to calculate the IC50 and IC90 of the compounds.

A scheme of a typical experiment is organized as follows:

- One set of control (CTR) to check the 100% clonogenicity in the absence of drug and solvent;
- One set of the negative control drug (NCTR, a single concentration of a drug without a toxic effect);
- One set of dose-response curve of the drug: D0 (only solvent, without drug); D1–D8 (eight decreasing drug concentrations).

Each set is performed in triplicate (33 cultures), and therefore to test three drugs, 99 cultures per each experiment are needed. Each drug is tested three times in three independent experiments that provide a number of data enough to confer to the results an acceptable statistical power.

An example of a typical complete trial (three experiments, one for a screening phase and two for IC determination) is reported in Table 6.1.

6.3 Materials

- Hemocytometer (e.g., Bürker, Neubauer)
- Inverted microscope (20–25× magnification)
- Freezing container, "Mr. Frosty" (Nalgene)
- Liquid nitrogen container
- Gauze tissue
- Surgical material: little pair of scissors, surgery tweezers
- Needle 23 or 25 Ga, 18 Ga, 19 Ga
- 0.22-μm disposable filter
- Cell strainer 100 μm (BD Falcon)
- Cryotubes 1.5 ml
- Petri dish ∅ 150 mm, ∅ 60 mm
- Petri dish ∅ 35 mm (Nunc)
- Gridded petridish ∅ 60 mm

Table 6.1 Complete Trial Layout

	Experimental Setup	Tubes	Dishes
Experiment 1, Day 1	Control (CTR)	3	9
	Negative Drug Control (NCTR)	3	9
	Solvent control (D0)	3	9
	Drug A (D1- D8)	8	24
	Drug B (D1- D8)	8	24
	Drug C (D1- D8)	8	24
			Total 99
Experiment 2, Day 2	Control (CTR)	3	9
	Negative Drug Control (NCTR)	3	9
	Solvent control (D0)	3	9
	Drug A (D1- D8)	8	24
	Drug B (D1- D8)	8	24
	Drug C (D1- D8)	8	24
			Total 99
Experiment 3, Day 3	Control (CTR)	3	9
	Negative Drug Control (NCTR)	3	9
	Solvent control (D0)	3	9
	Drug A (D1- D8)	8	24
	Drug B (D1- D8)	8	24
	Drug C (D1- D8)	8	24
			Total 99

- Disposable syringe 10 ml, 1-ml B-D insulin syringe
- 15-ml round bottom tubes (BD Falcon)

6.3.1 Reagents

- Antibiotics: penicillin 100 U/ml, streptomycin 100 μg/ml
- Iscove's Modified Dulbecco's medium 1X (IMDM)
- Fetal bovine serum (FBS)
- Dimethyl sulfoxide (DMSO)
- Dulbecco's Phosphate Buffer Saline without calcium and magnesium (PBS)
- Ficoll-Paque Research Grade (Pharmacia Biotech) (store at 4°C and protect from direct light)
- Anticoagulant suggested Citrate-Phosphate-Dextrose solution (CPD, Sigma-Aldrich)
- Human Serum Albumin (HSA) fraction V for clinical applications (Sigma)
- Dextran T 40: Rheomacrodex 10% (Dextran Mw 40,000) (Pharmacia Biotech)
- DNase I: RNase free, 10 U/μl
- Trypan Blue 0.4%
- Turk solution (Merck)
- Methylcellulose culture media (MCM): 1% Methylcellulose in IMDM, 30% fetal bovine serum, 1% bovine serum albumin, 2 mM L-glutamine, 10 ng/ml human or mouse recombinant GM-CSF (Stem Cell Technologies)
- Hank's balanced salt solution (HBSS)

6.4 Methods

6.4.1 Preparation of methylcellulose stocks

1. Thaw bottles of methylcellulose culture media (MCM) overnight at +4°C.
2. Homogenize the MCM by inverting the bottle several times.
3. With a 10-ml syringe and a 18-Ga needle, dispense all of the MCM in the bottle into 4-ml aliquots in 15-ml round bottom tubes.
4. The precision in dispensing MCM represents a critical point in order to obtain the correct final concentration of methylcellulose and toxicant at later steps. If the medium runs along the tube wall, centrifuge the tube 140 × g, 30 seconds to move it to the bottom.
5. Store aliquots at –20°C. Thaw and use as needed individual tubes.

6.4.2 Source of murine hematopoietic progenitors

Murine hematopoietic progenitors were collected from bone marrow of male BDF/1 (C57Bl/6 × DBA-2) mice, 8–12 weeks old. Use three mice per experiment and perform all the procedures in rigorous sterile conditions by using sterile reagents and materials and operating in a microbiological safety cabinet. Keep the cells at + 4°C (melting ice).

6.4.2.1 Isolation of murine bone marrow cells (mu-BMC)

1. Sacrifice the animals by cervical dislocation (without anesthesia). After animals have been killed, proceed as soon as possible. Wash the mice thoroughly with 70% ethyl alcohol and leave them for 1–2 minutes entwined into a gauze alcohol imbued.
2. Remove the skin of the legs and isolate the intact femora by cutting the muscle ligaments using a little pair of scissors and tweezers in order to clean the bones well.
3. Put the six femora in a 150-mm Petri dish containing 10-ml Iscove's modified Dulbecco's medium (IMDM) supplemented with antibiotics (Penicillin 100 U/ml – Streptomycin 100 μg/ml and 10 μg/ml Gentamicin) and maintain them at +4°C on ice.
4. Clean the knee from the articular cartilage and cut both ends from the femora just below the head.
5. Hold the femur shaft with surgery tweezers, insert a 23–25-Ga needle mounted on a 5-ml syringe into the knee end, and flush the marrow out of the end with IMDM without antibiotics. Twice use 1.5-ml flushings per femur. Collect the bone marrow cells from all femora in a single 15-ml round bottom or a 50-ml tube. Calculate 3 ml for each femur.
6. Disperse the mu-BMC with the syringe by repeated flushing (five times) and transfer the cell suspension with the syringe into a 50-ml sterile tube by filtering them through a 100-μm disposable cell strainer.
7. Wash the cells at 400× g, 10 minutes, discard the supernatant, and resuspend the pellet in IMDM (calculate 1 ml for each femur) supplemented with 30% FBS without antibiotics.
8. Dilute 50 μl of cells plus 450 μl of HBSS containing 0.04% Trypan Blue and evaluate the percentage of cell viability immediately after making the suspension in a hemocytometer. Thereafter, count the cells in the same way by using 10 μl of cells plus 90 μl of the Turk Solution.
9. Adjust the suspension at 0.59×10^6 viable nucleated cells/ml (viability must be greater than 95%) to have a final cell density of 40,000 cells/dish. The suspension must be constituted of single cells. If cell aggregates are present, disperse them by gentle pipetting before counting.

6.4.3 Source of human hematopoietic progenitors

Human hematopoietic progenitors are cryopreserved human cord blood mononuclear cells. A minimum of two frozen aliquots of cells from a single donor is required to complete the screening phase and the IC determination phase of testing.

6.4.3.1 Collection of human umbilical cord blood

1. Add 7 ml of a citrate phosphate dextrose (CPD) solution in a 50-ml tube. Keep the tube at room temperature.
2. Collect 43 ml of the cord blood after a normal delivery in sterile conditions if possible.
3. Mix CPD and the cord blood (ratio 1:7.1).

6.4.3.2 Isolation of human umbilical cord blood mononuclear cells (hu-UCB)

1. Maintain the Ficoll-Paque and the cord blood samples at room temperature prior to and during the density gradient.
2. Dilute 1 volume of cord blood with 1 volume of PBS (1:1 dilution). For each 10 ml of diluted cord blood, one gradient tube will be required.
3. Invert the Ficoll bottle several times to ensure thorough mixing.
4. Add 1 volume (i.e., 5 ml) of Ficoll-Paque in a 15-ml centrifuge tube and carefully layer two volumes (i.e., 10 ml) of the diluted cord blood sample. Do not mix the Ficoll with the diluted cord blood sample.
5. Centrifuge at 400× g for 30 minutes at 18–20°C, continuously.
6. Draw off the upper layer using a pipette, leaving the mononuclear cell layer undisturbed at the interface.
7. Transfer the mononuclear cell layer of the gradients to one clean, 50-ml centrifuge tube, using a pipette.
8. Add three volumes of PBS to the volume of mononuclear cells. Suspend the cells by gently drawing, using a pipette.
9. Centrifuge at 400× g for 10 minutes at 18–20°C, and draw off the supernatant with pipette.
10. Repeat the cell washing procedure (steps 7 and 8).
11. Resuspend cells in 0.5 ml of PBS and place on melting ice.
12. Dilute 50 μl of cells in 450 μl of HBSS containing 0.04% Trypan Blue and evaluate the percentage of viability immediately after making the suspension in a hemocytometer. Thereafter, count the cells in the same way by using 50 μl of cells plus 450 μl of the Turk solution.

6.4.3.3 Cryopreservation of hu-UCB

1. Prepare the freezing medium composed by IMDM supplemented with 20% DMSO and 40% FBS.
2. Adjust the cellular suspension to a concentration of 4×10^6–4×10^7 cells/ml in IMDM.
3. Dilute the cellular suspension 1:1 with the freezing medium to obtain a final concentration of 2×10^6–2×10^7 cells/ml.
4. Fill the cryotubes with 1 ml of the cell suspension immediately after step 3 and place them at –80°C for 24 hours in the Mr. Frosty container.
5. After 24 hours, move the cryotubes into liquid Nitrogen.

6.4.3.4 Thawing of hu-UCB

Perform all the procedures in rigorous sterile conditions operating in a Microbiological Safety Cabinet and using sterile reagents and materials.

1. Warm the thawing medium (IMDM containing 10% FBS and DNase I at 10 U/ml).
2. Quickly thaw the vial of frozen cells in a 37°C water bath. Wipe the outside of the vial with 70% ethanol.
3. Transfer a maximum of 2 ml of cell suspension to a 15-ml conical tube.

4. Rinse the vial with 1 ml of the medium. Add the rinse drop-wise to the cells while gently swirling the tube (~1 minute).
5. Slowly add enough medium drop-wise to the cells until the total volume is 5 ml, while gently swirling after each addition of the medium (~3 minutes).
6. Centrifuge the cell suspension at 200× g at room temperature for 15 minutes.
7. Carefully remove by pipette (and save in a second tube) most of the wash, leaving 2 ml behind so that the cell pellet is not disturbed. Gently resuspend the cell pellet in the remaining medium.
8. Add 5 ml of medium to the cell suspension and gently swirl.
9. Centrifuge the cell suspension at 200 × g at room temperature for 15 minutes.
10. Carefully remove by pipette all the surnatant (and save in the same tube as in step 7). Gently resuspend the cell pellet in 1 ml of IMDM+30% FBS. From now on, put the cells in melting ice (+4°C).
11. Count the cells by hemocytometer diluting 50 μl of cells into 450 μl of HBSS containing 0.04% Trypan Blue. If the cell count is lower than expected, centrifuge wash saved at a higher speed, count, and combine if necessary.
12. Adjust the suspension at 1.1×10^6 viable nucleated cells/ml (viability must be greater than 80%) to have a final cell density of 75,000 cells/dish.

6.4.4 Technical procedure for GM-CFU test

1. Mix methylcellulose, test article, and cells before plating the culture dishes.
2. Store aliquots of test article, paired solvent, and dilutions as directed by the supplier.
3. On the night before the testing day, completely thaw the aliquots of methylcellulose at +4°C. There are 11 tubes of cell culture mixture for each drug: CTR (100% clonogenicity), NCTR (negative drug control), D0 (solvent controls), and D1–D8 (drug curve). Each tube should contain 4.0 ml of the methylcellulose culture medium.
4. On an incubator tray, label 35-mm Petri dishes according to the experimental design.
5. Prepare the drug and solvent stocks immediately prior to use according to the specific SOP.
6. Add 100 μl of IMDM to CTR. Add 78 μl of IMDM to NCTR and to each of D0–D8
7. Prepare the toxicant dilutions at room temperature in sterile 1.5-ml Eppendorf tubes. For the screening phase, serial dilutions 1:10 are prepared in solvent (from 10^{-1} to 10^{-8}). For the IC determination phase the range of dilutions is calculated as reported in Section 4.4. Using certified pipettes with tips, add 22 μl of solvent or 22 μl of each toxicant dilution to the methylcellulose tubes in melting ice. Vortex each tube twice for 5 seconds. The final volume in each tube should be 4.1 ml after adding the toxicant.
8. Immediately add 0.3 ml of mouse or human MNC cell suspension to each correspondent tube, move the tube gently to mix, and then vortex vigorously three times for 8 seconds.
9. Let the tubes stand for 5 minutes on melting ice to release air bubbles. (If indicated by GLP, label the D8 and D4 dilution tube with toxicant name, dose level, date, and

test location. Store these two tubes at –80° for future analysis and record the stored samples on the Registration Form.)

10. Distribute 1 ml of the cell-medium mixture into each of three Petri dishes using 1-ml B-insulin syringe with 19-Ga needle or 1-ml tips with a certified pipette. Gently rotate the plate to spread the mixture evenly by allowing the meniscus to attach to the dish wall.
11. Incubate the cultures at 37°C in the air +5% CO_2 under saturated humidity for 7 days (murine assay) or 14 days (human assay).

Sufficient humidity during incubation is critical because the drying of cultures drastically reduces colony formation. If a humidified incubator is not available, saturated humidity can be achieved by incubating six culture dishes with one 60-mm dish (without a lid) containing water inside a 150-mm Petri dish covered with a lid.

6.4.5 Passing from screening phase to IC determination phase

1. From the GM-CFU results in the screening phase, identify the first drug dilution that completely inhibits GM-CFU (FCID) and the last drug dilution that does not inhibit GM-CFU (LNID).
2. Calculate the Log_{10} R as follows:

$$\text{Log}_{10} \Rightarrow \text{Log}_{10}[\text{FCID/LNID}]$$

3. Then,

If $\text{Log}_{10}\, R = 1 \rightarrow$ use 1:2 serial dilutions assigning to D_6 the value= 2 × FCID

If $\text{Log}_{10}\, R = 2 \rightarrow$ use 1:3 serial dilutions assigning to D_6 the value= 3 × FCID

If $\text{Log}_{10}\, R = 3 \rightarrow$ use 1:4 serial dilutions assigning to D_6 the value= 4 × FCID

If $\text{Log}_{10}\, R = 4 \rightarrow$ use 1:5 serial dilutions assigning to D_6 the value= 5 × FCID

In order to reduce the variability, it is suggested that the dilutions be performed starting from the same concentration of toxicant stock. When the volume required for making the dilution is too small, it is better to prepare the drug stock before making a working stock (for example, pipette 100 μl of a 1:100 dilution of toxicant stock, rather than 1 μl of undiluted toxicant stock).

6.4.6 Incubator humidity test

If an incubator is not self-testing for humidity, a simple method to evaluate if humidity incubator is acceptable is the following:

1. Prepare three Petri dishes of Ø 60 mm (Area = 28.27 cm^2) and fill each dish with 10 ml of distilled water (Vs = starting volume). Put the dishes in the incubator in the center of the middle plateau.

2. After 72 hours measure the volume of water in each dish (Vf = final volume). Calculate the evaporation rate (ER) of each one as follows:

$$E.R.(\mu l \times h^{-1} \times cm^{-2}) = \frac{Vs - Vf}{h \times A} = \frac{V_{72}}{2.035}$$

3. Calculate the mean ± S.D. of the triplicate.
   E.R.Values (ranges): 1.1–1.7 Very Good; 1.7–2.1 Good; 2.1–2.7 Acceptable; 2.7–3.0 Poor; >3.0 Unacceptable).

6.4.7 Scoring the colonies

Count colonies after 7 days (murine model) or 14 days (human model) of incubation as follows:

1. Place the culture dish inside a 60-mm gridded tissue culture dish.
2. Score CFU-GM colonies by scanning the whole Petri dish by using an inverted microscope (about 20–25× magnification). It is critical to use 20–25× magnification; do not use 40× magnification. Score first a D8 plate to determine what the minimal acceptable aggregate is considered as a colony. In this highest drug level, the colonies will be the smallest and most difficult to define because of toxicity. After scoring one D8 replicate, randomly count one replicate from the other experimental groups. Then score a second D8 replicate, and then randomly score the second replicate from the remaining groups. Repeat this sequence for the third replicates.

Aggregates containing 50 or more cells are defined as CFU-GM colonies. Aggregates with 20–50 cells are defined as clusters.

For a correct discrimination between colonies and clusters, carefully evaluate the number of cells for each aggregate. It is also important to carefully examine at the edge of the plate, where many colonies grow when a high cell density is seeded.

6.4.8 Criteria for colony counting

Applies to colony morphology at 20–25× magnification; do not use 40×.

- *Compact colonies:* with a central dense nucleus and a peripheral halo. These colonies are very easy to score.
- *Diffuse and spread colonies:* without apparent nucleus. Care must be taken with the magnification, since an excessive one (> 30×) can lead to lack of detection of this kind of colony. With high densities of colonies in the plate (> 150 colonies/plate), it is really hard (sometimes impossible) to score these colonies. This is one of the reasons for suggesting not to score very high number of colonies (although correlations could be good, the scoring is really hard).
- *Multicentric colonies:* with two or more dense nucleus nearby, with a common peripheral halo growing at the same depth in the plate. They should be considered as one colony.

- *Burst-forming units (BFUs):* multifocal colonies, with aggregates of several colonies or clusters, with or without a peripheral halo. These must be counted as one colony.

To help the colony identification, see their morphology in Figure 6.1.

6.5 Data Acquisition, Anticipated Results, and Interpretation

A statistical comparison between the absolute counts of colonies (GM-CFU) in CTR and D0 cultures gives preliminary essential information of the toxic activity of the solvent. In the presence of such a toxic effect it is imperative that the kinetics of inhibition of CFU-GM by a drug is normalized on D0 values instead of CTR counts. This means that the counts of colonies at each drug dose (from D1 to D8) must be expressed as a percentage of the counts in D0 (the culture containing the solvent specific for each different drug).

On such a curve of the kinetics, inhibition is possible to determine the inhibitory concentration of the drug tested by applying different methods suggested in Section 6.5.1.

As shown in the validation study, the drug concentration that inhibits CFU-GM by 90% (IC90) is a more predictive for the MTD in animals and in man than the IC50 [7] Therefore, according to our prediction model, the human in vivo MTD can by calculated by multiplying the actual in vivo murine MTD by the in vitro human and mouse IC90 ratio (IC90 hu/IC90 mu) by applying the following formula:

$$P-HuMTD = Actual\ murine\ LD_{10} \times \frac{Human\ IC_{90}}{Murine\ IC_{90}}$$

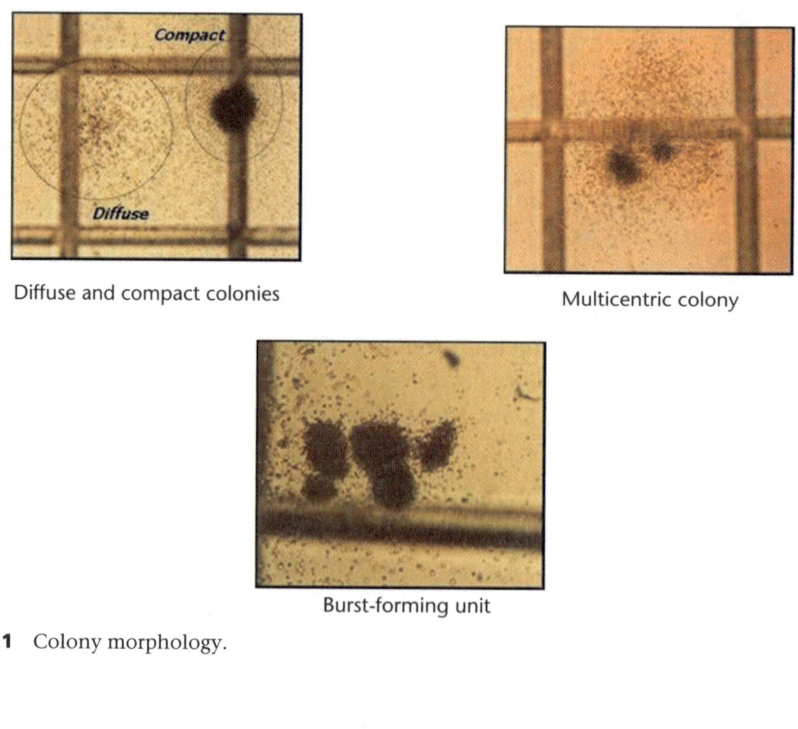

Diffuse and compact colonies

Multicentric colony

Burst-forming unit

Figure 6.1 Colony morphology.

This prediction model seems to be able to correlate the inhibition of CFU-GM in vitro and the depth of the absolute neutrophil count (ANC) nadir in vivo. As pharmacokinetic differences across species may contribute to as much as a fourfold difference in MTD, and the prediction model does not accommodate this source of variability. By this study we can consider as an accurate prediction a prediction of a human MTD that lies within fourfold of the actual human MTD value [3, 14–16].

The prediction model utilizes information from the in vitro analysis of toxic effects upon the actual human target cell and offers the advantage of being mechanism-naïve and would only fail to identify hematotoxicants that adversely affect myelopoiesis via indirect physiological mechanisms such as the induced release of inhibitory cytokines, the inhibited release of stimulatory cytokines, or the metabolic activation of protoxicants. It provides human toxicology and pharmacology information and an experimental basis for selecting the best animal models for investigating clinical hematotoxicity. The prediction model suggested here has been applied in a prevalidation study [8] to correlate the inhibition of CFU-GM in vitro and the depth of ANC nadir in vivo. The mouse MTD values used for validating the prediction model and the human MTD values to which they were compared were derived from studies with the exact same route of toxicant exposure (usually intravenous, but also oral) and similar dosing regimens. Given the schedule dependency of many marrow toxicants, it was very important to match route and regimen across species, so the true interspecies difference measured by the CFU-GM test, which is a difference in potency at the same target organ, could be detected accurately.

6.5.1 Statistical guidelines

The difference between the counts in CTR and D0 can be analyzed by applying a simple "t" student test to compare the means of the two groups and calculating a two-tail p value for evaluating statistical significance.

To estimate the Inhibitory Concentrations (Cs) (from IC1 to IC99) for each drug, the response variable can be transformed in Probit and then plotted versus the log10 concentration of each of the seven doses [17]. Next, by resorting to the Fieller theorem [18], it is also possible for each IC estimate the pertinent 95% confidence interval (95% CI). Finally, by resorting to approaches based on the analysis of variance [19], it is possible to study the variability for each drug, the estimates IC50 and the estimates IC90 within the laboratory.

For intricate experimental layouts with a large sample size, it is recommend to use SAS [20] for the analysis. For experimental layouts with small sample sizes, one could utilize a statistical package such as Graphpad InStat, wherein the IC values can be calculated using a simple method based on the interpolation of a regression curve as suggested by Reed and Muench [21], which in its simplest form is:

$$Log\ IC_{90} = Log\ D_x + Log\ \frac{D_y}{D_x} \times \frac{\% > 90\% - 10}{\% > 90\% - \% < 90\%}$$

D_x is the higher dose of xenobiotic correspondent to a clonogenicity % > 90% (% > 90%); and D_y is the lower dose of xenobiotic correspondent to a clonogenicity % < 90% (% < 90%).

6.6 Discussion and Commentary

The hematopoietic system is organized into three conceptually distinct levels of cellular development: the most primitive hematopoietic precursor cells sharing extensive proliferation and differentiation potential, the "committed" progenitor cells with lineage-restricted differentiation abilities, and the cells that have already display the morphological characteristics of the lineage to which they are committed. The "committed" population is easy to be assessed by standard clonogenic assays. The assay proposed here is concerning the myelocytic lineage commitment.

It is known that neutrophils are short-lived, circulating white blood cells that provide a critical line of defence against microbial infections. These cells are continuously produced by rapidly proliferating precursor cells in hematopoietic tissues called colony-forming units—granulocyte/macrophage (CFU-GM), so named because they form clonal colonies in semisolid cell cultures when stimulated with appropriate cytokines. These special cell types produce a new replacement of neutrophils and when in the blood they fall to abnormally low levels, the risk of infection increases, and the subject is said to be experiencing an episode of neutropenia. This condition can be caused by the toxic effects of xenobiotics and radiation upon hematopoietic progenitors that prevent their proliferation and differentiation into neutrophils. In the in vitro clonogenic assays for CFU-GM, this toxicity manifests as a loss of the colony-forming ability.

Toxicant-induced neutropenia can be described clinically by different hematology parameters but predicting neutropenia for a xenobiotic from in vitro testing requires an accurate prediction for each parameter [6, 22]. Currently, the clinical prediction models for the depth of the neutrophils' nadir published in literature are complex and differ for the amount of pharmacological information required to make predictions and the accuracy of those predictions [4, 6, 7, 22]. For the purpose of regulating exposure levels of thousands of xenobiotics, the most generally useful model requires the least amount of specialized pharmacological information.

Experiments with antineoplastic agents in animal models revealed a relationship between the reduction in CFU-GM and the decrease in absolute neutrophils count (ANC) [4]. These results suggested that in vitro toxicology tests using CFU-GM should be able to predict the exposure level of xenobiotics that would cause clinical neutropenia after acute exposure.

The protocol described here is based on essential modifications to existing methods for in vitro culture of CFU-GM that have been introduced to avoid complex and potentially misleading interactions between the reactive anticancer drugs and medium components:

- The absence of 2-Mercaptoethanol (2-ME), because of its possible reactivity with a lot of toxicants evaluated in the CFU-GM test;
- The absence of transferrin, because it carries ferrous ions that, in the presence of high oxygen tension existing in most CO_2 incubators, could react with xenobiotics producing free oxygen radicals;
- The presence of only one specific growth stimulant of colonies formation (recombinant GM-CSF), because cell-line conditioned medium and cytokine/growth factor cocktails can stimulate precursors other than the CFU-GM.

Based on these modifications, the standard operating procedure (SOP) reported in this protocol (for murine and human species) is able to generate reproducible IC values that express the direct toxic effect of the compound to GM-CFU [8, 9].

It is suggested to apply the assay according to two phases: one screening experiment to detect the range in which IC10, IC50, and IC90 values fall and then two experiments with narrow dose ranges. Details on the dilution protocol to apply in screening phase and how to pass to the IC determination are reported in Sections 6.3.3 and 6.3.4.

This protocol offers the advantage to work with a mechanism-naïve that misses hematotoxicants that adversely affect myelopoiesis via indirect physiological mechanisms, such as the induced release of inhibitory cytokines, the inhibited release of stimulatory cytokines, or the metabolic activation of protoxicants.

A further group of considerations concerns the application of the assay (by using the IC values obtained) to predict the human MTD, defined as the dose level found in clinical trials that causes severe toxicity in the majority of patients. For this purpose the formula reported in the protocol uses the IC90 values determined in vitro for murine and human CFU-GM. It is important to note that the predicted marrow MTD for the neutrophils lineage will equal the actual human MTD only when the myelopoietic tissue of the bone marrow is a primary target of toxicity. Therefore, the prediction model, based on in vitro toxicity, is a risk assessment tool for predicting exposure levels that will cause severe neutropenia. However, it is not useful for hazard classification because it predicts maximum tolerated exposure levels only for the bone marrow, but not other potential target organs. In this last case, however, the proposed methods can be of great utility because they allow the comparison of different metabolites of a drug if they are known.

A final important consideration concern the possibility to apply the assay according to a standardized protocol using 96-microwell cultures [11]. This protocol permits the assessment of hundreds of cultures in the same experiment with a significant reduction of costs and workload. Moreover, it is possible to reduce the amount of drug or xenobiotic necessary to prepare the starting solutions to test. This may be very important and sometimes crucial in testing new molecules, because during the first phase of screening, the pharmaceutical companies have a very small quantity of chemical on which to perform many tests. Furthermore, this miniaturization may facilitate the application of the clonogenic assays to other toxicants as microbial toxins, food contaminants, or environmental pollutants. The manual colony counting is both time-consuming and may give rise to a variability between operators. A major improvement may be the automation of the test by using plates readers. Unfortunately, satisfactory results have not yet used colorimetric techniques based on optical density. Image analysis offers a more promising approach for automatic colony scoring.

In an attempt to have a panel of tools for studying hematoxicity, we also have to remember here that two other tests have been optimized and/or prevalidated: one to evaluate the toxicity on megakaryocytes (Mk-CFU) [12], and one in which a CFU-GM assay is optimized for testing rat bone marrow, as this animal species is very important for toxicological studies that will allow the verification of the existing correlations between the in vitro toxicity and the values found by applying the in vitro tests [13].

As suggested by ECVAM directories, these techniques will contribute to the replacement, reduction, and refinement of laboratory animal procedures in accordance with the 3Rs concept of Russell and Burch [23].

Troubleshooting Table

Problem	Explanation	Potential Solutions
There is culture contamination (mycetes or bacteria).	There is a low sterility condition during the cells' collection.	Improve the sterility condition in BMC/UCB preparation. Verify the reagent's sterility. Add antibiotics in the first stage of in vivo cell explant.
No colonies in control dishes at all.	Cells do no growth at high MCM concentration or incorrect CO_2% and temperature. Some plastics or reagents can be toxic. Poor cell viability. GM-CSF has lost activity.	Check the evaporation rate, %CO_2 and T° of incubator. Check the plastics, reagents, and preparation steps of MCM. Test a different lot of GM-CSF.
There are no colonies but there is some degree of cell proliferation.	Dividing cells do not remain aggregate but they grow isolated and are dispersed if MCM concentration is too low. MCM disomogeneity (presence of cloths).	See above (especially if MCM is prepared starting from more concentrated stock).
Colonies are only in control dishes.	Cell growth may be influenced by toxicity of drug solvent. The drug concentration may be wrong (over than IC90).	Test the solvent toxicity. Select a less toxic solvent (for example, choose a hydrochloride drug form, soluble in water). Verify the drug preparation and dilutions.
It is difficult to score colonies.	Too many colonies are present that overlap. Cells were not homogeneously dispersed and produce colonies too close together.	Reduce the number of seeded cells. Mix the tubes more and better.
Fibroblasts adhere on the bottom of dishes.	Cell suspension is not well filtered. There is no good performance of dishes or plastic.	Reduce filter mesh size (70 or 40 μm). Change the lot or brand of dishes.

6.7 Application Notes

The method here proposed represents an important tool for increasing the safety in entering Phase I human treatment because refines the empirical prediction of MTD by reducing the risk for the volunteers (see Chapter 5 for a definition of an accurate prediction of human MTD). According to the validation study, the assay correctly predicts the human MTD for many drugs [9] and the percentage of successful is about 87% (near to 94% predictivity for nonnucleoside structures). Coupled with the reproducibility of the SOP application, this predictivity confirms that the SOP and the prediction model can be considered very satisfactory (Table 6.2).

6.8 Summary Points

To achieve the best outcomes using the method described earlier, some crucial points must be carefully considered.

1. Certified quality of reagents and plastic. Some of them can be toxic for hematopoietic cell progenitors.
2. Absolute condition of sterility. The most critical points are: cord blood collection, the following separation of UCB on Ficoll, and the collection of mouse bone

Table 6.2 Successful Prediction of Human Maximum Tolerated Dose (MTD) for Bleomycin, Fludarabine, Etoposide, and Adriamycin

	Bleomycin	Fludarabine	Etoposide	Adriamycin
IC90 hu/IC90 mu (determined by GM-CFU assay)	0.428	0.034	0.912	0.926
mu LD10 (mg/m^2/dose) (from preclinical data)	27.9	1008.9	23.1	11.1
hu MTD (mg/m^2/dose) (predicted by algorithm)	11.9	34.3	21.1	10.2
Actual hu MTD (mg/m^2/dose) (from the literature data)	15	25	54	22.5

marrow. Antibiotics can be helpful in this phase, but they must be absent in all the following steps.

3. High precision in preparing the drug dilutions. To reduce variability, it is suggested to perform dilutions starting from the same concentration of toxicant stock. When the volume required for making the dilution is too small, it is better to prepare the drug stock before making a working stock (for example, pipette 100 μl of a 1:100 dilution of toxicant stock, rather than 1 μl of undiluted toxicant stock).
4. Solvent concentration. It is important that all the drug dilutions have the same solvent concentration. In our experience, 0.5% DMSO (one of the most widely used solvents) represents a threshold of the toxic effect. Especially if poor information is available on solvent hematotoxicity, it is suggested that separate preliminary experiments be performed to determine the first nontoxic concentration of the solvent used.
5. Humidity of incubator. This factor is important especially for a human assay that needs 14 days of culture. If evaporation is too high, methylcellulose concentrates and cells die. It is important to set up a good humidification as described in the example.
6. Precision in preparing and dispensing MCM. This is critical to obtain the correct final concentration of methylcellulose and toxicant at later steps.
7. For a correct count of colonies, use a 20–25× magnification (do not use 40×). To discriminate between colonies and clusters, carefully evaluate the number of cells for each aggregate. It is also important to carefully examine the edge of the plate, where many colonies grow when a high cell density is seeded.
8. Passing from screening phase to IC determination phase. When the toxicity of the molecule under study is unknown, it is suggested that the assay be performed at least in two phases: one screening experiment to detect the range in which IC10, IC50, and IC90 values fall, and then another phase of experiments with narrower dose ranges for a precise IC determination. For details on the dilution protocol to apply in the screening phase and how to pass to the IC determination phase, a protocol that can be adapted is supplied. An essential point to consider is that the curve for IC determination must always contain a dose giving 0% toxicity and a dose giving 100% toxicity.

Acknowledgments

We thank Loredana Cavicchini for her excellent technical support in the standardization of the method here described.

References

[1] Lewis, I. D., et al., "Standardization of the CFU-GM Assay Using Hematopoietic Growth Factors," *Journal of Hematotherapy*, Vol. 5, No. 6, 1996, pp. 625–630.

[2] Gribaldo, L., et al., "The Use of In Vitro Systems for Evaluating Haematotoxicity. The Report and Recommendations of ECVAM Workshop 14," *Alternatives to Laboratory Animals*, Vol. 24, 1996, pp. 211–231.

[3] Parchment, R. E., M. Huang, and C. L. Erickson-Miller, "Roles for In Vitro Myelotoxicity Tests in Preclinical Drug Development and Clinical Trial Planning," *Toxicologic Pathology*, Vol. 21, No. 2, January 1993, pp. 241–250.

[4] Metcalf, D., "The Basic Biology of Hematopoiesis," in D. Metcalf, (ed.), *The Haematopoietic Colony Stimulating Factors*, Amsterdam, the Netherlands: Elsevier, 1984, pp. 1–26.

[5] Pessina, A., "The Granulocyte Macrophage Colony-Forming Unit Assay," in M. Clynes, (ed.), *Animal Cell Culture Techniques*, Berlin-Heidelberg, Germany: Springer-Verlag, 1998, pp. 217–230.

[6] Parchment, R. E., "Alternative Testing Systems for Evaluating Non-Carcinogenic, Hematologic Toxicity," *Environmental Health Perspectives*, Vol. 106, No. 2, April 1998, pp. 541–557.

[7] Parchment, R. E., et al., "Predicting Hematological Toxicity (Myelosuppression) of Cytotoxic Drug Therapy from In Vitro Tests," *Annals of Oncology*, Vol. 9, No. 4, April 1998, pp. 357–364.

[8] Pessina, A., et al., "Prevalidation of a Model For Predicting Acute Neutropenia by Colony Forming Unit-Granulocyte/Macrophage (CFU-GM) Assay," *Toxicology In Vitro*, Vol. 15, No. 6, December 2001, pp. 729–740.

[9] Pessina, A., et al., "Application of the CFU-GM Assay to Predict Acute Drug-Induced Neutropenia: An International Blind Trial to Validate a Prediction Model for the Maximum Tolerated Dose (MTD) of Myelosuppressive Xenobiotics," *Toxicological Sciences*, Vol. 75, No. 2, October 2003, pp. 355–367.

[10] Pessina, A., et al., "A Methylcellulose Microculture Assay for the In Vitro Assessment of Drug Toxicity on Granulocyte/Macrophage Progenitors (CFU-GM)," *Alternatives to Laboratory Animals*, Vol. 32, No. 1, 2004, pp. 17–23.

[11] Malerba, I., et al., "Inhibition of CFU-E/BFU-E and CFU-GM Colony Growth by Cyclophosphamide, 5-Fluorouracil and Taxol: Development of a High Throughput In Vitro Method," *Toxicology In Vitro*, Vol. 18, No. 3, June 2004, pp. 293–300.

[12] Pessina, A., et al., "Application of Human CFU-Mk Assay to Predict Potential Thrombocytotoxicity of Drugs," *Toxicology In Vitro*, Vol. 23, No. 1, February 2009, pp. 194–200.

[13] Pessina, A., et al., "Refinement and Optimisation of the Rat CFU-GM Assay to Use Cryopreserved Bone Marrow Cells for 'In Vitro' Toxicology Application," *Alternatives to Laboratory Animals*, 2009 (in press).

[14] Parchment, R. E., et al., "In Vivo-In Vitro Correlation of Myelotoxicity of 9-Methoxypyrazoloacridine (NSC-366140, PD115934) to Myeloid and Erythroid Hematopoietic Progenitors from Human, Murine, and Canine Marrow," *Journal of the National Cancer Institute*, Vol. 86, No. 4, February 1994, pp. 273–280.

[15] Erickson-Miller, C. L., et al., "Differential Toxicity of Camptothecin, Topotecan and 9-Aminocamptothecin to Human, Canine, and Murine Myeloid Progenitors (CFU-GM) In Vitro," *Cancer Chemotherapy and Pharmacology*, Vol. 39, No. 5, February 1997, pp. 467–472.

[16] Volpe, D. A., et al., "Myelotoxic Effects of the Bifunctional Alkylating Agent Bizelesin to Human, Canine, and Murine Myeloid Progenitor Cells," *Cancer Chemotherapy and Pharmacology*, Vol. 39, No. 1-2, November 1996, pp. 143–149.

[17] Finney, D. J., *Statistical Method in Biological Assay*, 2nd ed., New York: Hafner Publishing Co., 1964.

[18] Fieller, E. C., "The Biological Standardization of Insulin," *Journal of the Royal Statistical Society*, Vol. 7, No. 1, 1940, pp. 1–64.

[19] Winer, B. J., D. R. Brown, and K. M. Michels, *Statistical Principles in Experimental Design*, 3rd ed., New York: McGraw-Hill, 1991.

[20] SAS Institute Inc. SAS OnlineDoc, Version 8. Cary, NC, USA: SAS Institute Inc, 1999.

[21] Reed, L. J., and H. A. Muench, "A Simple Method of Estimating Fifty Percent Endpoints," *The American Journal of Hygiene*, Vol. 27, 1938, pp. 493–497.

[22] Parchment, R. E., and M. J. Murphy, Jr., "Human Hematopoietic Stem Cells: Laboratory Assessment and Response to Toxic Injure," in G. Sipes, C. A. McQueen, and A. J. Gandolfi, (eds.), *Comprehensive Toxicology*, New York: Pergamon, 1997, pp. 335–361.

[23] Russell, W. M. S., and R. L. Burch, *The Principles of Humane Experimental Technique*, London, U.K.: Methuen, 1959.

[24] European Centre for the Validation of Alternative Methods (ECVAM), http://ecvam-dbalm.jrc.it/ft_doc/ESA%20statement%20CFU-GM%20assay%2020060515.pdf.

CHAPTER

7

Epigenetically Stabilized Primary Hepatocyte Cultures: A Potential Sensitive Screening Tool for Nongenotoxic Carcinogenicity

Tatyana Y. Doktorova,[1] Tamara Vanhaecke,[2] Vera Rogiers[2], and Mathieu Vinken[2]

[1]Department of Toxicology, Vrije Universiteit Brussel (VUB), Laarbeeklaan 103, B-1090 Brussels, Belgium.
[2]Postdoctoral research fellows of the Fund for Scientific Research Flanders (FWO-Vlaanderen), Belgium,
Corresponding author: Tatyana Y. Doktorova, Department of Toxicology, Vrije Universiteit Brussel, Campus Jette, Building G, 2nd Floor, Laarbeeklan 103, 1090, Brussels, Belgium,
e-mail: tatyana.yordanova.doktorova@vub.ac.be, phone: +32 2 477 45 07, fax: +32 2 477 45 82

Abstract

Replacing, reducing, and refining the use of animals in xenobiotic toxicity testing are being gradually implemented in European legislation. Therefore, research in this field and development of alternative methods are becoming more important. As the liver and hepatocytes in particular are a main target for toxicity and carcinogenicity in the organism, a lot of attention is paid to the establishment of liver-based in vitro models. Primary hepatocytes are considered as the golden standard, but their cultures are prone to progressive dedifferentiation, thereby restricting their use to short-term purposes. In this chapter, we present a method to stabilize the differentiated phenotype of hepatocytes in primary culture by remodelling the chromatin structure by using the histone deacetylase inhibitor trichostatin A. After describing the setup of stabilized primary hepatocyte cultures and providing a troubleshooting guide, some results are shown with respect to their potential applicability for the testing of nongenotoxic hepatocarcinogenic substances.

Key terms carcinogenicity testing
nongenotoxic carcinogens
primary hepatocyte cultures
trichostatin A

7.1 Introduction

Historically, the assessment of potential adverse effects of chemicals on human health and environment has been performed mostly on experimental animals. Scientific and technological developments, ethical and economic constrains, and certainly also public concern about animal use have led to the regulatory acceptance of directives, in particular in Europe, according to which animal experimentation should be partly or totally replaced by nonanimal tests, the number of animals involved reduced, and their pain or distress minimized [1].

With the introduction of the new chemical legislation in Europe, known as REACH (Registration, Evaluation, Authorization and Restriction of Chemicals), approximately 30,000 substances currently on the market have to undergo toxicological testing [2]. In spite of many efforts in the field, a limited number of validated alternative methods still exist [3]. Consequently, the majority of the safety tests, in particular long-term tests such as repeated dose toxicity, developmental toxicity, and carcinogenicity, will have to be performed mainly on animals. The continuous search for alternatives and their validation are thus important challenges.

The identification of the carcinogenic potential of chemicals is one of the difficult areas. Based upon their mechanism of action, chemical carcinogens can be subdivided into genotoxic and nongenotoxic substances. Testing for potential genotoxicity is performed by a number of regulatory approved in vitro tests, such as the bacterial reverse mutation test and the in vitro mammalian chromosome aberration test [4–7]. Only when positive, these are followed by in vivo testing [8]. For the detection of nongenotoxic carcinogens, however, no suitable in vitro tests are currently available and the carcinogenicity nature of a compound is without pretesting directly assessed in vivo [9]. One of the reasons for this gap might be the diverse mechanisms through which nongenotoxic carcinogenic chemicals act, such as induction of cell proliferation, modulation of apoptotic activity, and inhibition of the gap junction intercellular communication (GJIC) [10, 11].

In 2007, a European project, called carcinoGENOMICS, was initiated with the aim to develop in vitro methods that would be able to discriminate between the genotoxic or nongenotoxic character of carcinogens. Since liver and, in particular, hepatocytes represent a main target of toxicity and are more specific of carcinogenicity, liver-based in vitro models are included in this project. Indeed, hepatocytes drive the majority of the xenobiotic biotransformation machinery and thus represent a first line of defense against toxic insults [12–14]. At present, primary hepatocyte cultures are considered to be the most appropriate experimental systems within the field of liver-based in vitro modeling, as they provide a good reflection of the hepatic in vivo situation. Cultured hepatocytes, however, suffer from a progressive loss of their differentiated phenotype, both at the morphological and functional levels [15]. A number of strategies have been introduced to counteract the dedifferentiation process, based mainly on mimicking the natural hepatocellular microenvironment. These include the addition of differentiation-promoting compounds to the cell culture medium, the restoration of cell-cell contacts and the reintroduction of the cell-extracellular matrix interactions. Although these methodologies or a combination of them, gave promising results, they were only successful to a limited extent [13–17].

Our group has introduced an innovative strategy to reduce dedifferentiation of primary cultured hepatocytes. It is based upon epigenetically interfering with gene expression by exposing the hepatocytes to inhibitors of histone deacetylase (HDAC) isoenzymes, of which trichostatin A (TSA; 7-[4-(dimethylamino)phenyl]-*N*-hydroxy-4,6-dimethyl-7-oxo-2,4-heptadienamide; Figure 7.1) is a prototypical example. The reversible acetylation of histones is a major epigenetic determinant of gene expression. The hyperacetylation of core histones results in the loosening of the histone-DNA contacts and thus chromatin decondensation. In turn, this facilitates the accessibility of transcription factors to DNA and enhances gene transcription. The inverse reaction is catalyzed by HDAC enzymes and is usually associated with gene silencing [18–20].

Thus far, we could demonstrate that TSA induces cell cycle arrest, reduces the occurrence of spontaneous apoptosis, increases liver-specific functionality, and maintains GJIC in primary hepatocyte cultures [19–23]. These outcomes commonly result in the stabilization of the differentiated phenotype of the hepatocytes.

In this chapter, we provide a detailed protocol of how to set up TSA-stabilized primary hepatocyte cultures and we give some practical tips taken from our own experience. Also, the potential of this experimental setting for the detection of nongenotoxic hepatocarcinogens is shown by some preliminary results.

7.2 Experimental Design

The isolation of hepatocytes from freshly removed rat liver and their subsequent cultivation are described. The isolation is performed by the use of the two-step collagenase perfusion technique of Berry and Friend [24] and is further modified by Seglen [25]. Briefly, the liver is first perfused with a Ca^{2+}-free buffer in order to weaken the Ca^{2+}-dependent contacts between cells, followed by a second perfusion with a buffer containing a Ca^{2+}-activated collagenase. Subsequently, the isolated hepatocytes are cultivated in a monolayer configuration under conventional conditions in the presence of TSA.

7.3 Materials

As a cellular source, male outbred Sprague-Dawley rats weighing 200–300g (Charles River Laboratories, Brussels, Belgium) are used. Rats are free of pathogens and kept under controlled environmental conditions (12 hours light/dark cycle) with free access to food (Animalabo A04) and water.

7.3.1 Reagents

All solutions are sterilized by passing through a 0.22-μm filter.

- 30-mM TSA stock solution in dimethyl sulfoxide (DMSO):
  - TSA (Sigma T8552, Belgium): 5 mg
  - DMSO (pro analyse, Merck 0983, Germany): 551 μl

  Store at –20°C.
- Phosphate-buffered saline (PBS) pH = 7.65:
  - KCl (Merck 4936, Germany): 1.0g
  - $KH_2PO_4.12H_2O$ (Merck 4873, Germany): 1.0g

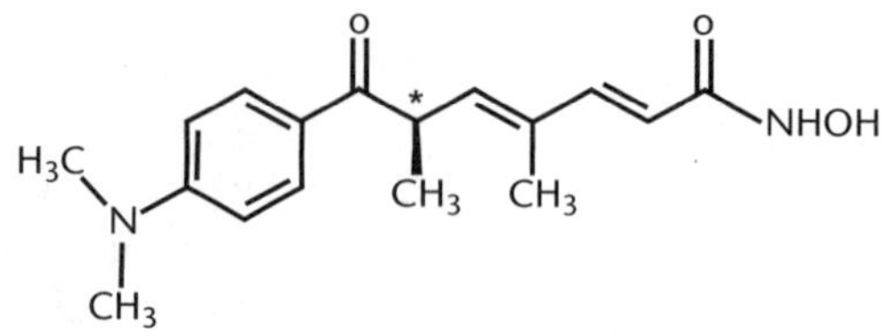

Figure 7.1 Chemical structure of trichostatin A (TSA; 7-[4-(dimethylamino)phenyl]-*N*-hydroxy-4,6-dimethyl-7- oxo-2,4-heptadienamide), containing a chiral center (*).

- NaCl (Merck 6404, Germany): 14.0g
- $Na_2HPO_4.12H_2O$ (Merck 6579, Germany): 15.5g
- milliQ-water (Millipore RO-4/45, United States): ad 5,000 ml

Store for 9 months at 4°C.

- N-[2-hydroxyethyl]piperazine-N′[2-ethanolsulfonic acid] (HEPES)buffer pH = 7.65:
  - HEPES (Sigma H3375, Belgium): 380 mg
  - KCl (Merck 4936, Germany): 200 mg
  - NaCl (Merck 6404, Germany): 8.0g
  - $Na_2HPO_4.12H_2O$ (Merck 6579, Germany): 100 mg
  - milliQ-water (Millipore RO-4/45, United States): ad 1,000 ml

Store for 6 months at 4°C.

- Krebs-Henseleit buffer (KHB) pH = 7.4:
  - KCl (Merck 4936, Germany): 361.6 mg
  - $KH_2PO_4.12H_2O$ (Merck 4873, Germany): 426.3 mg
  - $MgSO_4.7H_20$ (Sigma M7774, Belgium): 298.2 mg
  - NaCl (Merck 6404, Germany): 7.1g
  - $NaHCO_3$ (Sigma S5761, Belgium): 2.1g
  - milliQ-water (Millipore RO-4/45, United States): ad 1,000 ml

Must be saturated with carbogen for 60 minutes prior to use and can be stored for 6 months at 4°C.

- Krebs-Henseleit buffer supplemented with calcium (Ca^{2+}-containing KHB) pH = 7.4:
  - $CaCl_2.2H_20$ (Sigma C7902, Belgium): 73.5 mg
  - KHB pH = 7.4: ad 200 ml

Must be saturated with carbogen for 60 minutes, prior to use and can be stored for 6 months at 4°C.

- Leibovitz medium:
  - Leibovitz (Gibco 074-1300, Belgium): 14.7g
  - bovine serum albumin (Sigma A9647, Belgium): 2.0g
  - milliQ-water (Millipore RO-4/45, United States): ad 1,000 ml

Store for 6 months at 4°C.

- 10^6 IU/3 ml sodium benzyl penicillin prestock solution:
  - Sodium benzyl penicillin (Penicilline, Continental Pharma, Belgium): 10^6 IU
  - PBS pH = 7.65: ad 3 ml

Store at –20°C.

- 733 IU/ ml sodium benzyl penicillin stock solution:
  - sodium benzyl penicillin prestock solution (10^6 IU/3 ml): 1.1 ml
  - PBS pH = 7.65: ad 500 ml

Store at –20°C.

- 5 mg/ml streptomycin sulphate stock solution:
  - Streptomycin sulphate (Sigma S9137, Belgium): 2.5g
  - PBS = 7.65: ad 500 ml

Store at –20°C.

- 1 mg/ml sodium ampicillin stock solution:
  - Sodium ampicillin (Pentrexyl, Bristol-Myers Squibb, Belgium): 0.5g
  - PBS pH = 7.65: ad 500 ml

Store at –20°C.

- 5 mg/ml kanamycin monosulphate stock solution:
  - Kanamycin monosulphate (Sigma K4000, Belgium): 2.5g
  - PBS pH = 7.65: ad 500 ml

Store at –20°C.

- 1 mg/ml insulin stock solution:
  - Insulin (from bovine pancreas, Sigma I1882, Belgium): 10 mg
  - PBS pH = 7.65: 10 ml

After adjusting the pH with HCl 1N to pH = 2, store at –20°C.

- 3.5 μg/ml glucagon stock solution:
  - Glucagon (from porcine pancreas, Sigma G3157, Belgium): 350 μg
  - 0.001 N NaOH solution: ad 100 ml

Store at –20°C.

- 25 mg/ml hydrocortisone sodium hemisuccinate stock solution:
  - Hydrocortisone sodium hemisuccinate (Solu-Cortef, Pharmacia, Belgium): 100 mg
  - PBS pH = 7.65: ad 4 ml

Store at –20°C.

- Wash medium:
  - Kanamycin monosulphate stock solution (5 mg/ml): 10.0 ml
  - Sodium ampicillin stock solution (1 mg/ml): 10.0 ml
  - Sodium benzyl penicillin stock solution (733 IU/ml): 10.0 ml
  - Streptomycin sulphate stock solution (5 mg/ml): 10.0 ml
  - L-glutamine solution (29.22 mg/ml; Sigma G7513, Belgium): 10.0 ml
  - TSA stock solution (30 mM): 833.3 μl
  - William's E medium (Gibco 2551-032, Belgium): ad 1,000 ml

Store for 7 days at 4°C. Prior use, it should be placed in a thermostated bath (37°C) for 30 minutes. TSA stock solution should be added ex tempore.

- T0 medium:
  - Kanamycin monosulphate stock solution (5 mg/ml): 10.0 ml
  - Sodium ampicillin stock solution (1 mg/ml): 10.0 ml
  - Sodium benzyl penicillin stock solution (733 IU/ml): 10.0 ml
  - Streptomycin sulphate stock solution (5 mg/ml): 10.0 ml
  - L-glutamine solution (29.22 mg/ml; Sigma G7513, Belgium): 10.0 ml
  - Glucagon stock solution (3.5 μg/ml): 2.0 ml
  - Fetal bovine serum (Gibco 26140-087, Belgium): 100 ml
  - TSA stock solution (30 mM): 833.3 μl
  - William's E medium (Gibco 2551-032, Belgium) ad: 1,000 ml

Store for 7 days at 4°C. Prior use, it should be placed in a thermostated bath (37°C) for 30 minutes. TSA stock solution should be added ex tempore.

- T4 medium:
  - Kanamycin monosulphate stock solution (5 mg/ml): 10.0 ml
  - Sodium ampicillin stock solution (1 mg/ml): 10.0 ml
  - Sodium benzyl penicillin stock solution (733 IU/ml): 10.0 ml
  - Streptomycin sulphate stock solution (5 mg/ml): 10.0 ml
  - Hydrocortisone sodium hemisuccinate stock solution (25 mg/ml): 1.0 ml
  - L-glutamine solution (29.22 mg/ml; Sigma G7513, Belgium): 10.0 ml
  - Glucagon stock solution (3.5 μg/ml): 2.0 ml
  - Insulin stock solution (1 mg/ml): 0.5 ml
  - Fetal bovine serum (Gibco 26140-087, Belgium): 100 ml
  - TSA stock solution (30 mM): 833.3 μl
  - William's E medium (Gibco 2551-032, Belgium): ad 1,000 ml

Store for 7 days at 4°C. Prior use, it should be placed in a thermostated bath (37°C) for 30 minutes. TSA stock solution should be added ex tempore.

- 115 IU/ml collagenase solution:
  - Collagenase (*Clostridium peptidase A* type I; Sigma C0130, Belgium): 18400 IU
  - Ca^{2+}-containing KHB: 10 ml

Prepare ex tempore.

- 200 IU/ml sodium heparin solution:
  - Sodium heparin solution (5,000 IU/ml; Heparine Novo, S.A. Novo Nordisk, Denmark): 0.1 ml
  - Sterile physiological saline: 2.4 ml

Prepare ex tempore.

- 60 mg/ml sodium pentobarbital (Nembutal, Sanofi, Belgium)
- 1 mg/ml Trypan blue solution:
  - Trypan blue (Merck 11732, Germany): 100 mg
  - PBS pH = 7.65: 100 ml

Store at room temperature.

- Denaturated alcohol (Merck 983, Germany)
- Liquid nitrogen (Air Liquide, Belgium)
- Carbogen (95% O_2/5% CO_2, Air Liquide, Belgium)

7.3.2 Facilities/Equipment

- 63-μm perlon filter (Helene-Cavenaille, Belgium)
- 0.22-μm filter (Sterivex, Millipore, Belgium)
- Bile cannula (inner diameter 0.28 mm, outer diameter 0.61 mm; Intramedic, Belgium)
- Centrifuge (Heraeus Sepatech, Megafuge 1.0 R, Van der Heyden, Belgium)
- General sterile glassware (VEL, Belgium)
- Hemocytometer (area 1 mm^2/depth 0.1 mm; VEL 1502010, Belgium)
- Incubator (CO_2-water jacketed; Nuaire US autoflow, IKS, The Netherlands)
- Inverse-phase light microscope (Phase Contrast 1.25, Nikon, Optiphot, Belgium)
- Pasteur pipettes (Bilbate, United Kingdom)
- Perfusion apparatus (composed in the laboratory, Figure 7.2)
- Perfusion pump (Watson Marlow 503S, Belgium)
- Petri dishes (uncoated, diameters 15 cm, 10 cm, 6 cm, 35 mm; BD, Belgium)
- Pipetor (Pipetus, Flow Laboratories, Belgium)
- Sterile centrifuge tubes (50 ml, 25 ml; Falcon, BD, Belgium)
- Sterile volumetric pipettes (25 ml, 10 ml, 5 ml; BD, Belgium)
- Thermostated bath (37°C; GFL 1086, VEL, Belgium)
- Vertical laminar air flow cabinet (Gelaire Class 100, VWR, Belgium)

7.4 Methods

7.4.1 Isolation of hepatocytes from rat liver

Sterilization of the perfusion apparatus

1. Sterilize the perfusion apparatus with 400 ml of denatured alcohol, and circulate for 15 minutes.

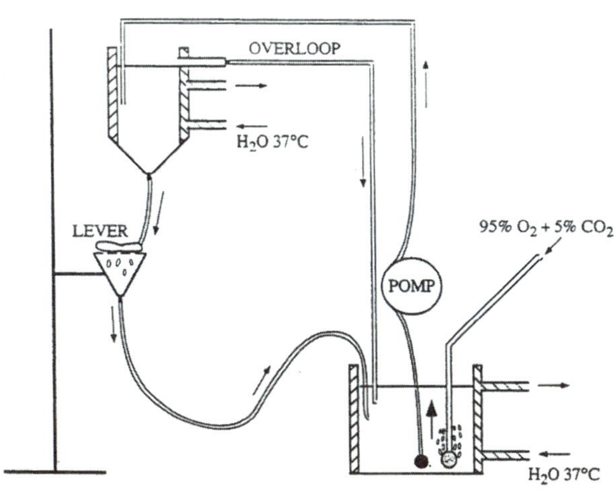

Figure 7.2 Schematic representation of the perfusion apparatus used to isolate hepatocytes from a freshly removed rat liver.

2. Rinse the apparatus three times with sterile double destillated water (10 minutes/circulation).

Isolation of the rat liver

1. Anesthetize the rat by injecting sodium pentobarbital solution (0.1 ml/100g body weight) intraperitoneally.
2. Shave the rat's abdomen, disinfect with denaturated alcohol, and immobilize the rat on a surgery table.
3. Open the abdomen through a U-shaped incision.
4. Ligate the bile duct twice, close the lower ligature, insert the bile cannula, and fix it by closing the upper ligature.
5. Ligate the vena cava once and the vena porta twice.
6. Inject 1 ml of freshly prepared heparin solution in the vena saphena medialis.
7. Close the lower ligature of the vena porta, insert the glass cannula in the vena porta, and close the upper ligature.
8. Close the ligature of the vena cava.
9. Remove the liver from the rat and rinse it with KHB.

Two-step perfusion of the rat liver

1. Circulate 250-ml KHB in the perfusion apparatus, thermostated at 37°C while carbogen is passed through.
2. Transfer the excised rat liver to the perfusion plate and connect the glass cannula, already inserted in the vena porta (perfusion rate 40–50 ml/minute, temperature 37°C). The connecting tubes have to be completely free of air bubbles.
3. For the first perfusion step, remove the blood from the liver with 100-ml KHB and circulate the remaining 150-ml KHB for 15 minutes. The liver should blanch within a few seconds; otherwise, it should be discarded.
4. For the second perfusion step, add the collagenase solution to the remaining KHB in the perfusion apparatus. Circulate for 25 minutes.
5. Disconnect the perfused rat liver from the perfusion apparatus and transfer it to a Petri dish filled with Leibovitz medium.

Purification of rat hepatocytes

1. Open the Glisson's capsule and suspend the cells in the Leibovitz medium.
2. Filter the resulting cell suspension through a sterile perlon filter.
3. Allow the cells to sediment for 15 minutes.
4. Remove the supernatant, wash the cells with 40 ml of the HEPES buffer, and dispense the cell suspension into two 50-ml sterile centrifuge tubes.
5. Centrifuge the cell suspension for 1 minute at $63 \times g$.
6. Remove the supernatant and wash the cell pellet with 20 ml of the HEPES buffer.
7. Centrifuge the cell suspension for 1 minute at $63 \times g$.
8. Remove the supernatant and wash the cell pellet with the wash medium.
9. Centrifuge the cell suspension for 1 minute at $63 \times g$.
10. Remove the supernatant and resuspend the cell pellet in 100 ml of the wash medium.

7.4.2 Cultivation of primary rat hepatocytes (Troubleshooting Table)

1. Seed the hepatocytes at a density of 4×10^5 cells/ml of T0 medium in 2 ml, 4 ml, 11 ml, and 25 ml in 35-mm, 6-cm, 10-cm, or 15-cm diameter Petri dishes, respectively. Work under sterile conditions (laminar air flow cabinet).
2. Incubate the cells for 4 hours at 37°C.
3. Four hours after seeding, renew the medium with 2 ml, 4 ml, 11 ml, and 25 ml T4 medium in the 35-mm, 6-cm, 10-cm, or 15-cm diameter Petri dishes, respectively.
4. Keep the cells in an incubator at 37°C and renew the medium every day thereafter with the same volume of the T4 medium.

7.5 Data Acquisition

Typically, $200–350 \times 10^6$ hepatocytes can be isolated from a single rat liver. It is highly recommended that freshly isolated rat hepatocytes should only be used when a minimum viability of at least 80% is obtained. Cell counting and the simultaneous assessment of cell viability are performed by a Trypan blue exclusion test, as discussed next.

1. Mix 600 μl of Ca^{2+}-containing KHB with 200 μl of Trypan blue solution and 400 μl of cell suspension.
2. Count the viable (white) and dead (blue) cells in a hemocytometer under a light microscope (10×20). Usually, four fields are counted per chamber.
3. Calculate the mean number of viable and dead cells according to the following equations:

$$viability(in\,\%) = (\#\, viable\ cells \times 100)/[(\#\, viable\ cells) + (\#\, dead\ cells)]$$

$$cell\ concentration(in\ cells/ml) = [(number\ of\ viable\ cells) \times 3 \times 1{,}000]/0.1 \times 10^6$$

7.6 Anticipated Results and Interpretation

In-depth characterization of our model showed that, despite the fact that primary hepatocytes usually dedifferentiate and undergo phenotypical changes in culture, expo-

Troubleshooting Table

Problem	Explanation	Potential Solutions
Fungus infection of the cultures	Reduced sterility	Add Amphotericin B (0.25 μg/ml) to the cell culture medium.
		Check the proper function of the laminar air flow cabinets.
Significant variation between repeats of the same biological replicate	Unequal seeding of the cells	Always stir the cell suspension before seeding to ensure equal distribution of the cells.
		Normalize the results by the determination of the total protein content.
No difference according to the morphological analysis between control cells and TSA-treated cells after 7 days of culture	TSA stored under inappropriate conditions	Check whether R-isomer is present. R(+)TSA is the biologically active isomer.

Culture conditions of hepatocytes are as described in Section 7.4.

sure to TSA leads to sustained liver-specific functions for a longer time. Indeed, research from our lab showed that TSA causes a cell cycle arrest at the G1/S border, as judged by drastically decreased DNA replication and the absence of cdk1 expression, an S/G2/M phase marker [21]. Furthermore, TSA counteracts the occurrence of spontaneous apoptotic cell death in primary cultures of hepatocytes, as evidenced by decreased caspase-3 processing and activity, upregulated expression of the antiapoptotic Bcl-xL protein, and reduced production of the proapoptotic Bid protein [22]. TSA also stabilizes liver-specific functionality, including albumin secretion and expression of cytochrome P450 (CYP450) isoenzymes (e.g., CYP 1A1, CYP 2B1 and CYP 3A2) (Figure 7.3). This is mediated, at least in part, by an increased presence of liver-enriched transcription factors [19]. Overall, these TSA-induced changes are associated with the enhancement of GJIC, which in turn results from an improved production of the main gap junction building subunit connexin32 (Figure 7.4) [23].

Abrogation of intercellular communication mediated by gap junctions and the sustained induction of biotransformation enzymes are two prominent mechanisms that underlie nongenotoxic hepatocarcinogenicity in vivo [10, 26, 27]. Since both features are abundantly present in TSA-optimized primary hepatocyte cultures, this in vitro model may potentially be used for the detection of nongenotoxic hepatocarcinogens. In order to test this hypothesis, the concordance and the sensitivity of the system after exposure to nongenotoxic hepatocarcinogens were investigated. In this context, concordance refers to the agreement between new in vitro test results and in vivo data. Earlier in vivo studies revealed that nongenotoxic carcinogens reduce hepatic GJIC in rodents, partly resulting from a decay of the connexin32 pool [25–28]. Preliminary experiments with the nongenotoxic hepatocarcinogen methapyrilene hydrochloride showed, after 72 hours of exposure, a dose dependent decrease of the connexin32 protein expression (Figure 7.5), which is in accordance with earlier observations done in rat liver [27].

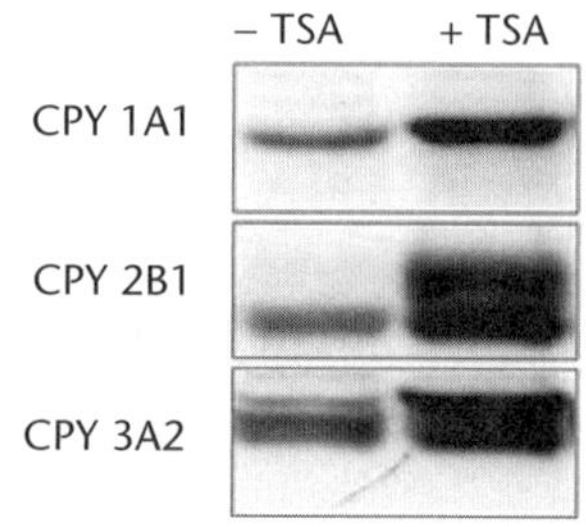

Figure 7.3 Effects of TSA on CYP450 protein expression after 7 days of cultivation. Culture conditions of hepatocytes are as described in Section 7.4. Western blotting and identification of CYP450 are done according to [19].

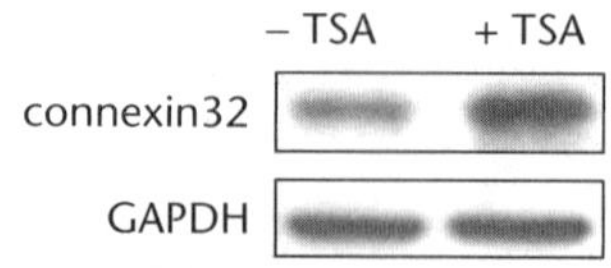

Figure 7.4 The inductive effect of TSA on connexin32 protein expression after 7 days of cultivation. Glyceraldehyde-3-phosphate dehydrogenase (GAPDH) is used as a loading control. Culture conditions of hepatocytes are as described in Section 7.4. Western blotting and identification of connexin32 is done according to [23].

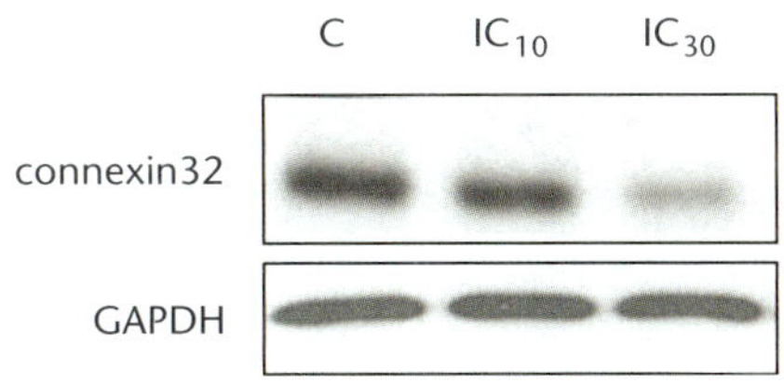

Figure 7.5 Inhibition of connexin32 expression after 72 hours exposure of the cells to two concentrations of methapyrilene hydrochloride (MP) (C = control; IC_{10} = concentration of MP at which 90% of the cells are viable; IC_{30} = concentration of MP at which 70% of the cells are viable). Glyceraldehyde-3-phosphate dehydrogenase (GAPDH) is used as a loading control.

The sensitivity of the system addresses the potential to detect changes at an earlier stage than would be the case in cultured hepatocytes not exposed to TSA. Indeed, when testing the cytotoxic potential of the prototypical nongenotoxic hepatocarcinogens phenobarbital sodium, piperonyl butoxide and methapyrilene hydrochloride by means of a conventional 3-(4,5-dimethylthiazol-2-yl)-2,5-diphenyltetrazolium bromide test (MTT test) in the TSA-stabilized primary hepatocyte culture system, IC_{10} values were lower in comparison with those obtained in the non-TSA-treated cultures (Table 7.1). This means that cytotoxic responses can be measured at lower concentrations in TSA-exposed hepatocytes in comparison with control cells. It is, however, recommended to check for potential interactions (chemical, physical or biological) between TSA and the compound under investigation. Also, an appropriate set of non-TSA-treated controls should be included.

The culture conditions of hepatocytes are as described in Section 7.4. Western blotting and identification of connexin32 is done according to Vinken et al. [23].

7.7 Discussion and Commentary

Innovative 3R tests (reduction, replacement, and refinement of animal experimentation) for hazard identification, especially of nongenotoxic hepatocarcinogens, are urgently needed, in particular in the light that the carcinogenic potential of new chemicals, or existing ones under REACH, needs to be assessed, and regulatory bodies, in particular, in the European Union, try to implement the 3Rs principle of Russell and Burch [29]. Currently, some regulatory approved in vitro and short-term in vivo tests for the detection of genotoxic events exist, but the assessment of the carcinogenic potential of chemicals causing tumors by mechanisms not involving genotoxicity is still performed in animals. The 2-year carcinogenicity study, usually applied for identification of carcinogens, not only requires a control group but also at least three doses of the chemicals to

Table 7.1 Assessment of TSA Treatment

Compound	IC_{10} in TSA-Treated Cultures	IC_{10} in TSA-Nontreated Cultures
Methapyrilene hydrochloride	$4.97 \pm 1.15\ \mu M$	$8.84 \pm 1.31\ \mu M$
Piperonyl butoxide	$11.31 \pm 2.43\ \mu M$	$13.44 \pm 0.74\ \mu M$
Phenobarbitol sodium	$0.60 \pm 0.09\ \mu M$	$0.66 \pm 0.04\ \mu M$

be tested. As such, at least 50 animals of each sex are consumed per dose [9]. As a lot of animals, time, and money are involved, the search for reliable alternatives is a must.

The role of the liver as a major organ of defense against toxicity in the organism makes liver-based in vitro models an attractive field of research. Primary hepatocyte cultures are frequently used to identify and quantify hepatotoxicity caused by a variety of chemical agents, but their main disadvantage is their progressive dedifferentiation and loss of functionality as a function of time [12–15]. Our research group has introduced a novel approach to partly maintain the liver-specific functions of the hepatocytes in culture for a longer time, using the HDAC inhibitor TSA. The exposure of the cultured hepatocytes to TSA led to cell cycle arrest, delayed spontaneous apoptotic cell death, better maintenance of the differentiated state, and sustained GJIC [19–23]. These characteristics are important for in vitro models applied for xenobiotic toxicity testing. In fact, our TSA-exposed system seems to detect cytotoxicity of chemical substances at lower concentrations than observed in conventional models. Furthermore, preliminary results also show concordance with in vivo data. Therefore, the in vitro model described represents a little step forward towards the achievement of the ultimate goal, namely the development of a reliable in vitro screening test system for nongenotoxic carcinogens.

7.8 Application Notes

The use of biomarkers in hazard identification is of general interest, since they could provide early information on exposure to toxic agents. The currently ongoing FP6 European project carcinoGENOMICS aims at identifying a set of biomarkers for the detection of carcinogenic events, anchoring on gene expression modulations, metabolic profiles, and mechanistic pathways. The major goal is to develop a battery of mechanism-based in vitro tests accounting for various modes of carcinogenic action in different organs, including liver, lung, and kidney. The model described in this chapter is actually part of this project and since biomarkers for the detection of genotoxic effects such as DNA adducts, chromosomal aberrations, and sister chromatid exchanges already exist, our focus is mainly on biomarkers for the detection of nongenotoxic carcinogens. It is known that this class of chemicals acts through a variety of mechanisms such as induction of proliferation, induction of oxidative stress, and inhibition of GJIC. Therefore, our efforts are focused on finding a group of biomarkers able to cover this variety.

7.9 Summary Points

1. Cultures of primary hepatocytes are considered to be the golden standard in the field of liver-based in vitro modeling, as they provide an acceptable reflection of the hepatic in vivo situation. Their use, however, is largely restricted to short-term purposes due to the progressive occurrence of hepatic dedifferentiation.
2. The HDAC inhibitor TSA stabilizes the in vivo–like hepatocellular phenotype in primary cultures of hepatocytes by inducing cell cycle arrest, reducing apoptotic activity, and enhancing liver-specific functionality and intercellular communication mediated by gap junctions.

3. This stabilized cell culture model may represent a promising tool for toxicity testing of xenobiotics and opens possibilities for screening of nongenotoxic carcinogens.

Acknowledgements

This work was financially supported by the FP6 European project carcinoGENOMICS (PL 037712).

References

[1] European Union, "Directive 2003/65/EC of the European Parliament and of the Council of 22 July 2003 Amending Council Directive 86/609/EEC on the Approximation of Laws, Regulations and Administrative Provisions of the Member States Regarding the Protection of Animals Used for Experimental and Other Scientific Purposes (Text with EEA Relevance)," *Official Journal of the European Union,* L 230, 2003, pp. 32–33.

[2] European Union, "Regulation (EC) No. 1907/2006 of the European Parliament and of the Council of 18 December 2006 Concerning the Registration, Evaluation, Authorisation and Restriction of Chemicals (REACH), Establishing a European Chemicals Agency, Amending Directive 1999/45/EC and Repealing Council Regulation (EEC) No 793/93 and Commission Regulation (EC) No 1488/94 as Well as Council Directive 76/769/EEC and Commission Directives 91/155/EEC, 93/67/EEC, 93/105/EC and 2000/21/EC," *Official Journal of the European Union*, L 396, 2006, pp. 1–849.

[3] Rogiers, V., and M. Pauwels, *Safety Assessment of Cosmetics in Europe (Current Problems in Dermatology),* Basel, S. Karger AG, 2008.

[4] European Union, B.13/14. "Mutagenicity—Reverse Mutation Test Bacteria. Commission Directive 2000/32/EC of 19 May 2000 Adapting to Technical Progress for the 26th Time Council Directive 67/548/EEC on the Approximation of the Law, Regulations and Administrative Provisions Relating to the Classification, Packaging and Labelling of Dangerous Substances," *Official Journal of the European Union*, L 136, 2000, pp. 57–64.

[5] European Union, "B.10. Mutagenicity—In Vitro Mammalian Chromosome Aberration Test. Commission Directive 2000/32/EC of 19 May 2000 Adapting to Technical Progress for the 26th Time Council Directive 67/548/EEC on the Approximation of the Law, Regulations and Administrative Provisions Relating to the Classification, Packaging and Labelling of Dangerous Substances," *Official Journal of the European Union,* L 136, 2000, pp. 35–42.

[6] Organisation for Economic Co-operation and Development, "OECD Guideline for Testing of Chemicals—Guideline 471: Bacterial Reverse Mutation Test," 1997.

[7] Organisation for Economic Co-operation and Development, "OECD Guideline for Testing of Chemicals—Guideline 473: *In vitro* Mammalian Chromosome Aberration Test," 1997.

[8] Eastmond, D.A., et al., "Mutagenicity Testing for Chemical Risk Assessment: Update of the WHO/IPCS Harmonized Scheme," *Mutagenesis*, Vol. 24, No. 4, 2009, pp. 341–349.

[9] European Union, "B.32. Carcinogenicity Test. Comission Directive 88/302/EEC of 18 November 1987 Adapting to Technical Progress for the Ninth Time Council Directive 67/548/EEC on the Approximation of Laws, Regulations and Administrative Provisions Relating to the Classification, Packaging and Labelling of Dangerous Substances," *Official Journal of the European Union,* L 133, 1988, pp. 37–42.

[10] Combes R. D., "The Use of Structure-Activity Relationships and Markers of Cell Toxicity to Detect Non-Genotoxic Carcinogens," *Toxicology In Vitro,* Vol. 14, No. 4, 2000, pp. 387–399.

[11] Yamasaki, H., et al., "Nongenotoxic Carcinogens: Development of Detection Methods Based on Mechanisms: A European Project," *Mutation Research*, Vol. 353, No. 1-2, 1996, pp. 47–63.

[12] Vanhaecke, T., and V. Rogiers, "Hepatocyte Cultures in Drug Metabolism and Toxicological Research and Testing," in I. R. Phillips and E. A. Shephard, (eds.), *Methods in Molecular Biology,* Totowa, NJ: Humana Press, 2006, pp. 209–227.

[13] Vinken, M., et al., "Involvement of Cell Junctions in Hepatocyte Culture Functionality," *Critical Reviews in Toxicology,* Vol. 36, No. 4, 2006, pp. 299–318.

[14] Papeleu, P., et al., "Cell Cultures as In Vitro Tools for Biotransformation Studies," in S. G. Pandalai, (ed.), *Recent Research Development in Drug Metabolism and Disposition*, Kerala, India: Transworld Research Networks, 2002, pp. 199–234.

[15] Elaut, G., et al., "Molecular Mechanisms Underlying the Dedifferentiation Process of Isolated Hepatocytes and Their Cultures," *Current Drug Metabolism,* Vol. 7, No. 6, 2006, pp. 629–660.

[16] Guguen-Guillouzo, C., et al., "Maintenance and Reversibility of Active Albumin Secretion by Adult Rat Hepatocytes Co-Cultured with Another Liver Epithelial Cell Type," *Experimental Cell Research*, Vol. 143, No. 1, 1983, pp. 47–54.

[17] Begue, J. M., et al., "Prolonged Maintenance of Active Cytochrome P-450 in Adult Rat Hepatocytes Co-Cultured with Another Liver Cell Type," *Hepatology*, Vol. 4, No. 5, 1984, pp. 839–842.

[18] Papeleu, P., T. Vanhaecke, and V. Rogiers, "Histone Deacetylase Inhibition: A Differentiation Therapy for Cultured Primary Hepatocytes?" *Current Enzyme Inhibition*, Vol. 2, No. 1, 2006, pp. 91–104.

[19] Henkens, T., et al., "Trichostatin A, a Critical Factor in Maintaining the Functional Differentiation of Primary Cultured Rat Hepatocytes," *Toxicology and Applied Pharmacology*, Vol. 218, No. 1, 2007, pp. 64–71.

[20] Vinken, M., et al., "Histone Deacetylase Inhibitors as Potent Modulators of Cellular Contacts," *Current Drug Targets*, Vol. 7, No. 6, 2006, pp. 773–787.

[21] Papeleu, P., et al., "Trichostatin A Induces Differential Cell Cycle Arrests But Does Not Induce Apoptosis in Primary Cultures of Mitogen-Stimulated Rat Hepatocytes," *Journal of Hepatology*, Vol. 39, No. 3, 2003, pp. 374–382.

[22] Vanhaecke, T., et al., "Effect of the Histone Deacetylase Inhibitor Trichostatin A on Spontaneous Apoptosis in Various Types of Adult Rat Hepatocyte Cultures," *Biochemical Pharmacology*, Vol. 68, No. 4, 2004, pp. 753–760.

[23] Vinken, M., et al., "Trichostatin A Enhances Gap Junctional Intercellular Communication in Primary Cultures of Adult Rat Hepatocytes," *Toxicological Sciences*, Vol. 91, No. 2, 2006, pp. 484–492.

[24] Berry, M. N., and D. S. Friend, "High Yield Preparation of Isolated Rat Liver Parenchymal Cell," *Journal of Cell Biology*, Vol. 43, No. 3, 1969, pp. 506–520.

[25] Seglen P. O., "Preparation of Isolated Rat Liver Cells," *Methods in Cell Biology*, Vol. 13, 1976, pp. 29–83.

[26] Chipman, J. K., A. Mally, and G. O. Edwards, "Disruption of Gap Junctions in Toxicity and Carcinogenicity," *Toxicological Sciences*, Vol. 71, No. 2, 2003, pp. 146–153.

[27] Cowles, C., A. Mally, and J. K. Chipman, "Different Mechanisms of Modulation of Gap Junction Communication by Non-Genotoxic Carcinogens in Rat Liver In Vivo," *Toxicology*, Vol. 238, No. 1, 2007, pp. 49–59.

[28] Kolaja, K. L., D. T. Engelken, and C. D. Klaassen, "Inhibition of Gap-Junctional-Intercellular Communication in Intact Rat Liver by Nongenotoxic Hepatocarcinogens," *Toxicology*, Vol. 146, No.11, 2000, pp. 15–22.

[29] Russell, W. M. S., and R. L. Burch, *The Principle of Humane Experimental Technique*, London, U.K.: Methuen, 1959.

CHAPTER

8

A Statistical Method to Reduce In Vivo Product Testing Using Related In Vitro Tests and ROC Analysis

Samuel S. Murray,[1] Elsa J. Brochmann,[2] Keyvan Behnam,[3] and Judith O. Harker[4]

[1]Geriatric Research, Education, and Clinical Center, VA Greater Los Angeles and Department of Medicine and Interdepartmental Program in Biomedical Engineering, University of California, Los Angeles [2]Geriatric Research, Education, and Clinical Center, VA Greater Los Angeles and Department of Medicine, University of California, Los Angeles, CA [3]Research Service, VA Greater Los Angeles and Lanx, Inc., Broomfield, CO [4]Research Service, VA Greater Los Angeles

Corresponding author: Samuel S. Murray, address: Geriatric Research, Education, and Clinical Center, VA Greater Los Angeles, GRECC (11-E); VA Medical Center; 16111 Plummer St.; Sepulveda, CA 91343 e-mail: Samuel.Murray@va.gov, fax: 818.895.9519

Abstract

Many biological materials, therapeutics, and reagents are tested in animals. It seems self-evident that an in vivo test would most reliably replicate the conditions of the relevant clinical or biological situation. However, in many cases precise and reproducible in vitro tests also exist. Quality control personnel and investigators rely excessively on in vivo gold-standard tests because of a sense of uncertainty as to how reliably the secondary test functions as a replacement for the primary in vivo test. In this chapter we present a statistical method in which the results of the secondary test are calibrated against the results of the primary test in such a way as to allow the investigator to employ the secondary test in the context of very precisely defined and controlled levels of uncertainty. This method will allow for a substantial reduction in the use of animal subjects.

Key terms alternatives to animal testing
cut-point analysis
ROC analysis

8.1 Introduction

The quality of many biological products is assessed by the use of a gold standard in vivo test. In this chapter we use the example of tests for the activity of demineralized bone matrix (DBM) products and bone morphogenetic proteins (BMPs) derived from our previous work in this area [1]. However, the methods can be applied to any product. For many of these products there exist precise and consistent in vitro testing models. The inability to precisely correlate the results of the in vivo test with those of the in vitro test introduces uncertainty into the testing process. Thus, in order to reduce uncertainty, an unnecessary reliance upon primary in vivo testing develops. Furthermore, there is often a tacit, but false, assumption that because the in vivo model more closely resembles the physiological or clinical situation, it is infallible. For example, in the ectopic bone formation assay for BMP activity, about 10% of samples will unexpectedly show no response whatsoever [personal communication with Ed King, 2008] The expectation of 0% uncertainty is unrealistic. In this chapter we will describe a relatively simple method of statistical analysis that allows investigators and quality control personnel to precisely relate the results of in vivo and in vitro tests in such a way as to allow them to exactly control the magnitude of uncertainty. By establishing control over uncertainty in product testing, in vivo testing can be greatly reduced and restricted to calibration, periodic recalibration, and structured confirmation.

The receiver operating characteristic (ROC) is a type of numerical analysis that was developed during World War II in an effort to optimize information recovery from radio signals of various types. As a signal is amplified, the magnitude of the useful information is increased, but the magnitude of information-canceling "noise" is also increased. Accuracy depended upon signal strength relative to noise and also on the operator's decision threshold in deciding what would be identified as a signal. Therefore, a method of optimizing the retrieval of information was developed. This type of analysis can be applied by analogy to biological tests in which a decision threshold or cutoff is set as indicating a positive result. The standard vocabulary of test evaluation is employed. The standard definitions used are presented in Table 8.1. The utility of the various descriptive statistics in the interpretation of numerical data has been presented adequately in many reviews [2]. As the threshold for "positive" is decreased in a diagnostic test, information reflecting the actual clinical reality (the number of true positives) is increased, but so is the magnitude of information-interfering data (the number of false positives).

Table 8.1 Definitions of Standard Parameters Used in the Evaluation of a Test

	Condition Present (In Vivo Score Pass)	Condition Absent (In Vivo Score Fail)	Total
Test Positive (assay = cut-point)	A	B	E
Test Negative (assay < cut-point)	C	D	F
Total	G	H	I

In parentheses, the definitions are applied to our analysis. A: True Positive. B: False Positive. C: False Negative. D: True Negative. A/G: Sensitivity (Se; True Positive Rate). D/H: Specificity (Sp; True Negative Rate). A/E: Predictive Power of a Positive Test (PPPT). D/F: Predictive Power of a Negative Test (PPNT). (A+D)/I: Accuracy. B/H: False Positive Rate = (1-Sp). C/G: False Negative Rate = (1-Se). (A/G)/(B/H): Likelihood Ratio for a Positive Test = (Se)/(1-Sp). (C/G)/(D/H): Likelihood Ration for a Negative Test = (1-Se)/(Sp)

The ROC curve presents the true positive rate (sensitivity) plotted against the false positive rate (1-specificity) at all possible decision thresholds. The decision threshold represented by the point on the curve furthest from the diagonal is the threshold that yields the best sensitivity while minimizing the false positive rate. This point is frequently regarded as the best *cut-point*. However, specific circumstances may require the selection of a different cut-point. For example, in the case of a screening test, a lower threshold will identify more true positives but assumes that a higher false positive rate is acceptable. Conversely, tests for which a positive result has serious consequences, such as surgery, a higher threshold will minimize false positives. Similarly, in product testing, the decision threshold may be adjusted to reflect considerations such as the cost of producing a lot of material and the need to ensure that each lot has a specified activity. ROC analysis allows the investigator or quality control personnel to set the definition of "positive" for the primary (gold standard) in vivo test and to set the degree of certainty required (the probability that the secondary in vitro test result reflects the actual clinical situation or the gold-standard surrogate for the actual clinical situation) and provides a cut-point for the secondary test result to achieve those exact expectations.

8.2 Experimental Design

Table 8.2 gives a hypothetical data set for 100 samples of DBM tested by in vivo testing and rated on a standardized semiquantitative scale (0, +, ++, +++, ++++) and also tested by ELISA for BMP-2 content. Our practice is to obtain three independent observations, convert the scores to integers (+ = 1, ++ = 2, and so forth), and use the mean of the three values as the final score. The value for sample size of 100 was selected arbitrarily. The range of values for the secondary test (BMP-2 ELISA) is based on our experience with a number of commercial products. The distribution of the results is completely fictitious and does not reflect any specific commercial product. It is to be emphasized that the nature of the product, DBM, is irrelevant. It is not necessary to know the variability within the secondary test though it is assumed that the test remains consistent. It is assumed that the in vivo test exactly reflects the true clinical situation and it is therefore assigned the appellation of gold standard, though in reality it is only a best approximation. Calculations of uncertainty are based on the empirical relationship between the two tests and, therefore, accurately reflect the results of the gold standard (the surrogate for the true situation) within the selected range of acceptable uncertainty.

8.3 Materials

A number of software programs that perform ROC computations are commercially available. Several of these programs have been critically examined and compared by other investigators [3]. For the calculations presented in this chapter, we have used Analyse-It, Methods Evaluation edition (version 2.12; Analyse-It, Leeds, United Kingdom). Analyse-It operates within Microsoft Excel (Microsoft, Redmond, Washington).

Table 8.2 A Hypothetical Data Set of In Vivo and In Vitro Testing of DBM

Subject Number	In Vivo Score	In Vitro Result	Subject Number	In Vivo Score	In Vitro Result	Subject Number	In Vivo Score	In Vitro Result	Subject Number	In Vivo Score	In Vitro Result
1	1.00	5.2	26	2.33	7.3	51	2.67	7.4	76	1.33	5.0
2	2.67	7.4	27	2.00	12.4	52	4.00	32.5	77	3.33	9.9
3	1.67	4.3	28	2.66	18.9	53	1.00	5.2	78	4.00	45.1
4	3.33	9.3	29	3.33	23.6	54	2.67	8.4	79	2.67	7.3
5	1.33	5.7	30	0.00	1.6	55	4.00	42.4	80	3.33	8.2
6	2.67	13.0	31	3.00	6.8	56	2.66	16.8	81	1.00	1.6
7	1.00	5.5	32	3.33	8.4	57	3.67	9.9	82	1.67	1.52
8	3.66	35.2	33	1.00	3.6	58	2.00	8.4	83	2.67	9.1
9	3.00	13.0	34	1.67	5.0	59	1.00	5.7	84	3.00	26.3
10	3.00	7.7	35	3.33	16.8	60	3.00	33.3	85	2.67	11.3
11	3.33	26.3	36	1.00	4.1	61	2.67	18.7	86	1.00	3.7
12	2.00	6.0	37	3.33	7.7	62	3.00	19.1	87	1.00	5.2
13	3.33	33.3	38	3.33	9.9	63	0.00	1.67	88	2.66	16.8
14	3.33	8.2	39	2.67	13.0	64	2.67	7.3	89	3.00	17.1
15	3.00	23.6	40	4.00	45.1	65	3.00	7.7	90	3.33	10.5
16	3.00	18.7	41	2.66	17.1	66	2.33	10.5	91	2.67	9.3
17	3.33	22.8	42	1.33	4.1	67	2.00	6.2	92	1.00	1.8
18	1.00	4.0	43	3.33	8.2	68	1.00	4.3	93	1.67	5.0
19	2.00	6.2	44	2.67	10.5	69	4.00	42.4	94	2.67	17.1
20	2.00	7.4	45	3.33	9.3	70	3.33	19.1	95	3.67	35.2
21	4.00	4.4	46	2.00	4.3	71	1.67	6.2	96	3.33	23.6
22	2.66	18.9	47	3.00	6.8	72	3.00	19.1	97	1.00	1.8
23	0.67	4.0	48	2.67	26.3	73	1.33	3.7	98	4.00	45.1
24	1.67	6.0	49	2.67	9.1	74	1.33	5.7	99	3.33	22.5
25	3.00	18.7	50	3.33	11.3	75	2.66	18.9	100	2.67	6.8

In vivo scores range from 0 to ++++. Final scores are the means of three independent observations (in vitro results are in ng of BMP-2 per g of DBM).

8.4 Methods

Before one undertakes computations using Analyse-It software, it is strongly recommended to view the instructive getting-started tutorial at (www.analyse-it.com/userguide/gettingstarted.aspx).

Presented next is a step-by-step protocol for the analysis of the data from our hypothetical data set (Table 8.3). Data from the gold standard in vivo test can be continuous (for example, the real number results of a determination; 1, 2.7, 3.25, and so forth), ordinal (1 = excellent, 2 = very good, 3 = good, and so forth), or nominal (pass versus fail). Prior to the computation, the data from gold standard are converted to a binomial nominal variable, that is, pass or fail. Note that the definition of pass can be changed in subsequent analyses. This allows for the construction of a more comprehensive decision-making reference table (Table 8.4). The following protocol gives an analysis with our data for the situation where "pass" is defined as an in vivo score = 3.0. We repeated the analysis employing definitions of pass as = 1.0 and = 2.0. The data from the secondary in vitro test can be continuous or ordinate as long as there is a sufficiently large series of categories to allow for meaningful discrimination. Our data set contains 100 sets of results. We have done this to illustrate the degree of precision that is possible. Obviously, copying this set is cumbersome. For the purpose of learning to use the soft-

Table 8.3 Cut-Point Analysis for the Data Set Given in Table 8.2

Cut-Point for BMP Assay Pass = 1.0	Pppt In Vivo Score Pass = 1.0	Cut-Point for BMP Assay Pass = 2.0	PPPT In Vivo Score Pass = 2.0	Cut-Point for BMP Assay Pass = 3.0	PPPT In Vivo Score Pass = 3.0
1.52	0.970	1.52	0.730	1.52	0.420
1.60	0.970	1.60	0.737	1.60	0.424
1.67	0.979	1.67	0.753	1.67	0.433
1.80	0.990	1.80	0.760	1.80	0.438
3.60	0.989	3.60	0.777	3.60	0.447
3.70	0.989	3.70	0.785	3.70	0.452
4.00	0.989	4.00	0.802	4.00	0.462
4.10	1.000	4.10	0.820	4.10	0.472
4.30	1.000	4.30	0.839	4.30	0.483
4.40	1.000	4.40	0.857	4.40	0.500
5.00	1.000	5.00	0.855	5.00	0.494
5.20	1.000	5.20	0.888	5.20	0.513
15.50	1.000	5.50	0.922	5.50	0.532
5.70	1.000	5.70	0.934	5.70	0.539
6.00	1.000	6.00	0.973	6.00	0.562
6.20	1.000	6.20	0.986	6.20	0.577
6.80	1.000	6.80	1.000	6.80	0.603
7.30	1.000	7.30	1.000	7.30	0.600
7.40	1.000	7.40	1.000	7.40	0.629
7.70	1.000	7.70	1.000	7.70	0.661
8.20	1.000	8.20	1.000	8.20	0.643
8.40	1.000	8.40	1.000	8.40	0.623
9.10	1.000	9.10	1.000	9.10	0.640
9.30	1.000	9.30	1.000	9.30	0.667
9.90	1.000	9.90	1.000	9.90	0.667
10.50	1.000	10.50	1.000	10.50	0.643
11.30	1.000	11.30	1.000	11.30	0.667
12.40	1.000	12.40	1.000	12.40	0.676
13.00	1.000	13.00	1.000	13.00	0.694
16.80	1.000	16.80	1.000	16.80	0.727
17.10	1.000	17.10	1.000	17.10	0.767
18.70	1.000	18.70	1.000	18.70	0.815
18.90	1.000	18.90	1.000	18.90	0.833
19.10	1.000	19.10	1.000	19.10	0.952
22.50	1.000	22.50	1.000	22.50	0.944
22.80	1.000	22.80	1.000	22.80	0.941
23.60	1.000	23.60	1.000	23.60	0.938
26.30	1.000	26.30	1.000	26.30	0.923
32.50	1.000	32.50	1.000	32.50	1.000
33.30	1.000	33.30	1.000	33.30	1.000
35.20	1.000	35.20	1.000	35.20	1.000
42.40	1.000	42.40	1.000	42.40	1.000
45.10	1.000	45.10	1.000	45.10	1.000

PPPT = Predictive Power of a Positive Test

ware, the readers can use the first 20 or 30 pairs of data in our set or their own data. However, if this is done, they will not be able to check their results against the final values in our results figures. The Excel file containing the data set can be obtained at the book's

Table 8.4 Decision Analysis Based on Cut-Point Analysis from Table 8.3

In Vivo Pass = (+)			In Vivo Pass = (++)			In Vivo Pass = (+++)		
Acceptable Uncertainty			Acceptable Uncertainty			Acceptable Uncertainty		
1% ("99% sure")	5% ("95% sure")	10% ("90% sure")	1% ("99% sure")	5% ("95% sure")	10% ("90% sure")	1% ("99% sure")	5% ("95% sure")	10% ("90% sure")
Cut-point used (ng/g)			Cut-point use (ng/g)			Cut-point used (ng/g)		
1.80	1.52	1.52	6.8	6.0	5.5	32.5	19.1	19.1

description page at http://www.artechhouse.com or via e-mail from the corresponding author.

8.4.1 Step-by-step protocol for the analysis of data using Analyse-It

1. Prepare the data spreadsheet. Data can be entered directly into the program, which works in Excel, or it can be entered into Excel separately and then pasted into the program. Open Excel and in row 1 enter six column titles in columns A through F as follows:

A	B	C	D	E	F
Subj #	in vivo score	Pass 1	Pass 2	Pass 3	BMP

2. Enter the subject numbers and test results. Enter subject number 1 through 100 in column A. Enter the results of the in vivo test in column B. Enter the results for the in vitro test (BMP ELISA) in column F (BMP). Use File, Save As to apply a name such as "my data 1" and save to the desktop.
3. Sort the data. Sorting the data at this point will reduce the chance of errors in the next step. Sorting is not mandatory, but it allows for misassignments of pass or fail to be readily spotted. First, highlight all cells in columns A through F and all rows 1 through 101 (your entire spreadsheet). On the top menu bar select "Data" and then "Sort." The dialog box will appear. Select "Sort By," "in vivo score" in the upper box. Click on "Ascending." The second two boxes remain empty. At the bottom find "My data range has" and click "Header row." Click "OK." The data will now be sorted beginning with subject numbers 30, 63, and 23 and with the corresponding in vivo scores of 0.00, 0.00, and 0.67 and BMP values of 1.60, 1.67, and 4.00 in rows 2, 3, and 4.
4. Assign "pass" or "fail" to the in vivo results. In this example we are defining "pass" as an in vivo score = 3.0. Our final decision-making table also contains results from analyses with other definitions of "pass." In column E ("pass 3"), place a "p" if the number in column 1 of the same row is = 3.0 and an "f" if the number is < 3.0. Save the results. In column E, rows 2–59 will be "f," whereas rows 60–101 will be "p." Complete the spreadsheet by assigning, in column D, "f" to in vivo scores that are < 2.0 and "p" to in vivo results that are = 2.0. Similarly, in column C, assign "f" to in vivo scores that are < 1.0 and "p" to in vivo results that are = 1.0. Save the file. The first five rows of the data should look like this:

A	B	C	D	E	F
Subject Number	In Vivo Score	Pass 1	Pass 2	Pass 3	BMP
30	0.00	f	f	f	1.60
63	0.00	f	f	f	1.67
23	0.67	f	f	f	4.00
1	1.00	p	f	f	5.20
7	1.00	p	f	f	5.50

5. Define this spreadsheet as an Analyse-It dataset. Open Analyse-It, Methods Development edition. The control panel for the most recent version of Excel is slightly different than that of previous versions. Both are very user-friendly. In the menu select "File," "open." Locate your Excel spreadsheet and open it in Analyse-It. In the top menu select "Analyze," then "Test Performance," and then "ROC Curve." "Welcome to the Dataset Wizard" will appear. Click "Next." Under "Data Layout" select "List" and click "Next." The "Dataset Format" window will appear and will automatically read the format of the data with the default of row 1 for the header. Make sure that there is a check in the "Apply Autoformat" box and click "Next." "Completing the Dataset Wizard" will appear. Click "Finish." The following message will appear: "Test Performance-ROC Curve requires a list dataset containing a dichotomous variable and 1 or more continuous variables." Click "OK."
6. Define the variables: Before an analysis can be undertaken, the variable for each column must be defined. Place the cursor in one of the cells of column A ("Subject Number") and then click on the "Variable" icon in the control panel. (The icon resembles a notepad with a pencil.) The "Variable Properties" menu appears. Select "Continuous" and click "OK." Place the cursor in one of the cells of column B ("in vivo score"). Click on the "Variable" icon. Select "Continuous," enter "2" for the number of decimal places, and click "OK." Place the cursor in one of the cells of column C ("Pass 1"). Click on the "Variable" icon and select "Nominal." Under "Categories: Id/Order," type "1" beside "f" and "2" beside "p," and then click "OK." Repeat the definition of "Pass 2" and "Pass 3" using exactly the same procedure as for "Pass 1." Finally, select a cell under column F ("BMP"), click the "Variable" icon, select "Continuous," enter "2" for the number of decimal places, and then click "OK." Save your assignments.
7. Perform a ROC analysis: Click on "Analyze" in the menu and select "Test Performance," and then select "ROC Curve." Fill in the spaces as follows in the "Test Performance ROC Curve" dialog box:

Prompt	Select or enter
True classification	Select "Pass 3"
Positive indicated by	Select "p"
Test	Select "BMP"
Positive	Select "greater than or equal"
Confidence interval	Enter "95" %
ROC plot	Select "1-Specificity v Sensitivity"
811144Decision plot	Select "Predictive value"

   Check the "Predictive value" box. Click "OK."
8. Transfer the results to a results spreadsheet. Open a new Excel spreadsheet and in row 1 label columns A through F as follows: A, "Cut-point"; B, "PPPT for P=1"; C,

"cut-point"; D, "PPPT for P=2"; E, "cut-point"; F, "PPPT for P=3." Name and save this file. From the ROC analysis data sheet, copy the figures for the cut-points ("Positive test >= cutoff") and paste them into column E of the new spreadsheet. Then copy and paste the results from the ROC analysis data sheet for PPPT ("Predictive value (+)") into column F. The analysis for the case where "pass" is defined as an in vivo score of = 3.0 is done. Examine the ROC curve and the descriptive statistics in the ROC analysis sheet below the curve and consider what conclusions can be drawn from them. To complete the analysis, return to the original data sheet, start at step 7 and change the designation of "True classification" to "Pass 2." Then repeat the process again for "Pass 1." Note that the cut-points are the same for all three groups.

8.5 Results

The descriptive statistic that is most useful in our analysis is the predictive power of a positive test (PPPT) for BMP ELISA with respect to predicting the result of the in vivo assay. This statistic gives the probability of correctly predicting the result of the in vivo test by any cut-point level of the in vitro assay. We have repeated the analysis for different "passing" levels of the in vivo test. These results are shown in Table 8.3. This arrangement allows great flexibility in decision-making because the decision-maker can set both the level of activity that is needed and the certainty of achieving that result. For example, if many lots were to be screened and any lot with activity subjected to another type of testing, passing would be set at = 1.0 and a relatively low certainty (90%) would be employed. On the other hand, if only material with high activity was desired and the level of activity needed to be assured, a passing of = 3.0 and a certainly of 99% would be used. All of these decisions are based on the cut-point used for the BMP ELISA. For example, say that a passing of = 2.0 is regarded as desirable and that a certainty of 95% is wanted. In Table 8.3 locate the heading "PPPT in vivo score pass = 2.0." Go down the column until you come to the first number = 0.95. In this case, the PPPT selected is 0.973 because 0.934 is less than 0.95. Then locate the corresponding cut-point in the column immediately to the left. In this case the cut-point is 6.0 ng/g. Thus, if a lot of DBM is tested with a BMP-2 ELISA and gave a result of = 6.0 ng/g, then there is a 95% probability that the lot would score (++) or better on the in vivo test. The ROC plot for the case where "pass" is set at = 3.0 is shown in Figure 8.1.

A summary of the cut-points for specific decision-making landmarks is given in Table 8.4.

8.6 Discussion and Commentary

8.6.1 Selecting the proper secondary test

Obviously, a secondary test that best predicts the results of the primary test is desired. There may, however, be other considerations such as the cost of reagents, the availability of commercial kits for the secondary test, sample preparation requirements for a specific biochemical analysis, propriety matters, and other concerns. In the situation of DBM, we and others assayed the materials using a battery of commercial kits for specific growth factors known to be important in bone formation and then applied regression

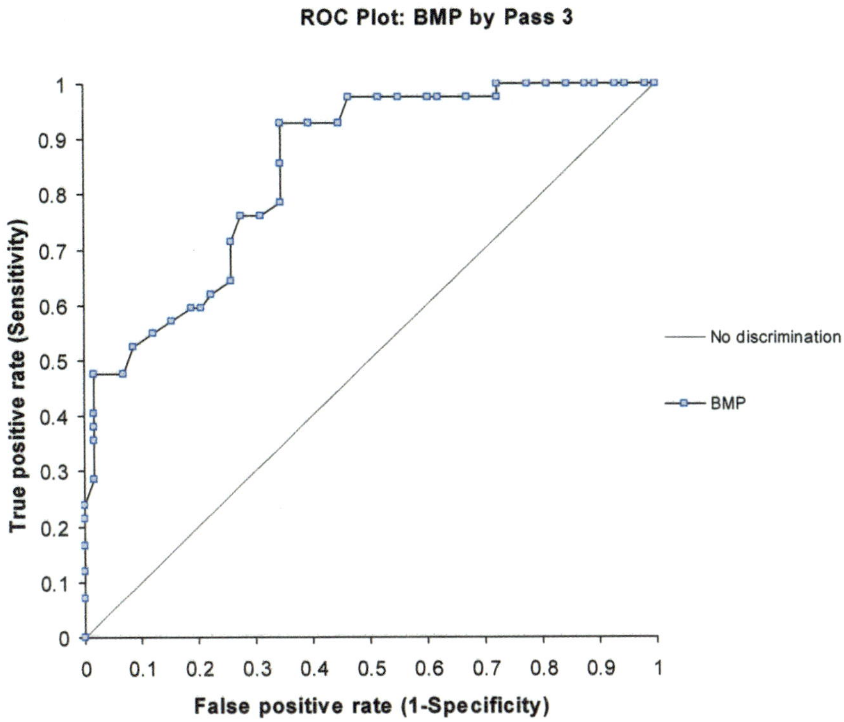

Figure 8.1 ROC plot from our example data using an in vivo result of = 3.0 as positive.

analysis to determine which test, or combination of tests, best predicted ectopic bone formation [1, 4]. It is to be emphasized that the best test need not be selected for the secondary test. For example, in our initial work, the BMP-7 content of DBM was nearly as predictive of bone formation as was the BMP-2 content. The use of BMP-7 content as the secondary test would be completely satisfactory. The predictive power of the combination of the BMP-2 and BMP-7 contents was more predictive than was that of BMP-2 alone, but the value added by the second test was minimal and therefore did not justify the added expense and effort. The TFG-beta content was also somewhat predictive, but that particular assay requires a cumbersome sample dilution step. Once selected, the relative limitations of the secondary test become less important because the level of uncertainty will always be revealed in the subsequent ROC analysis.

8.6.2 Determining the sample size for calibration and recalibration

In general, the performance of statistical tests improves as the sample size is increased. The situation with ROC analyses is somewhat more complex than the usual case. In the usual analyses of the usefulness of a test there is one cutoff value that is regarded to define "positive" and the condition, such as a disease to be detected, is either present or absent. Sample size calculations are then under taken based on this straightforward situation. However, in ROC analyses, one can sequentially set different levels of gold-standard bioactivity to correspond to the analogous situation of the presence of a clinical condition, and one needs to consider all possible definitions of positive derived from the secondary test.

Sample size affects the discriminating power of a test but not the accuracy. In the type of evaluation described in this chapter, the accuracy is intrinsic to the analysis and is given in the defined parameters of certainty. For example, in our illustrative data set, if the sample size had been greatly reduced, the ability to define a specific degree of certainty would remain, but the cut-point to achieve that certainty should have been shifted upward substantially. Thus, there is no harm in investigators undertaking a calibration study based on a "reasonable" number of samples based on considerations such as cost or the facilities available. One should make an effort to have several data points for each value over the entire range of the secondary test. We would recommend a sample size no smaller than 50. In our experience, employing a smaller sample size yields disappointing discriminating power. Precise calculations for sample size for the initial study are also made difficult because estimates of the variance for each test as well as advance knowledge of the relationship between the primary and secondary tests would be required.

For the purposes of recalibration, if no major changes in the testing conditions have occurred and one is satisfied with the discriminating power of the initial sample, then the same number would be used. In the more complex situation where changes in the testing situation had occurred and it was desired to confirm that the discriminating power was at least as good as the previous testing, then a larger sample size would be needed. Specifically, one would determine a sample size needed to test whether the area under the ROC curve in the recalibration sample is different from that of the initial sample. In supplementary material available at the book's description page at http://www.artechhouse.com, a discussion of the methods of Obuchowski [5] and Hanley and McNeil [6] for sample size determination is given.

8.6.3 Regulatory concerns

In the United States, the testing standards to which a product is to be held as well as the testing methods that will be employed are negotiated with the Food and Drug Administration (FDA) at the time of the approval process and are based on the claims of the manufacturer for the product. For example, if a claim is made that a product is "osteogenic," then testing that demonstrates that "osteogenicity" must be employed for approval and subsequent evaluation. Therefore, if it is desired to use a test such as a growth factor assay which indirectly predicts osteogenicity but does not measure osteogenicity per se, then evidence of the correlation between the two measurements would be required. If a producer of a product that was approved using a test of osteogenicity wished to switch to reliance on a growth factor assay as the primary means of quality control, the evidence of the correlation between the two testing methods as well as the protocol for the test and calibration would have to be agreed upon by the FDA.

8.6.4 Determining the frequency of recalibration

The reliability of this method of analysis is based upon the empirical relationship between the results of the in vivo and in vitro tests. Thus, anything that affects the material or the testing method will disrupt this relationship and influence the reliability. Examples of such factors include: a change in the manufacturing process, a significant change in the source of starting materials, a change in the supplier of testing reagents, and a change in analytic equipment. Such changes would require recalibration studies.

If there are no changes in the materials or testing methods, then the necessary interval for recalibration is arbitrary and can be decided on an administrative basis.

8.6.5 Determining the need for confirmatory testing

Confirmatory testing is, in reality, an abbreviated form of recalibration. There is no need for routine confirmatory testing if recalibration is done sufficiently to ensure the integrity of the empirical relationship between the two tests. This, however, assumes that all lots of product are of equal value. This may not be the case. For example, if product is easy to manufacture, there is a significant batch-to-batch variability in quality, and it is important to select lots of the highest quality for final use, it would not be unreasonable to employ in vitro testing of many lots to select the few lots that would be tested with in vivo testing. This should not, however, be undertaken for reassurance. If there is confidence in the integrity of the empirical relationship between the two tests, then no confirmatory testing should be done.

8.6.6 Statistical analysis

A discussion of the mathematical computations is not within the scope of this chapter. There are suitable reviews of this subject in the clinical literature [2, 7]. ROC curves such as the one presented in Figure 8.2 are plots of the probability of a true positive (sensitivity) versus the probability of a false positive (1-specificity) for all possible "positive test" decision thresholds.

When displayed in the manner presented in Figure 8.2, curves that are closer to the vertical axis on the left represent tests that are more accurate. The area under this curve can be calculated and used as an estimate of the relative overall accuracy of the test. In our analyses, the predictive power of a positive test (PPPT) assumes great importance. As can be seen from Table 8.1, this figure can also be derived from the values for true and false positives. In Table 8.1, true positives are represented by A and false positives are represented B. The PPPT = true positive/all positives = A/A+B. For our purposes, the PPPT gives the following information. If a sample has a positive in vitro test (that is, the assay result is equal to or greater than the decision threshold or cut-point), what is the proba-

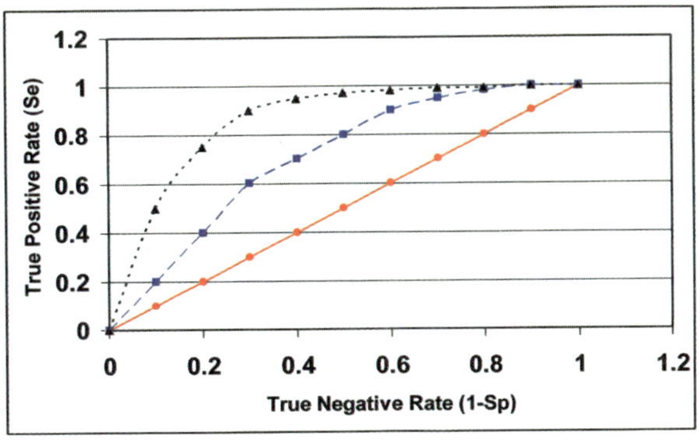

Figure 8.2 Examples of ROC curves. The diagonal line with equal values for x and y represents the curve for a test with no discriminating power above random probability. The curve to the far left would have greater predicting power than the curve closer to the diagonal. The area under these curves can be used as a measure of overall accuracy.

bility that the sample will be positive as determined by the gold-standard in vivo test? For example, a PPPT of 0.95 would indicate that there is a 95% chance that a sample with an in vitro score above the cut-point would have a positive in vivo score. Note that different levels of the gold standard can be set as positive or pass. The PPPT is calculated directly from the true positive and false positive values for each decision cut-point.

8.7 Summary Points

ROC curve analysis allows investigators to correlate the results of in vivo and vitro tests in such a way that they can rely much more on in vitro testing while maintaining the reliability of the in vivo test within precisely defined conditions of certainty. Several software programs are available that make this analysis easy to accomplish by general laboratory personnel. We present a step-by-step protocol to perform this type of analysis using Analyse-It software.

Acknowledgments

This chapter was supported by the Geriatric Research, Education, and Clinical Center, VA Greater Los Angeles Health Care System. This chapter is dedicated to my dear daughter Laura May Murray, whose respectful attitude towards animals has been an example to many.

References

[1] Murray, S. S., et al., "A Statistical Model to Allow the Phasing Out of the Animal Testing of Demineralised Bone Matrix Products," *ATLA*, Vol. 36, No. 4, 2007, pp. 1–5.

[2] Lang, T. A., and M. Secic, *How to Report Statistics in Medicine*, Philadelphia, PA: American College of Physicians, 1997.

[3] Stephan, C., et al., "Comparison of Eight Computer Programs for Receiver-Operator Characteristic Analysis," *Clinical Chemistry*, Vol. 49, No. 3, 2003, pp. 433–439.

[4] Blum, B., et al., "Measurement of Bone Morphogenetic Proteins and Other Growth Factors in Demineralised Bone Matrix," *Orthopedics*, Vol. 27, No. 1, Suppl., 2004, pp. s161–s165.

[5] Obuchowski, N. A., "Sample Size Calculations in Studies of Test Accuracy," *Statistical Methods in Medical Research*, Vol. 7, No. 4, 1998, pp. 371–392.

[6] Hanley, J. A., and B. J. McNeil, "The Meaning and Use of the Area Under a Receiver Operating Characteristic (ROC) Curve," *Radiology*, Vol. 143, No. 1, 1982, pp. 29–36.

[7] Obuchowski, N. A., "ROC Analysis," *Am. J. Roentgenology*, Vol. 184, No. 2, 2005, pp. 364–372.

CHAPTER

9

Application of the Benchmark Approach in the Correlation of In Vitro and In Vivo Data in Developmental Toxicity

Esther de Jong,[1,2] Wout Slob,[1,2] and Aldert H. Piersma1,2

[1]Institute for Risk Assessment Sciences (IRAS), Utrecht University, Utrecht, the Netherlands
[2]National Institute of Public Health and the Environment (RIVM), Bilthoven, the Netherlands
Corresponding author: Esther de Jong, address: National Institute for Public Health and the Environment (RIVM), Laboratory for Health Protection Research, Antonie van Leeuwenhoeklaan 9, 3720 BA Bilthoven, the Netherlands e-mail: Esther.de.Jong@rivm.nl, phone: +31 30 274 4510, fax: +31 30 274 4446

Abstract

The high number of animals and costs required for developmental toxicity testing has stimulated the development of in vitro alternative tests. Traditionally, to validate such alternatives, a comparison is made by correlating the in vitro results to no observed adverse effect levels (NOAELs) derived from in vivo studies. NOAELs have the disadvantage that their values depend on the design and quality of the specific study. In this chapter the alternative benchmark dose (BMD) approach will be discussed. In the BMD approach dose-response curves are fitted using all available data and used for estimating the dose (BMD) at which a certain change in the endpoint is reached. The main advantage of the BMD approach is that it is much less dependent on study design, resulting in more precise outcomes. This makes it possible to more precisely correlate in vitro benchmark concentrations (BMCs) to in vivo BMDs among different chemicals to investigate the predictability of in vitro assays.

Key terms
benchmark dose approach
BMC
BMD
developmental toxicity
in vitro–in vivo correlation

9.1 Introduction

Developmental toxicology has been identified as an area in toxicology that requires large numbers of animals for safety assessment. This is illustrated by an estimate of the number of tests and animals needed for each toxicological endpoint within the framework of the European legislation on chemical safety REACH (Registration Evaluation and Authorization of Chemicals). Although only 10% of the total number of chemicals to be tested within REACH would require developmental toxicity testing, over 25% of all animals used in REACH would be needed for this toxicological endpoint [1]. In terms of costs the percentage is even higher; out of the estimated overall cost of 1.5 billion euros, 32% would be attributed to developmental toxicity studies [1]. Furthermore, the awareness of ethical issues regarding animal experimentation has become more prominent over the years. Obviously, the search for in vitro alternatives to developmental toxicity studies deserves high priority.

The interest in alternative methods has led to the development of several in vitro test systems for developmental toxicity, such as the Embryonic Stem Cell Test (EST), the Whole Embryo Culture (WEC), and the rat limb bud MicroMass assay (MM) [2]. Of these three tests, the WEC (Figure 9.1) has the advantage that it offers the most complete in vitro method representing embryogenesis from cellular proliferation and differentiation to pattern formation.

The disadvantage of the WEC is that it still requires animals, although there is a significant reduction in numbers compared to in vivo developmental toxicity studies. Similarly, the MM also requires animals for isolating embryonic limb buds with the purpose of deriving primary cultures of dissociated mouse limb bud cells. These limb bud cells are used to study the effect of chemicals on the formation of chondrocyte foci. The EST uses the pluripotent nature of embryonic stem cells [Figure 9.2(a)] to study the effect of compounds on the differentiation of these stem cells into beating myocardial cells [Figure 9.2(b)] and has the advantage that it does not require the use of animals as continuous cell lines are used.

The EST, the WEC, and the MM have all three been part of a validation study coordinated by the European Centre for the Validation of Alternative Methods (ECVAM). The validation studies involved the testing of 20 chemicals in four independent laboratories in a double-blind protocol for each of the three in vitro assays [2–4]. The overall accuracy of each test was 70%, 79%, and 80% for the MM, the EST, and the WEC, respectively.

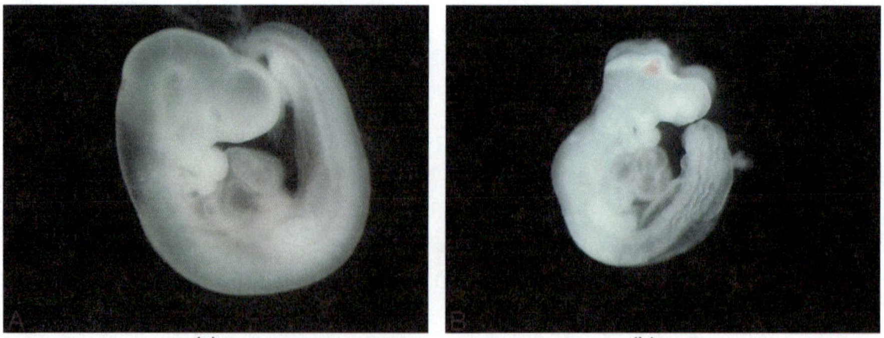

(a) (b)

Figure 9.1 Whole embryo culture (WEC): (a) normal embryo and (b) malformed embryo. (A. Verhoef, RIVM.)

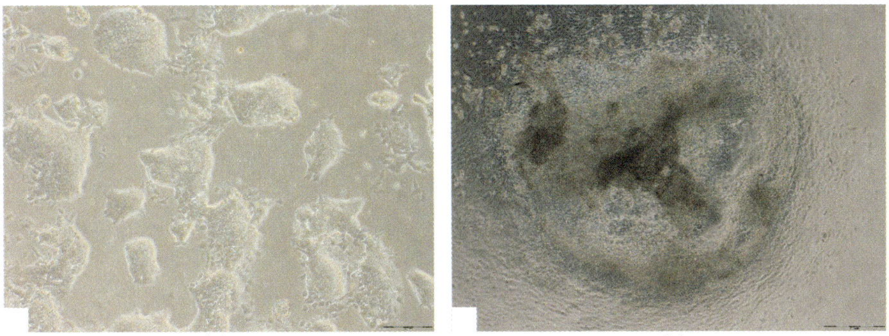

(a) (b)

Figure 9.2 Embryonic stem cell test (EST): (a) morphology of undifferentiated embryonic stem cells and (b) embryonic stem cells differentiated into cardiomyocytes. (E. de Jong, RIVM.)

Despite the good overall accuracy, the question arises whether 20 chemicals is sufficient to make a definitive statement on the predictability of these in vitro methods in general. In the ECVAM validation studies the in vitro potency of each compound was defined in terms of categories corresponding to nonteragonic and weak and strong teratogenic compounds. A limitation of this approach is that the definition of such categories is somewhat arbitrary. Furthermore, it does not take into account that chemicals that are in the low end of the strong teratogenic category and chemicals that are in the high end of the weak teratogenic category may actually have very similar potencies.

An approach that is often used to validate in vitro tests is by comparing in vitro results with no observed adverse effect levels (NOAELs) derived from in vivo studies. However, a disadvantage of NOAELs is that their values depend on the design and quality of the specific study, including group sizes, dose location, and scatter in the observed responses [5]. In vivo studies that are available in the literature may have differences in their study design, including differences in exposure timing and applied doses. Such differences may lead to different NOAELs even when the potencies of the compounds are in reality the same, as illustrated in Figure 9.3, which shows dose-response curves for two hypothetical studies.

In Figure 9.3(a) the NOAEL value is 200, while in Figure 9.3(b) this value lies substantially lower at 50 even though the underlying dose-response relationships are the same. Another method to derive a value to indicate the potency of a compound is the BMD approach. In the BMD approach a dose-response curve is fitted to determine the BMD for in vivo data, which is the dose at which a certain benchmark response (BMR) is reached [6]. Similarly, the benchmark concentration (BMC) can be determined for in vitro data. This method has the advantage over the classical NOAEL approach that is uses the full dose-response curve and is independent of the applied dose levels. In Figure 9.3 the dotted line indicates the BMD10s corresponding to a 10% decrease in the hypothetical endpoint. It shows that although there is a fourfold difference in the NOAEL values of the two studies presented, the BMD10 values are the same.

The goal of this chapter is to describe a methodology of how to optimally correlate in vitro data to existing in vivo data. An explanation will be given on how to derive equipotent concentrations/doses from in vitro and in vivo data. Critical aspects that should be taken into account in this type of analysis will also be discussed. For example, it will be discussed that the approach may be hampered by limitations in the in vivo

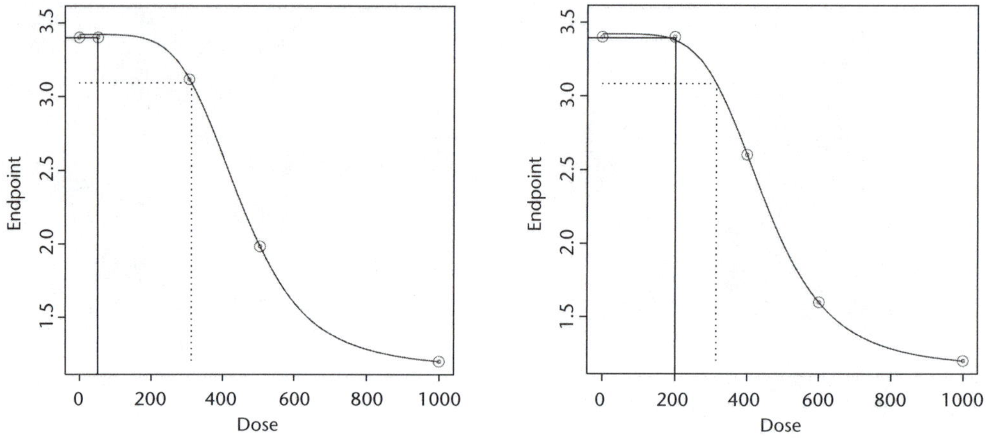

Figure 9.3 (a, b) NOAELs (straight lines) and BMD10 values (dotted lines) for the two studies with different applied doses, examining the same dose-response relationship.

database. Also, concerns regarding the in vitro method will be discussed, such as the difference in kinetics between the in vitro and the in vivo situation. While this chapter focuses on developmental toxicity the same approach is applicable to other toxicological endpoints.

9.2 Materials and Methods

The first step in designing a study with the purpose of correlating the performance of an in vitro test to in vivo studies is the selection of data to be used in the analyses. The data used should cover a wide range of embryotoxic potencies from highly embryotoxic to weak and nonembryotoxic compounds. The in vivo studies selected should meet certain selection criteria. For each individual compound the in vivo dose-response data on developmental toxicity should be suitable for estimating in vivo BMDs. Therefore, the study should preferably contain various dose levels to allow the estimation of a dose-response curve needed to derive BMDs. Furthermore, the study design (e.g., timing and duration of dosing) in the different in vivo studies should be as uniform as possible; otherwise, equipotent doses are not clearly defined. An important aspect of developmental toxicity is the timing of the exposure. The functionality of potential molecular pathways by which chemicals elicit their developmental effect may vary with time during development. For example, Ema et al. showed that for Wister rats exposed to a single dose of butyl-benzyl-phthalate the critical embryotoxic period lies between day 6 to 9 and on day 15 [7]. Exposures between days 6 to 9 resulted in an increased occurrence of skeletal malformations, while exposure on day 15 resulted in the induction of a cleft palate. Exposures between days 10 up to 14 did not result in any fetal malformations. So, if for this particular compound one study exposure was on day 7 and another study exposure was on day 11, the different exposure studies would result in distinct BMDs. Studies using the same range of exposure days are therefore preferred in particular if the sensitive period of the compound is not known. It is also important to select studies using the same animal species and preferably the same animal strain as well, as species might differ in sensitivity regarding developmental toxicity [8].

Compounds to be included in the dataset preferably have information on kinetic properties available, including data on absorption, distribution, metabolism, and elimination (ADME). Kinetic properties of compounds may have an important impact on the in vivo dose-response relationships. Due to differences in ADME rates between chemicals, equal external exposure levels can lead to very different internal concentrations (see Section 9.3). It is therefore preferable to select compounds for which information on kinetics is available to help interpret discrepancies found between the in vitro BMCs and the in vivo BMDs.

9.2.1 Derivation of in vitro BMC and BMD values

In the method described in this chapter BMC and BMD values are calculated to derive equipotent concentrations/doses to allow for comparing in vitro assays to in vivo studies. In the BMD approach, BMC and BMD values are calculated from a dose-response model fitted to the dataset. In contrast, the traditional NOAEL approach determines the highest experimental dose where no significant effect is detected using statistical analysis to compare each dose level with the control group. As a result, studies using small group sizes will need larger differences between the groups to become statistically significant, while studies with larger group sizes will detect smaller effects. Therefore, studies with small group sizes tend to have higher NOAELs. By interpolating between doses, the BMD is not directly dependent on the applied doses, and by using all the data it will result in a more precise estimate of the potency of compounds compared to the NOAEL.

To derive BMD (or BMC) values, it is important to find models that accurately describe the data. In the BMD approach the aim is not to select a single model that gives the best fit but to find all models with an acceptable fit. It is therefore recommended to run all available models, select the ones that give an acceptable fit, and determine the BMDs for these models. The differences in the BMD values associated with the acceptable models reflect the model uncertainty, which is to be interpreted as uncertainty in the data regarding the shape of the dose-response. For more information on how to apply dose-response curve fitting to derive BMDs (in the context of risk assessment), we recommend the EFSA report on this subject [9].

Examples of software developed with the purpose of deriving BMDs are the PROAST software, developed by the RIVM (www.rivm.nl/proast), and the benchmark dose software (BMDS) developed by the U.S. EPA (www.epa.gov/ncea). Both programs can be downloaded free of charge. In principle, the two software programs should give the same outcome when applied to the same dataset. Negligible differences may result from differences in the numerical algorithms employed by the programs. BMDS is a Windows-based package, while PROAST is written in the S language and requires the installation of the software package Splus or R. A license is needed for Splus, while the R software can be downloaded for free (www.cran.r-project.org). The advantage of using the Splus software is that it provides better graphical output in comparison to R.

Before a BMD (or BMC) can be derived, the BMR needs to be defined. The BMR is defined as the change in the response for which the BMD is determined. Preferably, it reflects the level below which there is no reason for concern [5]. While no adverse BMRs might be specified for in vivo studies, this may be more difficult for in vitro assays. However, since the objective of this chapter is to use the BMD approach as a tool in correlating in vitro results to in vivo data rather than for risk assessment, the issue of the relevance of the BMR plays less of a role. The choice of a BMR is not crucial as the ranking of the compounds should not change with different BMRs. When selecting a BMR, it should be taken into account that the BMR has to be observable within the dataset to prevent dependency on the model selected. It has been shown that model dependency of the BMC is more pronounced at lower BMRs [10]. Examples of BMRs that have previously been used for in vitro datasets are a BMR50 for the EST, corresponding to a 50% decline in differentiation into beating cardiomyocytes, and a BMR05 for the WEC, corresponding to a 5% decline in the total morphology score [11, 12]. For risk assessment purposes, the following BMR values have been recommended for in vivo data: for continuous data, a 5% change in the average response level [9, 13, 14], and for quantal data, an extra risk of 10% [9, 10, 15]. For the purpose of correlating in vitro to in vivo data, however, a 10% BMR for continuous in vivo data may be expected to result in more robust equipotent doses.

9.2.2 In vitro–in vivo correlation

The advantage of the BMD approach is that it increases the precision in a potential correlation between the BMC values derived from an in vitro assay and the in vivo BMDs. As stated in Section 9.2, all models found to have an acceptable fit should be applied to determine BMCs and BMDs. Usually more than one model is accepted, leading to multiple BMCs and BMDs for a single compound/endpoint. For the correlation test only one value can be used. In a risk assessment context, one option is to select the lowest BMC and BMD for each compound as these are the most conservative values. However, in attempting to establish a correlation, it is better to use the "best" estimates of the BMDs and BMCs. An obvious measure for that would be the geometric mean of the BMDs (or BMCs) associated with the different models.

To examine the correlation, the BMDs can be plotted against the BMCs (e.g., on a double logarithmic scale as shown in Figure 9.4). A straight line can then be fitted by minimizing the sum of products of the horizontal and vertical distances of the data to the line [16]. Note that the formula of the fitted line ($\log_{10}y = a + b \log_{10}x$) is equivalent with the formula $y = \tilde{a}x^b$ on the original scale, where $\tilde{a} = 10^a$. The slope (b) of this line will represent the power of the relationship and indicates whether the relationship is linear or nonlinear. If the slope equals 1, than the relationship between the BMCs and the BMDs is linear on the original scales. If the slope is smaller than 1, the curve will bend off at higher values on the original scales.

While in Figure 9.4 point estimates of BMDs and BMCs are plotted, it may be helpful to include confidence intervals of both estimates in the plots (resulting in crosses at each point). In this way it is better possibly to judge if an outlying point is just an uncertain point (i.e., the confidence intervals in both directions overlap with the fitted line), or if the estimation errors associated with that point do not explain that it deviates from the line (i.e., the confidence interval in one or both directions do not overlap with the fitted line). Confidence intervals can be calculated by the BMD software (PROAST or BMDS), but only for a given model that was fitted. As discussed above, all models that resulted in an acceptable fit should be taken into account, so that for each BMD (or BMC) several confidence intervals may become available, each related to another acceptable model. How to combine these separate intervals into a single interval is still subject of investigation. For quantal endpoints, an approach of "model averaging" has been proposed [17].

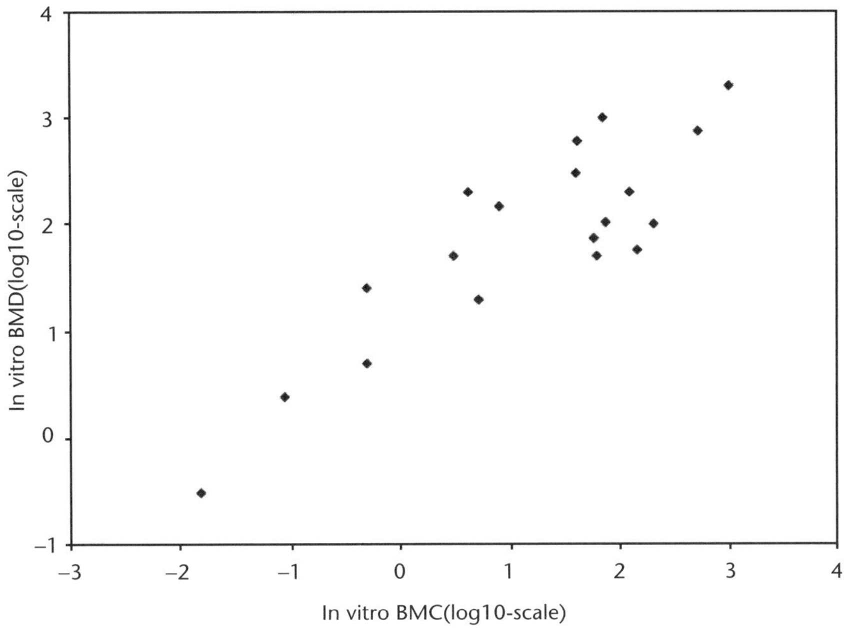

Figure 9.4 In vitro BMC values plotted against BMD values on a double logarithmic scale.

A simpler method would be to just take the lowest lower and the highest upper confidence bounds, leading to a single interval that is "conservative" (i.e., it is wider than the one that would have resulted from approaches like model averaging).

As mentioned in Section 9.1, differences in the kinetic properties between compounds can have a large impact on the correlation plot between BMCs and BMDs. Due to differences in the ADME parameters, equal external doses of different compounds can lead to very different internal concentrations. For example, if a compound has a relatively low absorption rate, a lower fraction of the exposure level will enter the circulation and end up at the target site compared to a compound with a high absorption rate. By strictly looking at the doses administered in the study while determining the in vivo potencies of these two compounds, the BMDs for the developmental effects will be the same leading to the conclusion that the potency of these two compounds is the same. However, if the internal concentrations are taken into account then the first compound would be more potent. An example of a study where differences in absorption rates of compounds may have played a part in the in vitro-in vivo correlation is the study by Janer et al. [18], where the effect of a number of phthalates metabolites in the WEC were tested and compared with the in vivo embryotoxic potency of their parent compounds. In the study one of the metabolites, monoethyl hexyl phthalate (MEHP), showed a relatively high in vitro embryotoxicity in comparison to the in vivo data [12]. The parent compound of MEHP has a lower absorption rate in comparison to the other phthalates tested, resulting in lower internal concentrations for MEHP and thereby reducing the in vivo potency in terms of a higher BMD.

Similarly, differences in elimination rates may also play a role in the level of developmental effects found in the in vivo studies. The internal exposure to a compound with a relatively long half-life will tend to persist during a longer period of time. Furthermore, compounds with a slow elimination rate can accumulate over time leading to a higher internal dose if a dose regime with multiple exposures is used in the in vivo study. An example of a possible role of elimination differences is given in a study where the effect of a number of glycol ether metabolites in the EST was determined and compared to the in vivo embryotoxicity of the parent compounds [11]. The study showed that, although the ranking of the potency of the compounds was similar in the EST as compared to the in vivo situation, the spread in the in vivo BMDs between the compounds was larger than between the BMCs in the in vitro studies. In this study, the most potent in vitro metabolite (methoxy acetic acid) also happened to have the lowest elimination rate, which enhances the in vivo potency of the associated parent compound (EGME) in terms of a lower BMD.

Apart from the fact that in vivo BMDs are expected to vary more than in vitro BMCs due to larger differences in experimental conditions, these examples show that differences in ADME rates may affect the correlation between BMCs and BMDs. They illustrate how knowledge of the kinetic properties of compounds might help interpret discrepancies found between the in vitro BMCs and the in vivo BMDs. In short, if a compound has a relatively high absorption rate or a long half-life, the internal exposure of that compound will tend to be higher and persist during a longer period of time. As a result, such compounds would tend to have a relatively low BMD value as compared to the ranked BMCs from the in vitro assay. Therefore, the corresponding BMDs would have been higher if the compounds had a half-life and absorption comparable to the other compounds. The latter is illustrated in Figure 9.5 where the arrows are located at observed BMDs for compounds with a relatively long half-life. In this way, one might imagine that the correlation between the two endpoints could be improved by taking the kinetic properties of the compounds into account and that the scatter in the plotted BMD versus BMC values could be decreased.

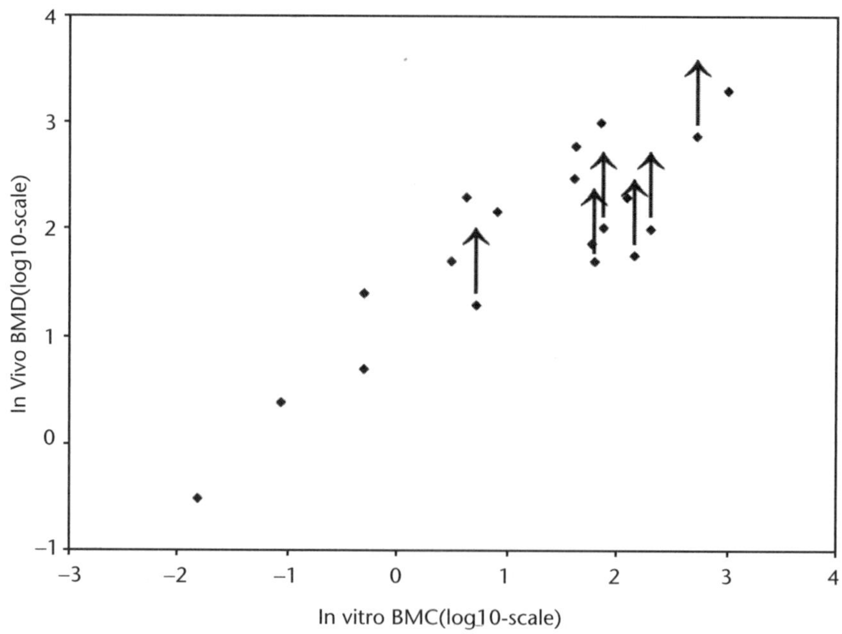

Figure 9.5 In vitro BMC values plotted against BMD values on a double-logarithmic scale with arrows indicating compound with a low elimination rate.

9.3 Discussion and Commentary

We discussed the BMD approach as a quantitative tool in analyzing and correlating in vitro and in vitro test results. This approach does not have some of the limitations of the NOAEL method such as the dependency on the applied doses and the study size. One of the concerns that have been raised against the BMD approach is the selection of the BMR. This has primarily to do with how to define a BMR that is protective for the human situation. However, as the goal is to correlate in vitro results with in vivo data to determine the applicability of an in vitro system, this is of less concern. The aim is to derive equipotent in vitro concentrations and in vivo doses for the embryotoxic potency instead of deriving doses that are relevant for risk assessment. The BMD approach gives the opportunity to reliably compare the potency among different chemicals and to correlate in vitro BMCs to in vivo BMDs to investigate the predictability of in vitro assays.

Some aspects should be kept in mind in using the BMD approach for in vitro–in vivo correlation. BMDs depend on the quality of the in vivo data used. It is difficult to find in vivo studies for all tested compound that fit all the selection criteria as described in Section 9.1. While comparing the in vitro results with the in vivo BMDs, possible differences in the experimental conditions of the in vivo studies, such as the animal species used, and the effect that this may have on the BMDs should be taken into account. Even if all the selected in vivo studies have exactly the same study protocol, intralaboratory variation may still lead to differences in the BMD values. It should be noted that the issue of variability among studies also applies to NOAELs, while, as illustrated in Figure 9.3, NOAELs are even sensitive to the study design (i.e., applied doses and group sizes). Another aspect that should be considered in BMD derivation for developmental toxicity is the issue of maternal toxicity. Some compounds (e.g., butoxyethanol) induce malformations but only at a maternally toxic dose [19]. Maternal toxicity and developmental effects cannot always be measured independently. The observed developmental effect can either be caused by direct action from the compound itself or be secondary to the maternal toxicity. This hampers the comparison between the in vitro and in vivo potency of compounds. If the developmentally toxic effect occurs as a consequence of maternal toxicity the BMD derived for that compound could be underestimated compared to the in vitro BMCs as maternal toxicity does not play a role in the in vitro assays.

Besides the concerns that can be raised about the in vivo data, limitations in the in vitro model should be taken into account as well. First, the issue of free versus nominal concentrations of compounds should be taken into account. Usually, the nominal concentrations, based on the amounts directly added in the in vitro system, are reported, for being the fastest and cheapest method. The concentrations can in reality change, for example, due to the evaporation from the medium or by binding to proteins in the medium. If, for example, the free concentration in the medium is lower than the nominal concentration, the BMC for that compound will be relatively high. A solution for this problem would be to measure the free concentrations of the chemicals, for example, by using solid phase microextraction (SPME) methods such as high performance liquid chromatography (HPLC). This would increase the reliability of the correlation between the BMCs and the BMDs. Second, in vitro methods in general are often simple in comparison to the complexity of the in vivo system. As discussed in Section 9.3, knowledge on kinetic properties of compounds may enhance the comparison between BMCs and BMDs. An example of how data on the kinetic properties of compounds can be incorporated into the in vitro–in vivo correlation is with the use of physiologically based kinetic (PBK) modeling. A PBK model consists of compartments that correspond to the different tissues of the body which are connected by the circulating blood system. The model describes the pharmacokinetic processes of compounds, including ADME properties. It provides the opportunity to calculate in vivo doses corresponding to a certain internal concentration. Verwei et al. [20] used PBK modeling to extrapolate in vitro concentration to in vivo doses for five compounds tested in the EST. They were able to adequately predict in vivo effect levels for four out of five of the tested compounds by using the PBK model and the in vitro effect levels.

In short, the BMD approach is the appropriate method for correlating an in vitro system to in vivo data in determining the predictability of an assay. It has clear advantages above the classical NOAEL method and leads to more precise estimates of the developmental toxic potencies of compounds. Issues such as differences in kinetics properties of compounds do influence the correlation and should be taken into account.

References

[1] Van der Jagt, K., et al., *Alternative Approaches Can Reduce the Use of Test Animals Under REACH*, Report EUR 21405, 2004.

[2] Genschow, E., et al., "The ECVAM International Validation Study on In Vitro Embryotoxicity Tests: Results of the Definitive Phase and Evaluation of Prediction Models," *European Centre for the Validation of Alternative Methods, Altern. Lab. Anim.*, Vol. 30, 2002, pp. 151–176.

[3] Spielmann, H., et al., "Validation of the Rat Limb Bud Micromass Test in the International ECVAM Validation Study on Three In Vitro Embryotoxicity Tests," *Altern. Lab. Anim.*, Vol. 32, 2004, pp. 245–274.

[4] Piersma, A. H., et al., "Validation of the Postimplantation Rat Whole-Embryo Culture Test in the International ECVAM Validation Study on Three In Vitro Embryotoxicity Tests," *Altern. Lab. Anim.*, Vol. 32, 2004, pp. 275–307.

[5] Dekkers, S., C. de Heer, and M. A. Rennen, "Critical Effect Sizes in Toxicological Risk Assessment: A Comprehensive and Critical Evaluation," *Environ. Toxicol. Pharmacol.*, Vol. 10, 2001, pp. 33–52.

[6] Crump, K. S., "A New Method for Determining Allowable Daily Intakes," *Fundam. Appl. Toxicol.*, Vol. 4, 1984, pp. 854–871.

[7] Ema, M., E. Miyawaki, and K. Kawashima, "Developmental Effects of Plasticizer Butyl Benzyl Phthalate After a Single Administration in Rats," *J. Appl. Toxicol.*, Vol. 19, 1999, pp. 357–365.

[8] Hurtt, M. E., G. D. Cappon, and A. Browning, "Proposal for a Tiered Approach to Developmental Toxicity Testing for Veterinary Pharmaceutical Products for Food-Producing Animals," *Food Chem. Toxicol.*, Vol. 41, 2003, pp. 611–619.

[9] EFSA, "Guidance of the Scientific Committee on a Request from EFSA on the Use of the Benchmark Dose Approach in Risk Assessment," *The EFSA Journal*, Vol. 1150, 2009, pp. 1–72.

[10] Sand, S., A. F. Filipsson, and K. Victorin, "Evaluation of the Benchmark Dose Method for Dichotomous Data: Model Dependence and Model Selection," *Regul. Toxicol. Pharmacol.*, Vol. 36, 2002, pp. 184–197.

[11] de Jong, E., et al., "Relative Developmental Toxicity of Glycol Ether Alkoxy Acid Metabolites in the Embryonic Stem Cell Test as Compared with the In Vivo Potency of Their Parent Compounds," *Toxicol. Sci.*, Vol. 110, 2009, pp. 117–124.

[12] Janer, G., et al., "Use of the Rat Postimplantation Embryo Culture to Assess the Embryotoxic Potency Within a Chemical Category and to Identify Toxic Metabolites," *Toxicol. In Vitro*, Vol. 22, 2008, pp. 1797–1805.

[13] Slob, W., and M. N. Pieters, "A Probabilistic Approach for Deriving Acceptable Human Intake Limits and Human Health Risks from Toxicological Studies: General Framework," *Risk Anal.*, Vol. 18, 1998, pp. 787–798.

[14] Slob, W., "Dose-Response Modeling of Continuous Endpoints," *Toxicol. Sci.*, Vol. 66, 2002, pp. 298–312.

[15] Sand, S., et al., "Identification of a Critical Dose Level for Risk Assessment: Developments in Benchmark Dose Analysis of Continuous Endpoint," *Toxicol. Sci.* Vol. 90, 2006, pp. 241–251.

[16] Bokkers, B. G. and W. Slob, "A Comparison of Ratio Distributions Based on the NOAEL and the Benchmark Approach for Subchronic-to-Chronic Extrapolation," *Toxicol. Sci.*, Vol. 85, 2005, pp. 1033–1040.

[17] Wheeler, M. W., and A. J. Bailer, "Model Averaging Software for Dichotomous Dose Response Risk Estimation," *J. Stat. Soft.,* Vol. 26, 2008, pp. 1–15.

[18] Janer et al., 2008.

[19] Wier, P. J., S. C. Lewis, and K. A. Traul, "A Comparison of Developmental Toxicity Evident at Term to Postnatal Growth and Survival Using Ethylene Glycol Monoethyl Ether, Ethylene Glycol Monobutyl Ether and Ethanol," *Teratog. Carcinog. Mutagen,* Vol. 7, 1987, pp. 55–64.

[20] Verwei, M., et al., "Prediction of In Vivo Embryotoxic Effect Levels with a Combination of In Vitro Studies and PBPK Modeling," *Toxicol. Lett.,* Vol. 165, 2006, pp. 79–87.

CHAPTER

10

Three-Dimensional Cell Culture of Canine Uterine Glands

Katharina Stadler, Cordula Bartel, and Ingrid Walter

Department of Pathobiology, University of Veterinary Medicine, Vienna, Austria
Corresponding author: Ingrid Walter, address: Institute of Micro- and Macroanatomy, Department of Pathobiology, University of Veterinary Medicine, Veterinaerplatz 1, 1210 Vienna, Austria
e-mail: Ingrid.Walter@vu-wien.ac.at, phone: 43 1 25077 3406, fax: 43 1 25077 3490

Abstract

We present a method to isolate complete canine uterine glands and keep them differentiated in a Matrigel-based three-dimensional cell culture system. To imitate the complexity of the in vivo endometrium as much as possible, endometrial stromal cells are cocultured together with the uterine glands. In this three-dimensional (3D) system the uterine glands keep their morphology and characteristics such as cytokeratin expression, steroid hormone receptors, and lectin binding patterns. We are convinced that this organotypic culture system of the canine endometrium is suitable for experimental approaches to study physiological and pathological processes on the molecular level.

Key terms
dog
endometrium
extracellular matrix
three-dimensional cell culture
uterine glands

10.1 Introduction

The endometrium is one of few organs that are subjected to recurrent extensive growth and tissue remodeling processes during hormonal driven cycles. The different cell types situated in the endometrium such as surface epithelial cells, glandular epithelial cells, and stromal cells and endothelial cells express steroid receptors to response to these hormonal signals [1, 2]. Therefore, they can be individually influenced by steroid hormones leading to significant cyclic morphological alterations that also include extracellular matrix components. Beside these striking physiological remodeling processes, the canine endometrium and in particular the uterine glands are affected by several, even life-threatening, pathological conditions with a mostly unknown etiology. To study these physiological and pathological events on the molecular level in the canine endometrium, we were in need of an in vitro model, as experiments on living dogs are problematic due to ethical and practical reasons. The living animal is so complex that monitoring the effect of individual factors on specific cells is impractical. A monolayer culture in contrast is not sufficient to study the complex molecular events in the endometrium as the cells tend to dedifferentiate and not to maintain their in vivo capabilities [3].

Attempts have been made to develop 3D cell culture models to close this gap. Three-dimensional systems of the human endometrium have been constructed with polarized epithelial cells, separated by the basement matrix from underlying stromal cells [4, 5], or as outgrowths from organ explants [6]. Cell culture models for the bovine endometrium were developed using multicellular spheroid cultures comprising epithelial and stromal cells [7]. These systems are definitely superior to monolayer cultures and are able to imitate some physiological processes. However, one of the most important structures that is subjected to remodeling processes during physiological and pathological processes in the endometrium—the uterine gland—is not present in these 3D systems. Three-dimensional glandular culture systems have been developed mainly for breast tissue to study the alterations during cancer development [8]. Mostly, isolated epithelial cells were seeded into suitable ECM components that supported the secondary lumen formation of cell aggregates [8–11]. In our experimental attempt we aspired to conserve the original uterine gland structure and keep it differentiated in the 3D cell culture system. To obtain an in vitro system with fully differentiated cells, we designed a method to cultivate canine uterine glands surrounded by stromal cells imitating the in vivo endometrium. The stromal-epithelial interactions were studied before in cocultures of epithelial cells or tumor cells and stromal cells in 3D culture systems [12, 13]. The interactions are based on the production of soluble factors, signaling by direct cell contact and cell-matrix interactions [14], and regulate growth, differentiation, and responsiveness of epithelial cells to steroid hormones [15]. To maintain cell differentiation, the presence of a suitable extracellular matrix (e.g., basement membrane components such as laminin, collagen type IV, fibronectin) that provides a specific microenvironment is essential [8, 16]. Important signals are given from the surrounding stromal cells (ECM, soluble factors) to the epithelial cells in healthy and pathologically altered tissues [17, 18]. These signals are assumed to be responsible for gene expression determining the cell phenotype [19]. Therefore, we tested a series of commercially available extracellular matrices (collagen IV, laminin, fibronectin, Matrigel) for their adequacy to support normal uterine gland differentiation [20]. Cultivation of uterine glands in Matrigel was most successful conserving the characteristics of glandular epithelial cells such as polarization, secretion, and expression of steroid hormone receptors. A main focus is given to the expression of estrogen and progesterone receptors in the cultivated glands to guarantee their reactivity to hormonal stimulation. The differentiation level and morphological condition of the cells during the culture time was controlled by methods of light and electron microscopy.

10.2 Materials

Canine uteri are obtained from routine castration surgery. The dissected uteri are transported to the laboratory in cool, sterile 500-ml PBS supplemented with 2.5-ml Gentamicin and 7.5-ml Nystatin.

10.2.1 Cell culture

All cell culture plasticware and other tools have to be sterile.

- Sterile working bench (Gelaire Flow Labaratories)
- Phase contrast light microscope (Leitz Wetzlar, Germany)
- Incubator (37°C, 5% CO_2; Forma Scientific)
- Centrifuge (Eppendorf)
- Pipette tips, 10 μL, 100 mL, 1,000 mL (Sarstedt)
- Pipettes, 10 mL, 25 mL, 50 mL (Sarstedt)
- Surgical blades (Swann-Morton)
- Forceps
- Glass beaker, 50 mL
- Magnetic stirrer
- Centrifuge tubes, 15 mL (Sarstedt)
- Petri dishes, size 35 mm × 10 mm (Sarstedt)
- Petri dishes, size 100 mm × 20 mm (Falcon)
- Cellector Complete, 85 ml, stainless steel, pore size 280 μm (Bellco Glass)
- Cell sieve, pore size 40 μm (Falcon)
- Collagenase type I (Sigma) 1 mg/mL
- Cell culture plates (6er, 24er, Becton Dickinson)
- Matrigel Inserts/Wells (Becton Dickinson)
- DMEM (Sigma)
- Fetal calf serum FCS (Gibco)
- Pen/strep stock solution (Gibco)
- Nystatin stock solution (Sigma)
- Bottle filter system, pore size 0.22 μm Cellulose Nitrate Filter (Corning Costar)

10.2.2 Histological preparation for light microscopy

- Fume hood
- Formaldehyde, 4% (Pharmacy)

- Embedding equipment (Sanova)
- Biopsy embedding cassettes (Sanova)
- Histocomp (Vogel)
- Metal molds (Sanova)
- Mircotome blades (Feather)
- Water bath (Medax)
- Glass slides (Stölzle—Oberglas)
- Mounting Medium (Fluka)
- Coverslips (Merck)

10.2.3 Histological preparation for electron microscopy

- Fume hood
- Glutardialdehyde 3% (Sigma)
- Soerenson buffer pH 7.4; 0.1M
- Propylenoxid (Sigma)
- Epon (Serva)
- Plastic molds (Gröpl)
- Oven (Heraeus)
- Ultramicrotome (Ultracut S, Leica)
- Diamond knife (Gröpl)
- Copper grids (Gröpl)
- Transmission electron microscope (Zeiss, EM 900)
- Negative films (Kodak)

10.3 Methods

10.3.1 Cell culture

10.3.1.1 Preparation of washing solution and media

1. For the preparation of a washing solution, mix 2.5-mL Gentamicin, 7.5-mL Nystatin with sterile 500-mL PBS, and pH 7.4, at room temperature. Be aware that Nystatin does not dissolve completely; ignore the remnants of crystals that will be visible.
2. Warm all components for the culture medium (DMEM, FCS, Pen-Strep, Nystatin) to 37°C in a sterile water bath.
3. Pipette 176-mL DMEM into a sterile glass beaker, add 20-mL FCS, 2-mL Pen-Strep, 2-mL N-Glutamin, and 2-mL Nystatin. Assemble a bottle filter system, connect to the vacuum pump, and filter the supplemented medium over the filter membrane into the flask. After finishing, close the flask with the sterile screwtop.
4. For the tissue disintegration solution, weigh 10-mg collagenase into a sterile beaker and dilute in a 10-mL supplemented medium.

10.3.1.2 Tissue preparation and cultivation

1. Rinse the uterus in PBS supplemented with 0.5% Pen/Strep stock solution and 1.5% Nystatin. Dissect the uterus from all adhering tissue such as fat [Figure 10.1(a, b)]. Cut the uterus in pieces of about 3 mm with sterile scalpel blades and put all the pieces in a Petri dish containing PBS with antibiotics and antimycotics.
2. Move the cut pieces into new Petri dish with fresh PBS/Pen/Strep/Nystatin at least three times and rinse thoroughly to remove as many red blood cells as possible.
3. Put all pieces into glass beaker with prewarmed collagenase I solution add sterile magnetic stirrer and incubate at 37°C and 5% CO_2. Stir the mixture every 20 minutes for 5 minutes and put back in incubator. Tissue integration takes between 3 and 6 hours depending on the tissue structure of the obtained uterus (e.g., density of stromal collagen). Therefore, the declaration of a standard incubation time for the collagenase solution is not possible and has to be identified during the procedure by direct observation. After 3 hours, take about 100 mL of the collagenase-tissue-solution with a blue pipette tip into a Petri dish and evaluate for the presence of isolated uterine glands under a phase contrast light microscope. If you find a sufficient number of free-floating uterine glands within the solution, proceed with the filtering steps. Otherwise, prolong the collagenase incubation time.

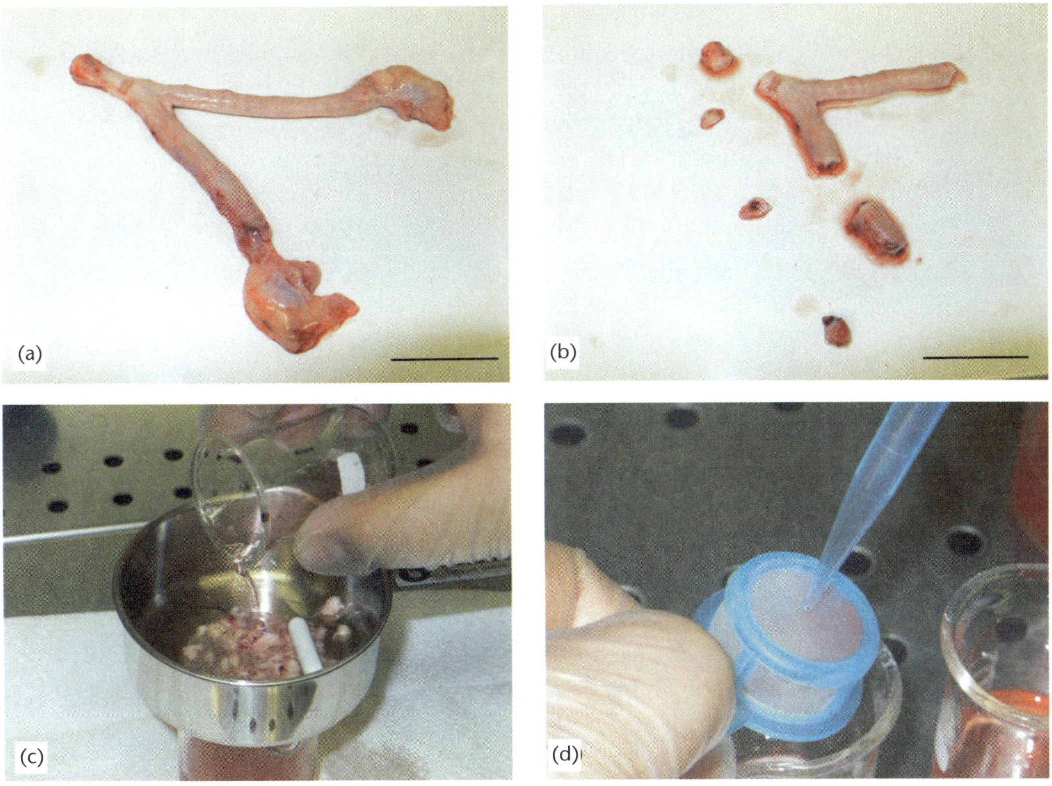

Figure 10.1 (a) Uterus and ovaries as obtained from routine castration freed from surrounding fat tissue. (b) Before cutting the organ into small pieces for cell culture, take one piece of the uterus close to the ovary, middle of the horn, and bifurcation for histological preparation. Scale bar = 4 cm. (c) Filtering of the disintegrated tissue in the collagenase solution to remove large tissue pieces. (d) Washing of the uterine gland fraction from the 40-μm cell filter.

4. When the disintegration is completed, prepare Matrigel cell culture plates by loading with DMEM and put in incubator for the equilibration of temperature and pH.
5. Filter the collagenase-solution through the stainless steel filter (pore size 280 μm) to remove tissue pieces that are not disintegrated [Figure 10.1(b)]. Discard the tissue pieces on the mesh and proceed with the filtered solution, which contains now epithelial cell sheets, uterine glands, and stromal cells.
6. Remove stromal cells filter through a cell filter (pore size 40 μm) into a sterile Petri dish. Isolated stromal cells will pass through the filter, whereas uterine glands will stay on the filter mesh. The filter mesh will be clogged easily; do not overload.
7. Flip the cell filter and rinse the uterine glands from the filter mesh surface [Figure 10.1(d)] into a beaker or a Petri dish with the supplemented DMEM (isolated glands) (see Figure 10.2).
8. Pool the stromal cell fraction and the uterine gland fraction into separate 10-mL tubes and centrifuge for 1 minute at 2,000 rpm. Dilute and resolve each pellet in 1-mL DMEM and aliquot cell suspensions to the cell culture dish as needed.
9. Change the medium the next day. Take care not to aspire the Matrigel membrane. Afterwards the medium has to be changed every second day.

10.3.2 Histological preparation for light microscopy

10.3.2.1 Fixation

1. To harvest cultured uterine glands for histological procedures, take the cell culture plates from the incubator and remove DMEM from the inserts carefully. Be sure not to aspirate the Matrigel with the uterine glands.
2. Wash the cell culture with PBS once and fix uterine glands by pipetting the fixation solution (4% formaldehyde) into the culture dish wells or inserts. Leave the fixative for 15 to 20 minutes at room temperature. Warning: Formaldehyde is toxic. Use the fume hood to handle this substance.

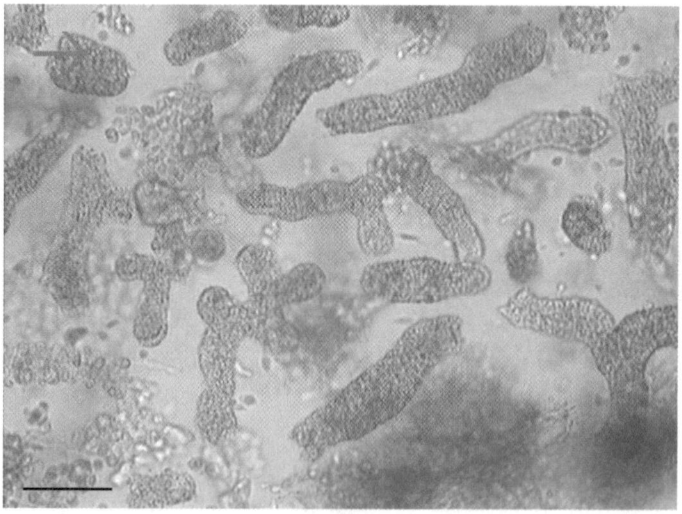

Figure 10.2 Phase contrast micrograph showing freshly isolated canine uterine glands. Note the characteristic tubular and branched morphology. Scale bar = 50 μm. (*From:* [20]. © 2009 Society for In Vitro Biology, formerly the Tissue Culture Association. Reprinted with permission.)

3. Remove the fixation solution and wash uterine glands with sterile distilled water.
4. Cut the Matrigel membrane from the insert using a surgical blade.

10.3.2.2 Embedding and cutting

1. Put the Matrigel membrane containing uterine glands into the biopsy embedding cassette.
2. Load the embedding machine with filled embedding cassettes and start the overnight paraffin embedding program.
3. The next day remove the cassettes from the embedding device and cast the tissue with metal molds. Let it solidify on cold plate.
4. Cut the cooled paraffin block with microtome and put the sections (3-μm thickness) in a water bath (35°C). Stick the section on coated glass slides and let it dry overnight at 37°C. The tissue sections are now ready for the staining procedures (H&E, immunohistochemistry, lectin histochemistry).

10.3.3 Histological preparation for electron microscopy

10.3.3.1 Fixation

1. Harvest the cultured uterine glands as stated in 10.3.2.1.
2. Wash the cell culture with the Soerensen buffer twice and fix the uterine glands by pipetting a cooled (4°C) fixation solution (3% glutaraldehyde) into the culture dish wells. Leave the fixative for at least 24 hours at 4°C. Warning: Glutaraldehyde is toxic. Use the fume hood to handle this substance.
3. Remove the fixation solution and wash the uterine glands three times with the Soerensen buffer. Post fixation with osmium (1%, 0.1M). Warning: Substance is toxic. Use the fume hood and gloves while handling.
4. Rinse with the Soerenson buffer three times for 10 minutes.
5. Cut the Matrigel membrane from the insert using a surgical blade.

10.3.3.2 Embedding and cutting

1. Put the Matrigel membrane containing uterine glands into a glass snap cap vial (5 ml, 40 × 19 mm).
2. Use a series of increasing alcohols (two times for 5 minutes 50%, two times for 60 minutes 70%, two times for 30 minutes 96%, two times for 30 minutes 100%) at room temperature to dehydrate the specimens.
3. Remove the 100% alcohol, add pure propylene oxide for 20 minutes, and then substitute by propylene oxide:Epon (ratio 1:1) for 1 hour. Warning: This is toxic. Replace the solution with propylene oyxde:Epon (ratio 1:3) overnight.
4. Remove the propylene oxide and substitute with pure Epon resin, put the samples into plastic molds, and orientate for the requested sectioning direction. Leave under the fume hood for 4 hours.
5. Epon blocks are incubated for 48 hours at 60°C for polymerization.
6. Sectioning is done on an ultramicrotome by using a diamond knife. The 60-nm-thick sections are put on copper grids and air dried.

7. Sections mounted on grids are contrasted by uranyl acetate and lead citrate rather then air dried. It is ready for evaluation in the transmission electron microscope.

10.3.4 Imaging

1. To obtain phase contrast microscopic images, we use a light microscope equipped with a Sony camera and AnalySiS software.
2. For images of histological sections, a Polyvar light microscope equipped with a digital camera combined with imaging software (e.g., Eclipse Net) is used. Images can be achieved in various established formats (jpeg, tiff, bmp, and so forth).
3. Electron micrographs are made on Kodak films and developed on photo paper.

10.4 Anticipated Results

The use of explanted uterine glands cultured in a suitable extracellular matrix together with stromal cells provides a complex model of the canine endometrium. The cells, as cultured on Matrigel-coated inserts, can easily be fixed, embedded, and processed as described for light and electron microscopic examinations. All staining techniques can be applied on paraffin or frozen sections; however, whole-culture stainings [8] should be possible. When epithelial cell sheets originating from the endometrial surface or from destroyed glands are present in the culture system, we observed that they form aggregates that develop a secondary lumen [20]. The contamination of the culture with blood vessels is not completely avoidable, however; those are identifiable by morphology and disintegrate within 1 day.

10.5 Discussion and Commentary

There is a need for 3D cell culture systems that reflect the organization and complexity of an organ to allow experimental intervention and therefore to avoid experiments on living animals. In case of the canine endometrium, experimental approaches are necessary to study the physiological endometrial biology (cyclic differentiation and remodeling processes) as well as pathological events such as pyometra and cystic endometrial hyperplasia at the molecular level.

The way of culturing canine endometrial cells in a 3D model highlighted in this chapter has been proven to result in differentiated, growth-arrested uterine glands [20]. We were able to show that the glandular epithelial cells keep their expression of cytokeratin (Figure 10.3) and showed in vivo–like morphological characteristics as demonstrated by transmission electron microscopy [20]. Lectin histochemistry of cultured glands verified a comparable carbohydrate pattern in the glycocalyx and secretions with the in vivo situation.

Cells of organs involved in reproduction such as the mammary gland and the endometrium depend on hormonal signals. Estrogen and progesterone receptors have been reported as markers for endometrial or mammary gland differentiation [21, 22]. A prerequisite for the applicability of a 3D model of the endometrium is the cellular responsiveness to steroid hormone signals via their receptors [23]. The persuability of

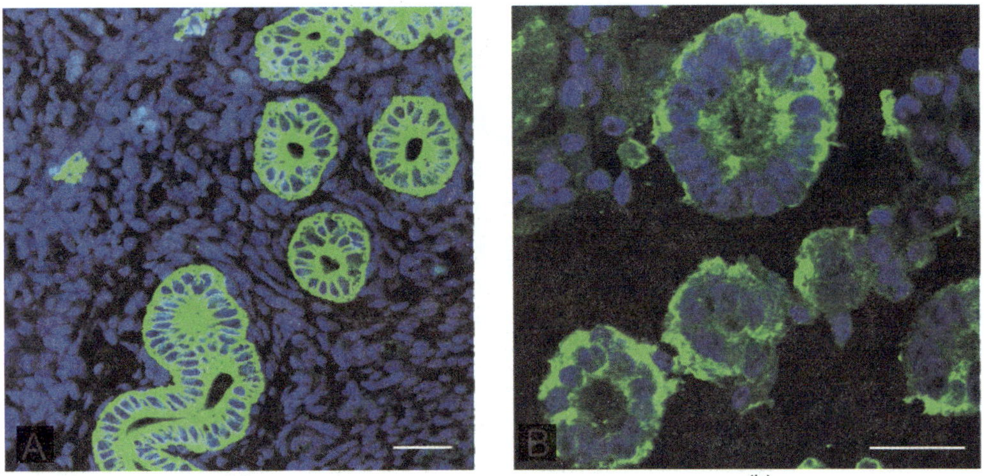

(a) (b)

Figure 10.3 Confocal laser scanning images. Immunofluorescence for cytokeratin of (a) canine in vivo endometrium compared to (b) isolated uterine explants. Scale bar = 20 μm. (*From:* [20]. © 2009 Society for In Vitro Biology, formerly the Tissue Culture Association. Reprinted with permission.)

cultured canine endometrial cells by steroid hormones has been proven for monolayer cultures [24]; however, the effects on the uterine glands can not be studied in such a 2D model. Although both epithelial cells and stromal cells express estrogen and progesterone receptors, stromal cell showed a higher responsiveness to steroid treatment [24]. The importance of stromal cells as mediators and signal transducers for the epithelial cells has been supposed [1, 2, 5]. Therefore, in our opinion, the addition of stromal cells to the uterine gland culture is essential to imitate the in vivo situation.

ECM-mediated signals are essential for epithelial differentiation far beyond the physical support of the cellular layer [8, 11, 21]. We have tested several different ECMs in 3D culture and found that the differences were minimal [20]. Although Matrigel gave the best support of glandular morphology, the type of matrix is probably not so important when the original basement membrane is preserved as in our culture system (Figure 10.4).

A limitation of the presented canine 3D endometrium model is caused by the variation of the endometrial tissues obtained from animals, even if clinically classified in the same estrous cycle stage. Therefore, it is recommended to control cycle stage and morphology by histological examination of each uterus. This morphological variability is also the reason for the observed needed differences in the duration of the collagenase digestion procedure. Optionally, a piece of the uterus could be snap frozen and a cryosection made for rapid diagnosis of the endometrial condition. Due to the relatively small size of the canine uterus, the number of cell cultures that can be established from one animal is limited. An option to overcome the limited material quantity is the isolation of epithelial and stromal cell fractions, expanding them in vitro and seeding them into Matrigel. We observed that aggregates of epithelial cells differentiate and develop a secondary lumen lined by cytokeratin expressing cells [20]. We have not tested these secondary glands in detail until now, but we assume that they will be keepable in culture for a prolonged time period. An improvement of the system by the addition of growth factors with the aim to reach long-term cell and tissue viability of complete uterine

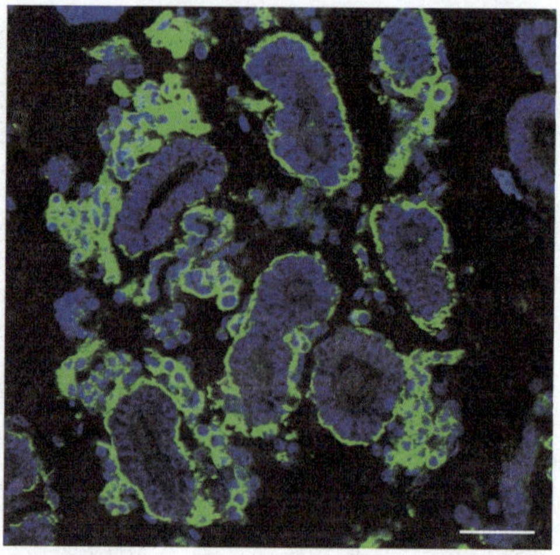

Figure 10.4 Immunohistochemical demonstration of laminin shows the preservation of the basement membrane in a freshly isolated glandular fraction. Scale bar = 20 μm. (*From:* [20]. © 2009 Society for In Vitro Biology, formerly the Tissue Culture Association. Reprinted with permission.)

glands will be studied in the near future. Moreover, a way to quantitate the glandular fraction has to be determined to better standardize culture conditions.

Taking the characteristics and markers of the cultured endometrial cells together, we are convinced that this 3D model is promising and applicable to experimental studies. Moreover, the method should be transferable to other species and most probably to other organs as well.

10.6 Application Notes

The 3D cell culture model of the canine endometrium as outlined in this chapter represents a useful tool for experimental studies of individual factors on the molecular level. In response to estrogen and progesterone, this organ is extremely variable concerning the surface epithelium, glandular epithelium, glandular morphology, stromal conditions, and more. The canine estrous cycle has some peculiarities such as the progesterone production of the preovulatory luteinized ovarial follicles that are reflected by the endometrial morphology [25]. Moreover, several pathologic conditions such as cystic endometrial hyperplasia are assumed to at least involve alterations of steroid receptor expression and distribution [26]. Therefore, a functional canine endometrial model has to ensure the presence of steroid hormone receptors as in our canine 3D system, indicating a susceptibility of cultivated glands and stromal cells for steroid treatment. With this model it is possible to directly assess the effect of various steroid hormone concentrations on the proliferation, secretory activity, and apoptotic events in the uterine glands. Moreover, the role of the stromal cells might be elucidated when culturing the glands with and without the surrounding fibroblasts. Information gained during the establishment of this 3D culture system can be assigned to other organotypic in vitro models helping to avoid animal experiments as much as possible.

Troubleshooting Table

Problem	Explanation	Potential solutions
Tissue does not react to collagenase digestion.	Wrong concentration, temperature, or duration.	Control collagenase concentration and temperature; try a longer incubation with collagenase.
Tissue including glands is disintegrated.	Incubation with collagenase was too long.	Try a shorter incubation time, and make light microscopic control by taking a sample of the solution.
Cell sieve is clogged.	Filter is overloaded.	Use a fresh filter.
Uterine glands are short and small.	Depends on condition of the original uterine tissue.	Compare with a histologic sample of the respective uterus.
Contamination of culture with blood vessels.	Blood vessels are released from the uterine stroma and are retained by the cell sieve.	Blood vessels disintegrated rapidly in standard cell culture medium and are distinguishable easily by morphology.
Contamination with erythrocytes.	It is not possible to remove all erythrocytes during the washing steps.	Remaining erythrocytes diminish within 1 day from the culture.
Uterine glands in culture are not surrounded by Matrigel.	Matrigel was not allowed to undergo moisture expansion.	Prolong equilibration time.

10.7 Summary Points

1. We have developed a 3D cell culture model of the canine endometrium using total uterine glands obtained from routine castration surgery.
2. Uterine glands together with stromal cells cultivated on Matrigel keep their morphology and differentiation.
3. The cultivated cells express steroid hormone receptors and can therefore be used for experimental approaches.
4. After the experiment the organotypic cell culture can be fixed and processed for light and electron microscopic histology.

References

[1] Vermeirsch, H., et al., "Immunohistochemical Detection of Estrogen Receptors in the Canine Uterus and Their Relation to Sex Steroid Hormone Levels," *Theriogenology*, Vol. 51, 1999, pp. 729–743.

[2] Vermeirsch, H., et al., "Immunohistochemical Detection of Progesterone Receptors in the Canine Uterus and Their Relation to Sex Steroid Hormone Levels," *Theriogenology*, Vol. 53, 2000, pp. 773–788.

[3] Bissell, M. J., "Modelling Molecular Mechanisms of Breast Cancer and Invasion: Lessons Learned from the Normal Gland," *Biochem. Soc. Trans.*, Vol. 35, Pt. 1, 2007, pp. 18–22.

[4] Bentin-Ley, U., et al., "Isolation and Culture of Human Endometrial Cells in a Three-Dimensional Culture System," *J. Reprod. Fertil.*, Vol. 101, 1994, pp. 327–332.

[5] Arnold, T. J., et al., "Endometrial Stromal Cells Regulate Epithelial Cell Growth In Vitro: A New Co-Culture Model," *Human Reprod.*, Vol. 16, No. 5, 2001, pp. 836–845.

[6] Esfandiari, N., Z. Nazemian, and R. F. Casper, "Three-Dimensional Culture of Endometrial Cells: An In Vitro Model of Endometriosis," *Am. J. Reprod. Immunol.*, Vol. 60, 2008, pp. 283–289.

[7] Yamauchi, N. et al., "A Three-Dimensional Cell Culture Model for Bovine Endometrium: Regeneration of a Multicellular Spheroid Using Ascorbate," *Placenta*, Vol. 24, 2003, pp. 258–269.

[8] Lee, G. Y., et al., "Three-Dimensional Culture Models of Normal and Malignant Breast Epithelial Cells," *Nat. Methods*, Vol. 4, No. 4, 2007, pp. 359–435.

[9] Debnath, J., and J. S. Brugge, "Modelling Glandular Epithelial Cancers in Three-Dimensional Cultures," *Nat. Rev. Cancer*, Vol. 5, No. 9, 2005, pp. 675–688.

[10] Krause, S., et al., "A Novel 3D In Vitro Culture Model to Study Stromal-Epithelial Interactions in the Mammary Gland," *Tissue Eng. Part C Methods*, Vol. 14, No. 3, 2008, pp. 261–271.

[11] Johnson, K. R., J. L. Leight, and V. M. Weaver, "Demystifying the Effects of a Three-Dimensional Microenvironment in Tissue Morphogenesis," *Methods Cell Biol.*, Vol. 83, 2007, pp. 547–583.

[12] Pierro, E., et al., "Stromal-Epithelial Interactions Modulate Estrogen Responsiveness in Normal Human Endometrium," *Biol. Reprod.*, Vol. 64, 2001, pp. 831–838.

[13] Sadlonova, A., et al., "Breast Fibroblasts Modulate Epithelial Cell Proliferation in Three-Dimensional In Vitro Co-Culture," *Breast Cancer Res.*, Vol. 7, No. 1, 2005, pp. R46–R59.

[14] Skibinski, G., J. S. Elborn, and M. Ennis, "Bronchial Epithelial Cell Growth Regulation in Fibroblast Co-Cultures: The Role of Hepatocyte Growth Factor," *Am. J. Physiol. Lung Cell Mol. Physiol.*, Vol. 293, 2007, pp. L69–L76.

[15] Arnold, J. T., et al., "Endometrial Stromal Cells Regulate Epithelial Cell Growth In Vitro: A New Co-Culture Model," *Human Reprod.*, Vol. 16, No. 5, 2001, pp. 836–845.

[16] Benton, G., and J. G. George, "Defining 3D Culture for Investigating Breast Cancer Progression," *Bio. Sci. Techol.*, Vol. 1, 2005, pp. 50–52.

[17] Bissell, M. J., et al., "The Organizing Principle: Microenvironmental Influences in the Normal and Malignant Breast," *Differentiation*, Vol. 70, 2002, pp. 537–546.

[18] Rønnov-Jessen, L., and M. J. Bissell, "Breast Cancer by Proxy: Can the Microenvironment Be Both the Cause and the Consequence?" *Trends Mol. Med.*, Vol. 15, No. 1, 2008, pp. 5–13.

[19] Ghajar, C. M., and M. J. Bissell, "Extracellular Matrix Control of Mammary Gland Morphogenesis and Tumorigenesis: Insights from Imaging," *Histochem. Cell Biol.*, Vol. 130, 2008, pp. 1150–2118.

[20] Stadler, K., et al., "A Three-Dimensional Culture Model of Canine Uterine Glands," *In Vitro Cell Dev. Biol. Anim.*, Vol. 45, 2009, pp. 35–43.

[21] Roskelley, C. D., A. Srebrow, and M. J. Bissell, "A Hierarchy of ECM-Mediated Signalling Regulates Tissue-Specific Gene Expression," *Curr. Opin. Cell Biol.*, Vol. 7, 1995, pp. 736–747.

[22] Classen-Linke, I., et al., "Establishment of a Human Endometrial Cell Culture System and Characterization of Its Polarized Hormone Responsive Epithelial Cells," *Cell Tissue Res.*, Vol. 287, 1997, pp. 171–185.

[23] Bläuer, M., et al., "Effects of Tamoxifen and Reloxifene on Normal Human Endometrial Cells in an Organotypic In Vitro Model," *Eur. J. Pharmacol.*, Vol. 592, 2008, pp. 13–18.

[24] Galabova-Kovacs, G., et al., "Steroid Receptors in Canine Endometrial Cells Can Be Regulated by Estrogen and Progesterone Under In Vitro Conditions," *Theriogenology*, Vol. 61, 2004, pp. 963–976.

[25] Concannon, P., W. Hansel, and K. McEntee, "Changes in LH, Progesterone and Sexual Behaviour Associated with Preovulatory Luteinization in the Bitch," *Biol. Reprod.*, Vol. 17: 1977, pp. 604–613.

[26] De Bosschere, H., R. Ducatelle, and M. Thsamala, "Uterine Oestrogen and Progesterone Receptor Expression in the Experimental Pyometra in the Bitch," *J. Comp. Pathol.*, Vol. 128, 2003, pp. 99–106.

CHAPTER

11

Markers for an In Vitro Skin Substitute

Danielle Larouche,[1,2] Jessica Jean,[1,3] François Berthod,[1,2] Lucie Germain,[1,2] and Roxane Pouliot[1,3]

[1]Laboratoire d'Organogenèse Expérimentale (LOEX), Centre de recherche FRSQ du CHA universitaire de Québec, Hôpital du Saint-Sacrement, 1050, chemin Sainte-Foy, Québec, Canada, [2]Département de Chirurgie, Faculté de Médecine, [3]Faculté de Pharmacie, Université Laval, Québec, Québec, Canada. Corresponding author: Roxane Pouliot, address: Laboratoire d'Organogenèse Expérimentale (LOEX), Centre de recherche FRSQ du CHA, Universitaire de Québec, Hôpital du Saint-Sacrement, 1050, chemin Sainte-Foy, Québec, Canada, G1S 4L8, phone: 418-682-7663, fax: 418-682-8000, e-mail: roxane.pouliot@pha.ulaval.ca

Abstract

The tissue engineering self-assembly approach allows the production of skin substitutes comprising both the dermis and epidermis, using methods promoting the secretion and organization of a dense extracellular matrix by skin cells. In a reconstructed epidermis, all cellular layers of the native tissue are present. An evaluation of the expression and localization of a number of specific protein markers revealed that the self-assembled, tissue-engineered skin substitute shares some common features with normal human skin, such as the expression of Ki-67, keratins 10 and 14, filaggrin, involucrin, transglutaminase, DLK, α3-integrin subunit, laminin-5, and collagens I, II, IV, and VII. At the ultrastructural level, many differentiation markers can be observed, including desmosomes, as well as an organized basement membrane presenting hemidesmosomes, lamina densa, and lamina lucida. In this chapter, protocols to generate skin substitutes by the self-assembly approach will be presented and the methods including the labeling of the principal skin differentiation markers by immunofluorescence will be examined.

Key terms collagens, ibronectin, filaggrin, involucrin, keratins, laminin, Loricrin, markers, skin substitute, transglutaminase

11.1 Introduction

Multiple kinds of artificial skin substitutes are now available. They have several prospective applications including the treatment and closure of skin wounds, models of skin biology and pathology, alternatives to animals for safety testing, and drug delivery [1]. Each of these prospective applications has distinct requirements for validation of skin substitutes. These categories of applications represent, respectively, the fields of surgery, investigative dermatology, toxicology, and pharmacology [1].

The various methods used for the reconstruction of living skin substitutes give different epidermal-cell phenotypes depending on the culture conditions [2]. However, the final goal in the development of alternative in vitro models, for cytotoxicity studies as well as for extensive burn coverage, is to create an in vitro model that enables extrapolation to in vivo results. In this regard, the morphological equivalence of the tissue architecture with native tissue is not a sufficient criterion of the quality of the reconstructed tissue [3]. The expression of various specific protein markers have to be monitored and evaluated as well. An appropriate cultured skin substitute should thus exert morphological, biochemical, and functional features that approach those of the native tissue [4].

The human epidermis is a stratified squamous epithelium characterized by a high keratin (K) content and by the ability to make cornified envelopes. When keratinocytes leave the basal compartment and progress upward in the epidermis, they sequentially undergo several differentiation changes and stop dividing [5, 6]. Four layers can be distinguished morphologically in healthy epidermis: stratum basale, spinosum, granulosum, and corneum. The function of stratum basale is to proliferate and to keep the epidermis firmly anchored to the basement membrane [7]. The function of stratum spinosum is to start the differentiation process by producing new cell-cell attachments, cytoskeleton, and so forth. The function of stratum granulosum is to produce cornified envelope proteins, crosslinking enzymes and appropriate lipids. In the stratum corneum, proteins are crosslinked, lipids are extruded, and the epidermal barrier is formed [7]. The cohesion between the dermis and the epidermis is ensured by the dermo-epidermal junction, which is also required for control of epidermal growth and differentiation [8, 9].

In this chapter, protocols to generate the self-assembled, tissue-engineered skin substitute model [10, 11] will be described in detail and the quality of the skin substitutes will be examined using tissue morphology and the presence and distribution of various specific protein markers. Morphological and ultrastructural markers of epidermal differentiation include individualization of the epidermal layers such as the stratum basale, stratum spinosum, stratum granulosum, and stratum corneum and the presence of desmosomes, hemidesmosomes, and a basement membrane at the dermo-epidermal junction. Proliferation markers such as Ki-67, as well as epidermal differentiation markers like α3-integrin subunit, keratin patterns (K14 and K10), involucrin, transglutaminase, the dual leucine zipper-bearing kinase (DLK), and filaggrin have additional specific biochemical characteristics. Other components of the basement membrane (collagens IV and VII, laminin-5) and of the dermis (collagen I) will also be examined.

11.2 Experimental Design

The concept of the self-assembly approach is to reconstruct an organ in a fashion resembling its formation in vivo, in which the use of appropriate culture and mechanical conditions induces cells to secrete a significant amount of extracellular matrix as during organogenesis [12–14]. The skin substitutes reconstructed in a 3D environment, as those produced by the self-assembly approach presented in this chapter, exhibits a well-developed epidermis that expresses differentiation markers and a well-organized basement membrane [10, 11]. Among the key developments in this procedure were the use of autologous cells and the capacity of mesenchymal cells, such as fibroblasts, to create their own extracellular matrix in vitro [15]. The model supports the hypothesis that keratinocytes and fibroblasts of individuals possess inherent properties of the native tissues as observed with skin substitutes reconstructed from psoriatic cells. The study showed that psoriatic skin reconstructed in vitro partially displays a psoriatic phenotype [16]. The immunohistochemical analyses performed to evaluate the quality of the skin model in vitro require biopsies from the native tissues as relevant controls. To insure reproducibility, at least three subjects per independent trial must be compared and the experiment performed three times.

11.3 Materials

11.3.1 Human tissue-engineered skin substitute reconstructed by the self-assembly approach

- Confluent human fibroblasts between their second and eighth passages (for a detailed protocol concerning the extraction from human skin and the culture of fibroblasts, refer to [17])
- Human keratinocytes at 80% confluence between their third and fifth passages (cocultured with iS3T3 cells) (for a detailed protocol concerning the extraction from human skin and the culture of keratinocytes, refer to [17])
- DMEM (Invitrogen, Burlington, Ontario, Canada)
- Ham's F12 medium (Invitrogen)
- 0.22-μm low-binding disposable filter (Milipore, Billerica, Massachusetts)
- $NaHCO_3$ (Fisher Scientific, Ottawa, Ontario, Canada)
- Adenine (Sigma, St. Louis, Missouri)
- Fetal calf serum (HyClone, Scarborough, Ontario, Canada)
- Fetal clone II serum (HyClone)
- Penicillin G (Sigma)
- Gentamicin (Sigma)
- Insulin (Sigma)
- Hydrocortisone (Calbiochem, Gibbstown, New Jersey)
- Cholera toxin (Sigma)
- Epidermal growth factor (Sigma)
- Ascorbic acid (Sigma) (light-sensitive)
- Ingots (16-G stainless steel grade # 316, Denmar, Quebec, Canada)

- Merocel sponge, Medtronic (Instruments Ophtalmiques INNOVA, Laval, Quebec, Canada)
- Seeding ring (stainless steel grade # 316, Denmar); dimension: 3-cm diameter, 7/8-inch wide, 1/8-inch height, sterilized with autoclave
- Dissecting curved forceps (Fisher Scientific) sterilized with autoclave
- Tissue culture flask, 25 cm^2 (BD Biosciences, Mississauga, Ontario, Canada)
- Cell culture dish (BD Biosciences)
- Petri dish, size: 100 mm × 15 mm (Fisher Scientific)
- Weller Universal Dual Heat Soldering Gun (available in local hardware)
- Anchoring paper: Cut a circle with a 60-mm diameter in a Whatman sheet (Fisher Scientific); remove the concentric inside disk of 25 mm diameter after cutting; and sterilize with autoclave
- Air-liquid stand (sterile homemade stand, 1/16-inch height)
- Standard culture materials (sterile pipettes, droppers, tips, test tubes)
- Sterile laminar flow hood cabinet (class II cabinet)
- Laboratory equipment (pipetman, micropipettes, water bath, autoclave, pH meter, incubator, preparative centrifuge, magnetic stirrer, stir bar, freezer, refrigerator, vaccum)
- Laboratory glassware (graduate cylinder, beaker, pipettes, bottles)
- Immunofluorescence
- Dissecting jeweler microforceps (Fisher Scientific)
- Dissecting curved scissors (Fisher Scientific)
- OCT compound (Miles Inc., Monrovia, California)
- Small container with liquid nitrogen
- Tissue holder
- Forceps (Fisher Scientific)
- Cryostat
- Freezers (–80°C and –20°C)
- Superfrost glass slides (Fisher Scientific)
- Slide warmer (Fisher Scientific)
- Diamond pencil (Fisher Scientific)
- Distilled water (H_2Od)
- $NaH_2PO_4{\cdot}H_2O$ (Fisher Scientific)
- Acetone
- NaCl (Fisher Scientific)
- KCl (Fisher Scientific)
- Na_2HPO_4 (Fisher Scientific)
- KH_2PO_4 (Fisher Scientific)
- $MgCl_2.6H_2O$ 2.8M solution kept at –20°C (Sigma)
- $CaCl_2.2H_2O$ (Sigma)
- Bovine serum albumin (BSA) (Sigma)
- Primary antibodies (see Table 11.1)

Table 11.1 Antibodies

Antibody Name	Antigen	Host	Isotype	Supplier	Working Dilution
Primary Antibody					
Antirecombinant human Ki-67 peptide mAb (B56)	Ki-67	Mouse	IgG1	BD Biosciences	1:400
Anti-K14 C-terminal peptide pAb	Keratin 14	Rabbit		Dr. Normand Marceau, Centre de recherche de l'Hôtel-Dieu de Québec, Université Laval, Quebec, Canada [18]	1:2,000
Anticytokeratin 10 mAb (RKSE60)	Keratin 10	Mouse	IgG1	Cedarlane	1:100
Antifilaggrin mAb	Filaggrin	Mouse	IgG1	Abcam, Cambridge, Massachusetts	1:800
Anti-involucrin mAb	Involucrin	Mouse	IgG1	Sigma	1:400
Antitransglutaminase mAb	Transglutami nase 1	Mouse	IgG2a	Biomedical Technologies Inc., Stoughton, Massachusetts	1:400
Anti-N-terminal portion of recombinant mouse DLK pAb	Dual leucine zipper-bearing kinase (DLK)	Rabbit		Dr. Richard Blouin, Département de Biologie, Université de Sherbrooke, Quebec, Canada [19]	1:400
mAb antidesmosomal protein (ZK-31)	Epitope on desmosome	Mouse	IgG1	Sigma	1:400
Anti-a3-integrin (VM-2) mAb (HB-8530) conjugated to Alexa Fluor 488 (Invitrogen)	a3-integrin subunit	Mouse	IgG1	ATCC, Manassas, Virginia	1:500
Anticollagen type IV pAb	Collagen IV	Rabbit		Abcam	1:100
Anticollagen type VII mAb	Collagen VII	Mouse	IgG	Millipore	1:400
Antilaminin-5 mAb	Laminin 5	Mouse	IgG1	Dr. Patricia Rousselle, Institut de Biologie et Chimie des Protéines, UMR 5086, CNRS, Université Lyon [20]	1:1,000
Anticollagen type I pAb	Human collagen type I	Rabbit	IgG	Cedarlane	1:400
Anticollagen type III mAb	Human collagen type III	Mouse	IgG1	Millipore	1:100
Isotype Control Antibody					
IgG1		Mouse	IgG1	Dako, Mississauga, Ontario, Canada	#
IgG2a		Mouse	IgG2a	Dako	#
IgG1a isotype control conjugated to Alexa Fluor 488		Mouse	IgG1a	Dako	#
Secondary Antibody					
Fluorescein conjugated antimouse IgG/IgM	Mouse IgG-IgM	Goat		Millipore	1:100
FITC congugated antirabbit IgG, H+L	Rabbit IgG, H+L	Goat		Millipore	1:200

mAb: monoclonal antibody, pAb: polyclonal antibody, #: Dilute at the same concentration than the associated primary antibody.

- Secondary antibodies (see Table 11.1)
- Hydrogen peroxide 30% (v/v) (Sigma)
- Microscope slide mailer (Somagen Diagnostics Inc., Edmonton, Alberta, Canada)
- Microscope slide rack (Somagen Diagnostics Inc.)
- NaN_3 (Sigma)
- Gelatin (Fisher Scientific)
- pH paper (EMD Chemicals Inc., Gibbstown, New Jersey)
- Glycerol (MP Biomedicals, Solon, Ohio)
- Absorbent paper

- Cover slips 24 × 50 mm, #1 (Fisher Scientific)
- Hoechst 33258 (cat. no. B2883, Sigma-Aldrich)
- Microscope equipped with epifluorescence

11.4 Methods

11.4.1 Preparation of solutions and materials for the in vitro fabrication of human skin substitutes by the self-assembly approach

- *DMEM-Ham.* DMEM : Ham's F12 medium, 3:1, 3.07 g/L $NaHCO_3$ (36.54 mM), 24.3 mg/L adenine (0.18 mM), 312.5 μL/L 2N HCl. Dissolve in apyrogenic ultrapure water. Adjust pH to 7.1. Sterilize by filtration through a 0.22-μm low-binding disposable filter. Aliquot and store at 4°C.
- *Fetal calf serum and fetal clone II serum.* Thaw in cold water. Inactivate in hot water (56°C) for 30 minutes. Distribute in single-use aliquots and store at –20°C.
- *Insulin.* Dissolve 250 mg in 50-mL 0.005N HCl (125 μL 2N HCl/50 mL apyrogenic ultrapure water) to make a 1,000× stock solution (0.87 mM). Sterilize by filtration through a 0.22-μm low binding disposable filter, distribute in single-use aliquots, and store at –80°C.
- *Hydrocortisone.* Dissolve 25 mg in 5 mL of 95% ethanol (4.8 mL 99% ethanol/0.2 mL apyrogenic ultrapure water). Complete to 125 mL with DMEM-Ham to make a 500× stock solution (0.53 mM). Sterilize by filtration through a 0.22-μm low binding disposable filter, distribute in single-use aliquots, and store at –80°C.
- *Cholera toxin.* Dissolve 1 mg in 1 mL of apyrogenic ultrapure water. Complete to 118.18 mL with DMEM-Ham supplemented with 10% (v/v) fetal clone II to make a 1,000× stock solution (10^{-7} M). Sterilize by filtration through a 0.22-μm low binding disposable filter, distribute in single-use aliquots, and store at –80°C.
- *Epidermal growth factor.* Dissolve 500 μg in 2.5 mL of 10 mM HCl. Complete to 50 mL with DMEM-Ham supplemented with 10% (v/v) fetal clone II to make a 1,000× stock solution. Sterilize by filtration through a 0.22-μm low binding disposable filter, distribute in single-use aliquots, and store at –80°C.
- *Penicillin G and Gentamicin.* Dissolve 50,000 IU/mL of Penicillin G and 12.5 mg/mL of Gentamicin sulfate in apyrogenic ultrapure water to make a 500× stock solution. Sterilize by filtration through a 0.22-μm low-binding disposable filter, distribute in single-use aliquots, and store at –80°C.

Preparation of culture media

- Human fibroblast culture medium (fDMEM). To make 1L, refer to Table 11.2. Thaw all components at 4°C. Store at 4°C for a maximum of 10 days.
- Complete human keratinocyte culture medium (complete hkDMEM-Ham). To make 1L, refer to Table 11.2. Thaw all components at 4°C. Store at 4°C for a maximum of 10 days.
- Complete air-liquid human keratinocyte culture medium (complete alhkDMEM-Ham). To make 1L, refer to Table 11.2. Thaw all components at 4°C. Stored at 4°C for a maximum of 10 days.

Table 11.2 Medium Reconstitution

Human Fibroblast Culture Medium (fDMEM)

Component	Quantity	Final Concentration
DMEM	900 mL	90% (v/v)
Fetal calf serum	100 mL	10% (v/v)
Penicillin G-Gentamicin 500×	2 mL	Penicillin G 100 IU/mL Gentamicin 25 μg/mL

Complete Human Keratinocyte Culture Medium (complete hkDMEM-Ham)

Component	Quantity	Final Concentration
DMEM-Ham	950 mL	95% (v/v)
Fetal clone II	50 mL	5% (v/v)
Insulin 1,000×	1 mL	5 mg/mL
Hydrocortisone 500×	2 mL	0.4 μg/mL
Cholera toxin 1000×	1 mL	10^{-10} M
Epidermal growth factor 1,000×	1 mL	10 ng/mL
Penicillin G-Gentamicin 500×	2 mL	Penicillin G 100 IU/mL Gentamicin 25 μg/mL

Air-Liquid Human Keratinocyte Culture Medium (alkDMEM-Ham)

Component	Quantity	Final Concentration
DMEM-Ham	950 mL	95% (v/v)
Fetal clone II	50 mL	5% (v/v)
Insulin 1,000×	1 mL	5 μg/mL
Hydrocortisone 500×	2 mL	0.4 μg/mL
Cholera toxin 1,000×	1 mL	10-10 M
Penicillin G-Gentamicin 500×	2 mL	Penicillin G 100 IU/mL Gentamicin 25 μg/ml

11.4.2 In vitro fabrication of human skin substitutes by the self-assembly approach

All further manipulations are performed under a sterile laminar flow hood cabinet.

11.4.2.1 Assembly of fibroblast sheets for dermal reconstruction

A schematic drawing of the method is presented in Figure 11.1(a).

1. Seed 2×10^5 fibroblasts per 25 cm^2 culture flask in 5 mL of fDMEM containing 50 μg/mL ascorbic acid. Incubate in 8% CO_2, 100% atmospheric humidity at 37°C. Change culture medium three times a week.
2. After about 28–35 days, open the top of the flask with a soldering iron.
3. Using curved forceps, detach carefully one fibroblast sheet. Transfer it into a 100-mm cell culture dish.
4. Anchor peripherally the fibroblast sheet with ingots. Move the ingots one by one towards the tissue periphery to flatten the fibroblast sheet.
5. Detach a second fibroblast sheet and transfer it onto the first fibroblast sheet. Repeat step 4.
6. Place a Merocel sponge (which has been cut to fit within the ingots) on top of the superimposed fibroblast sheets. Keep the Merocel sponge in place with ingots.

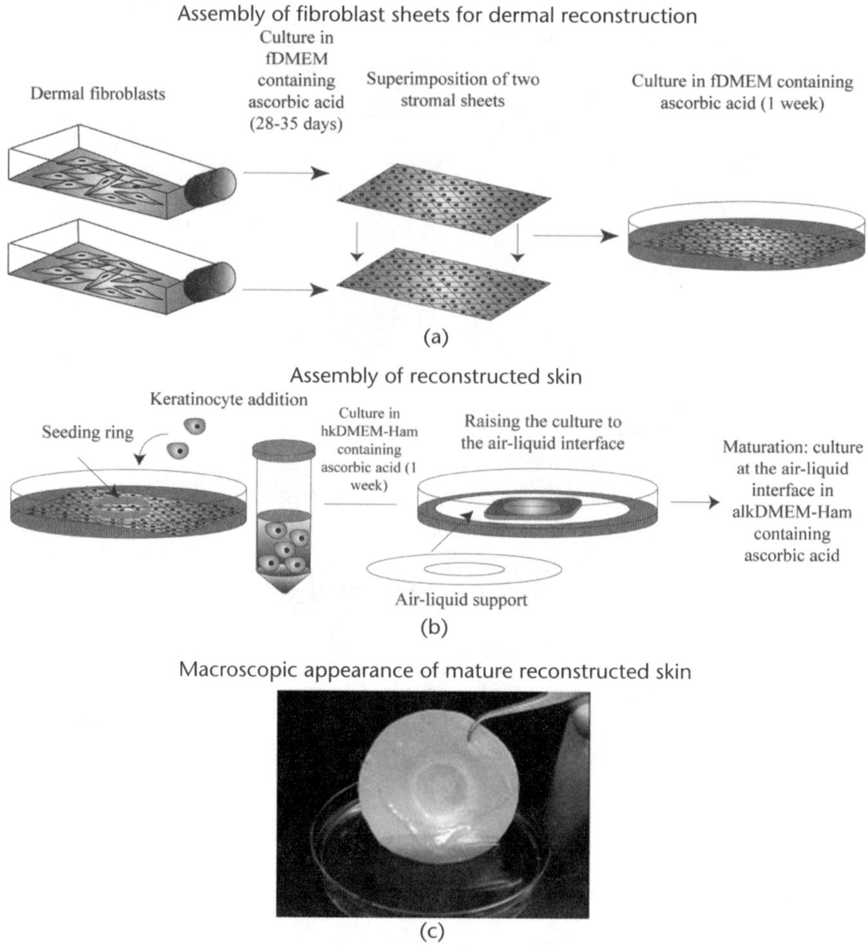

Figure 11.1 Schematic representation of the in vitro production of reconstructed skin. (a) Production of stromal sheets for dermal reconstruction. Dermal fibroblasts are cultivated in fDMEM containing 50 μg/mL of ascorbic acid for 28–35 days in order to produce stromal sheets. Then two stromal sheets are detached and superimposed; ingots (small weights) are placed around the edges to keep the sheets in contact. A Merocel sponge, which has been cut to fit within the ingots, is placed on top of stromal sheets, and maintained in place with ingots (after about 2 days, Merocel sponge is removed). The resulting reconstructed dermal substitute is cultivated for one more week in fDMEM containing 50 μg/mL of ascorbic acid. (b) Assembly of reconstructed skin. Keratinocytes are seeded on top of reconstructed dermis within a seeding ring. After 1 more week under submerged conditions in hkDMEM-Ham containing 50 μg/mL of ascorbic acid, the tissue is detached carefully from the bottom of the cell culture dish and placed on an anchoring paper in order that the area with keratinocytes is on top of aperture. The reconstructed tissue together with the anchoring paper are lifted and transferred onto an air-liquid stand, and then cultured at the air-liquid interface in the maturation medium (alkDMEM-Ham containing 50 μg/mL ascorbic acid). (c) Macroscopic aspect of reconstructed skin after 21 days of maturation at the air-liquid interface.

7. Add 25 mL of the fDMEM containing 50 μg/mL ascorbic acid. Incubate in 8% CO_2, 100% atmospheric humidity at 37°C.
8. After 2 days, remove the medium and then carefully remove the ingots (those on the Merocel sponge) and the Merocel sponge.
9. Put a seeding ring in the center of the fibroblast sheets (the seeding ring can be placed between 1 and 7 days before keratinocyte seeding).

10. Add 25 mL of the fDMEM containing 50 μg/mL ascorbic acid. Incubate in 8% CO_2, 100% atmospheric humidity at 37°C. Change the culture medium three times a week.

11.4.2.2 Assembly of reconstructed skin

A schematic drawing of the method is presented in Figure 11.1(b).

11. One week after the stacking of fibroblast sheets, remove culture medium and seed 8 × 10^5 keratinocytes within the seeding ring (from a suspension of cells in a complete hkDMEM-Ham). Incubate in 8% CO_2, 100% atmospheric humidity at 37°C.
12. About 2 to 4 hours later, add 25 mL of complete hkDMEM-Ham containing 50 μg/mL ascorbic acid. Incubate in 8% CO_2, 100% atmospheric humidity at 37°C. Change the culture medium three times a week.
13. Twelve hours later, the seeding ring can be removed.

12.4.2.3 Maturation of the tissue-engineered skin by culturing at the air-liquid interface

14. One week after the keratinocyte seeding, remove culture medium and ingots.
15. Using curved forceps, detach carefully the reconstructed skin from the bottom of the cell culture dish.
16. Place the reconstructed skin on the anchoring paper, centering it in order that the area with keratinocytes is on top of the aperture.
17. Put an air-liquid stand in a Petri dish.
18. Lift reconstructed skin together with the anchoring paper and transfer it onto the air-liquid stand.
19. Add 25 mL of alkDMEM-Ham containing 50 μg/mL ascorbic acid. Incubate in 8% CO_2, 100% atmospheric humidity at 37°C. Change the culture medium three times a week.

Note that tissue-engineered skin can be cultivated for more than 28 days at the air-liquid interface.

11.4.3 Tissue preservation and sectioning

1. Cut the tissue-engineered skin into small pieces (5 mm × 5 mm).
2. Sponge the excess tissue fluid with absorbent paper.
3. Completely coat the tissue with OCT without creating bubbles.
4. Put liquid OCT on a tissue holder and use the forceps to immerse it for 2–3 seconds in liquid nitrogen and then remove it before the OCT completely solidifies.
5. With forceps, take the OCT-coated piece of tissue and place it on top of the liquid OCT on the tissue holder.
6. Add OCT on top of the tissue sample to completely recover the tissue and prevent it from being freeze-dried.
7. Take the tissue holder with the forceps and immerge it in liquid nitrogen for approximately 15 seconds. The OCT should be white, but should not crack.

8. Store OCT blocks at –80°C until use.
9. Cut the tissue in a cryostat (4–6-μm-thick sections) and place the tissue sections on the superfrost glass slides.
10. Dry the sections on a warm plate at 37°C for 30 minutes.
11. Process slides for immunohistochemistry or store the slides at –20°C (they can be kept for few weeks).

11.4.4 Preparation of solutions and materials for immunofluorescence

- *Phosphate-buffered saline-calcium (PBS-Ca).* To make 1L, dissolve 8g NaCl (137 mM), 0.2g KCl (2.7 mM), 0.92g Na_2HPO_4 (6.5 mM), 0.2g KH_2PO_4 (1.5 mM), and 0.17 mL $MgCl_2.6H_2O$ (0.5 mM) of a 2.8M solution kept at –20°C, and 0.131g $CaCl_2{\cdot}2H_2O$ (0.9 mM) in dH_2O. Complete to 1L with dH_2O. Verify that the pH is 7.4. Note that a five-times concentrated PBS solution can be prepared and kept at 4°C, but the $CaCl_2$ should be omitted and added only after dilution of PBS to 1×. A ten-times concentrated solution can also be prepared and kept at 4°C.
- *PBS-Ca-BSA.* PBS-Ca containing 1% BSA. Dissolve 1g bovine serum albumin in 100 mL PBS-Ca.
- *Primary antibodies.* Dilute in PBS-Ca-BSA as specified in Table 11.2.
- *Secondary antibodies.* Dilute in PBS-Ca-BSA as specified in Table 11.2.
- *Mounting media.* To make 160 mL, dissolve 3.2g gelatin in 100-ml hot dH_2O. Add 0.896g NaCl (137 mM), 0.0224g KCl (2.7 mM), 0.1032 g Na_2HPO_4 (6.5 mM), 0.0224g KH_2PO_4 (1.5 mM,, and 0.16g NaN_3 (22 mM). Adjust pH to 7.6 using pH paper. Complete to 160 mL with dH_2O. Add 48 mL glycerol.
- *Hoechst solution.* Dissolve 25 mg of Hoechst 33258 in 1 mL of dH_2O to make stock 1 solution (25 mg/mL). Dilute 100 μL of stock 1 solution in 49.9 mL dH_2O to make 100× stock 2 solution (50 μg/mL). To dilute 100× stock 2 solution to 1× (0.5 μg/mL), add dH_2O.

11.4.5 Immunofluorescent labeling of human skin substitutes

1. Take the tissue sections from the freezer and leave them at room temperature for about 5 minutes to remove excess humidity. If the tissue sections are small, they can be circled by etching the slide with a diamond pencil.
2. Acetone fixation: Fix the tissue sections by immersion in cold acetone for 10 minutes at –20°C.
3. PBS-Ca washes: Put the slides in a rack and rinse three times in PBS-Ca for 5 minutes. Repeat twice.
4. First primary antibody incubation: Aspirate the liquid surrounding the tissue sections. Work rapidly to avoid drying the tissue. Cover each tissue section with 25 to 50 μL of the primary antibody. For control slides, use an unrelated antibody of the same isotype and then the monoclonal primary antibody used or PBS-Ca BSA. In the case of a polyclonal antibody, put PBS-Ca-BSA. Incubate for 45 minutes at room temperature.
5. Do three PBS-Ca washes.

6. Secondary antibody incubation: Aspirate the liquid surrounding the tissue sections. Process rapidly to avoid drying the tissue. Cover each tissue section with 25 to 50 μL of the secondary antibody. Incubate for 30 minutes at room temperature and protect from the light.
7. Do three PBS-Ca washes.
8. The tissue sections or cells can be counterstained with Hoechst (nuclei staining) as follows. Put the slides in a rack and immerse it in two successive baths of dH_2O (2 minutes by wash). Cover with 25 to 50 μL of the 1× Hoechst solution. Incubate for 10 minutes in the dark. Do three dH_2O washes.
9. Mount the slides as follows. Warm the solid mounting media in warm water or in the microwave to liquefy it. Place one drop of the mounting media on each tissue section. Put a cover slip gently on top while trying to avoid bubbles. Remove the excess of the mounting media by draining onto an absorbent paper.
10. Store in the dark at 4°C to avoid any bleaching of the fluorescence.
11. Examine the sections under a microscope equipped with epifluorescence.

11.4.6 Histological analysis

Cut the skin substitute into pieces of about 4 mm × 15 mm. Fix in the Histochoice MB fixative (Amresco) overnight at room temperature. Histologically process the tissue and embed in paraffin. Stain 6-μm-thick sections with Masson's Trichrome according to standard procedures. Examine the section under a light microscope.

11.4.7 Transmission electron microscopy

Cut the skin substitute into pieces of about 1 mm^3. Fix tissue with 2.5% glutaraldehyde in 0.1M sodium cacodylate buffer for a minimum of 4 hours or overnight at 4°C. Wash in 0.1M sodium cacodylate buffer, post fix in 2% osmium tetroxide, and embed in Epon 812 according to standard procedures. Stain thin sections with uranyl acetate and lead citrate.

11.4.8 Statistical analysis

- Although variations within batches are low, there are some interbatch variations which merit that the experiment should be repeated with cells from other donors.
- Immunohistochemistry is a qualitative analysis. No statistics are usually performed except if an evaluation of the percentage of cells (for example, the proportion of Ki-67-labeled cell) or the measurement of a morphological aspect (for example, the measurement of the epidermal thickness). However, it is necessary to be sure that the noticed phenomenon is reproducible (makes at least three independent experiments).
- Since there is a set of dependent variables between experiments, a normal distribution could not be assumed because it involves an in vitro environment.
- The variation between the experiments is usually low and does not necessitate large samples. A nonparametric test is suggested to perform statistical analysis.
- Western Blot or FACS analyses could be performed to complete the study in a more quantitative manner.

11.5 Anticipated Results

As highlighted in this chapter, the development of in vitro skin substitute by tissue engineering must include the monitoring of the expression of various specific protein markers. A series of coordinated morphological and biochemical changes that result in a highly specialized and organized stratified squamous epithelium takes place during terminal differentiation and is associated with typical markers. Thus, fully developed reconstructed skin must expressed a broad range of protein products typically found in native skin [21], as observed in the tissue-engineered skin substitutes produced by the self-assembly approach (the similarities and differences are summarized in Table 11.3).

The culture at the air-liquid interface exerts a powerful effect on skin epithelial cells [22]. Macroscopically, the gross aspect of the tissue-engineered skin substitute changes in appearance from shiny to an opaque white, indicating the formation of the stratum corneum, which is induced by the contact with air [for a macroscopic picture, see Figure 11.1(c)]. Human keratinocytes cultured on an appropriated stromal substitute under submerged conditions proliferate until confluence and then form an epidermis composed of one to three cell layers [Figure 11.2(a)]. The formation of a well-organized epidermis with all histological layers present in the normal epidermis is achieved in vitro by culturing the substitute at the air-liquid interface for about 14 days [Figure 11.2(d)]. At the beginning of culture, the most differentiated layers are not present [Figure 11.2(b)].

Table 11.3 Comparison Between Native Skin and the Air-Exposed Tissue-Engineered Skin Substitute

Characteristics	Native Skin	Air-Exposed Tissue-Engineered Skin Substitute
Tissue architecture		
Stratification	All epidermal strata SB, SS, SG, SC	All epidermal strata SB, SS, SG, SC (Figure 11.2)
Desmosomes	Present	Present (Figures 11.5 and 11.6)
Hemidesmosomes	Present	Present (Figure 11.6)
aspnumKeratohyalin granules	Present	Present
Cornified cell envelope	Present	Present
Expression of epidermal differentiation-specific protein markers		
Keratins	SB, SPB: K14	SB, SPB: K14 (Figure 11.3)
	SPB: K10	SPB: K10 (Figure 11.3)
Ki-67	SB	SB (Figure 11.5)
VM-2 (a3 integrin subunit)	SB	SB (Figure 11.5)
Involucrin	SG	SS, SG (Figure 11.3)
Transglutaminase	SG	SS, SG (Figure 11.3)
DLK	SG	SG (Figure 11.4)
Filaggrin	SG	SS, SG (Figure 11.4)
Extracellular matrix component		
Collagen I	D	RD (Figure 11.5)
Collagen III	D	RD (not shown)
Collagen IV	BM	BM, RD (Figure 11.5)
Collagen VII	BM	BM (Figure 11.5)
Laminin-5	BM	BM (Figure 11.5)

BC:, basal cells; SPB:, suprabasal cells; SB:, stratum basale; SS:, stratum spinosum; SG:, stratum granulosum; SC:, stratum corneum; K:, keratin; BM:, basement membrane; D:, dermis; RD:, reconstructed dermis.

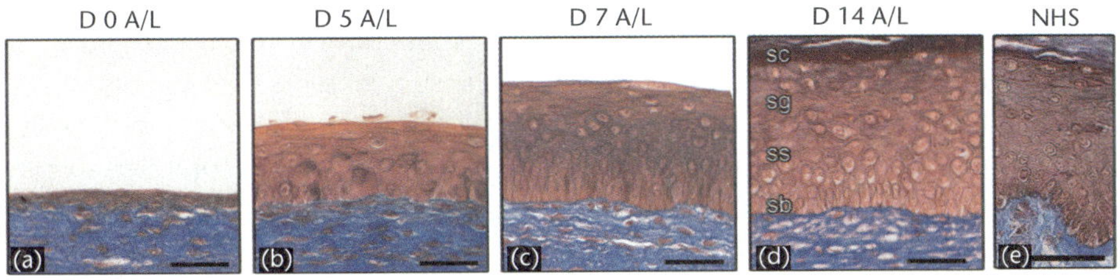

Figure 11.2 Histological modifications of cultured human tissue-engineered skin over time. Masson's trichrome staining of tissue-engineered skin substitutes produced by the self-assembly approach and cultured (a) 0, (b) 5, (c) 7, and (d) 14 days at the air-liquid interface and normal human skin (NHS, D). sb: stratum basale, ss: stratum spinosum, sg: stratum granulosum, sc: stratum corneum. Bars = 50 μm. (Reproduced and adapted from [28] according to the copyright policy of the publisher. © 2009 Nature Publishing Group.)

The stratum granulosum appears after about 7 days [Figure 11.2(c)]. With time, the stratum corneum usually becomes thicker during the culture than normal skin because of the absence of desquamation. Thus, the tissue-engineered skin substitute is a dynamic tissue in which keratinocytes are induced to differentiate by prolonging the culture period. The distribution of keratinocyte differentiation-associated markers is going to vary according to the differentiation state of the skin substitute.

During the keratinocyte differentiation process, one of the major changes is the induction of unusually large keratins, which occurs as a cell leaves the basal layer [23]. Basal cells express mainly the keratins K5 and K14, whereas suprabasal cells express keratins K1 and K10 [7]. In a tissue-engineered skin substitute, K14 expression appears in cells of the stratum basale [Figure 11.3(a)] while K10 expression begins from the stratum spinosum [Figure 11.3(b)].

Involucrin, a marker of terminal differentiation, is among the first cornified envelope precursors to be cross-linked during keratinocyte differentiation [24]. It is expressed from the upper spinous layer of the epidermis in normal skin [25, 26]. In the mature tissue-engineered skin substitute, involucrin appears directly above the basal cells [Figure 11.3(c)], while the transglutaminase expression begins to be significant in upper spinous layers [Figure 11.3(d)]. The dual leucine zipper-bearing kinase (DLK) is a mitogen-acti-

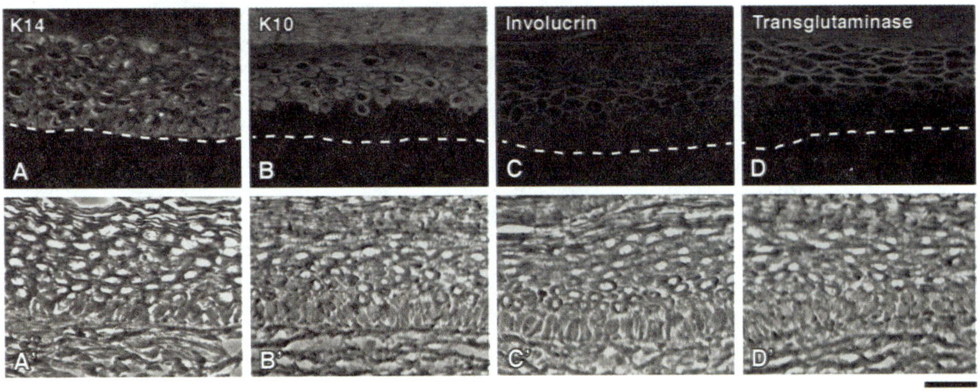

Figure 11.3 Expression of epidermal differentiation markers in the tissue-engineered skin substitute. Immunofluorescent labeling of (a) K14, (b) K10, (c) involucrin, and (d) transglutaminase in the tissue-engineered skin substitutes produced by the self-assembly approach. Phase contrast microscopy is also provided (e–h, respectively). The dotted line indicates the dermo-epidermal junction. Bar = 50 μm.

vated protein kinase kinase kinase that has an active role in the induction of terminal differentiation [27–29]. DLK and filaggrin are not expressed in the first phases of epidermal cell differentiation [Figure 11.4(a, d)]. These proteins are induced when the stratum granulosum develops at 7 days of culture at the air-liquid interface and increase with epidermal differentiation [Figure 11.4(b, c, e, f)].

Epidermis is a self-renewing stratified epithelium. In order to localize cycling cells or to determine the proliferating index of a skin substitute, detection of Ki-67, a nuclear antigen expressed by proliferating cells in all phases of the active cell cycle (G1, G2, S, and M) but absent in resting cells (G0) can be performed [30, 31]. Typically, Ki-67 appears in the nucleus of some keratinocytes dispersed within the stratum basale of the tissue-engineered skin substitute [arrows in Figure 11.5(a)]. Also, the monoclonal anti-

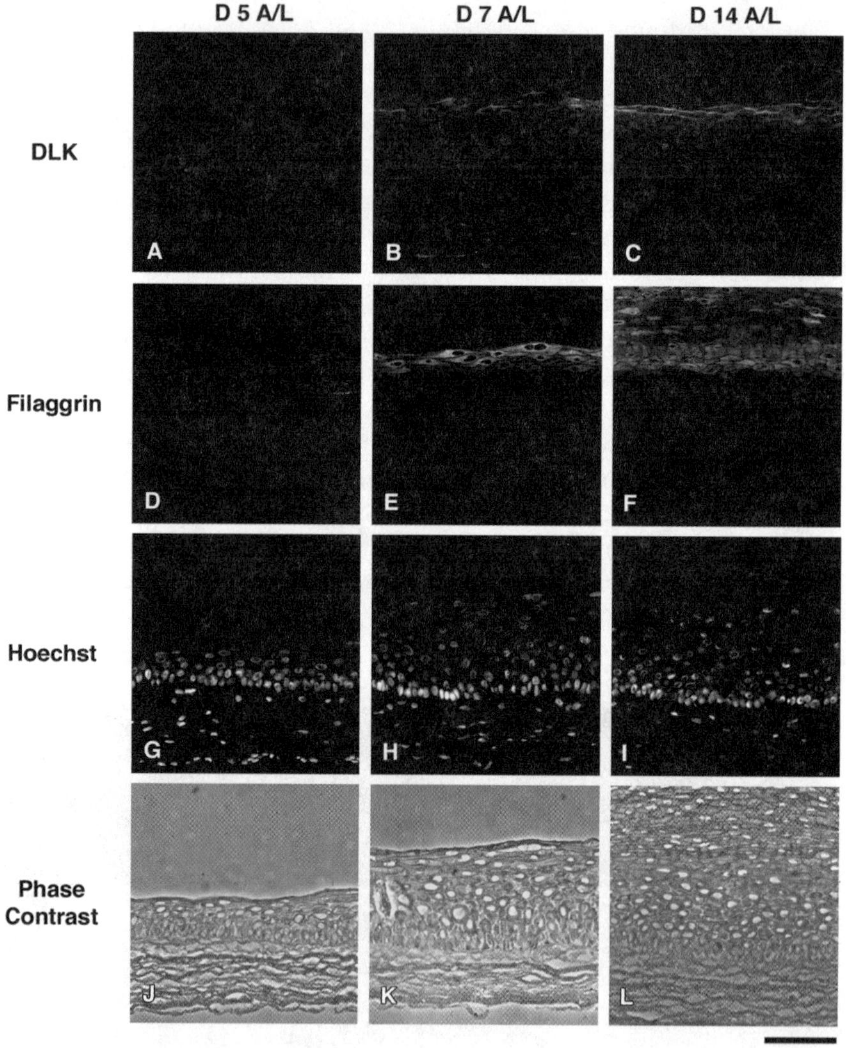

Figure 11.4 Expression of DLK and filaggrin with the culture time. Immunofluorescence labeling of (a–c) DLK and (d–f) filaggrin in tissue-engineered skin substitutes produced by the self-assembly approach cultured (a, d) 5, (b, e) 7, and (c, f) 14 days at the air-liquid interface. Nuclei were stained with Hoechst dye (g, h, and i, respectively). Phase contrast microscopy is also provided (j, k, and l, respectively). Bars = 50 μm. (Reproduced and adapted from [28] according to the copyright policy of the publisher. © 2009 Nature Publishing Group.)

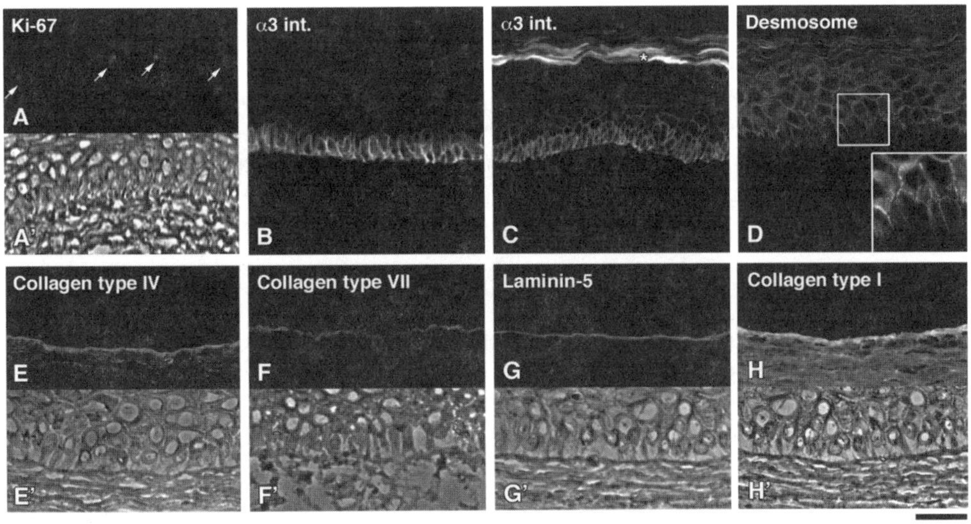

Figure 11.5 Expression of skin markers in the tissue-engineered skin substitute. Immunofluorescent labeling of sections from tissue-engineered skin substitutes produced by the self-assembly approach with antibodies raised against (a) Ki-67, (b, c) α3-integrin subunit, (d) desmosomal protein (pan desmocollin), (e) the basement membrane-associated collagen type IV (E), (f) collagen type VII and (g) laminin-5 proteins, and (h) collagen type I. (a', e', f', g', and h') are contrast microscopy corresponding to (a), (e), (f), (g), and (h), respectively. Note that the basal keratinocyte marker VM-2 also stains the first suprabasal cells of the epidermis because the section was not a perfect cross-section. *: background corresponding to the nonspecific binding of the antibodies. Bars = 50 μm.

body called VM-2, another basal keratinocyte marker, which is expressed in basal cells of normal skin [32], recognizes the α3 subunit of the $\alpha 3\beta 1$ integrin in the basal keratinocytes of the tissue-engineered skin substitute [Figure 11.5(b)].

Desmosomes are important structures in cell-to-cell adhesion. They serve as anchoring plates for keratin filaments [33]. In the tissue-engineered substitute, the mouse monoclonal ZK-31 antibody formed against the desmosomal protein (pan desmocollin) reveals a dot-like pattern on the keratinocyte cell membrane [Figure 11.5(d)]. The presence of desmosomes is also easily recognizable by transmission electron microscopy [Figure 11.6(a)].

A bilayered skin substitute comprising both living fibroblasts and keratinocytes presents advantages over other models. In addition to synthesizing ECM, fibroblasts secrete many components of the basement membrane such as collagen IV and VII and laminin-5 [34, 35]. In skin, the presence of a basement membrane prevents epidermal detachment from the dermis and also participates in stem cell regulation and anchoring [36, 37]. In the tissue-engineered skin substitute, collagen IV, VII as well as laminin-5 are all present at the dermo-epidermal junction [Figure 11.5(e–g)]. Hemidesmosomes, which are specialized junctions consisting of dense plaques associated to bundles of keratin filaments, are also present between the basal keratinocyte and the lamina densa in the mature tissue-engineered skin substitutes [Figure 11.6(b)]. The lamina lucida, principally composed of type IV collagen, is also apparent [Figure 11.6(b)].

The mechanical stability is an important challenge in the development of biological tissues by tissue engineering. The structure of the dermis confers considerable mechanical strength to skin. The dermis is composed mainly of collagen type I and type III fibers, which are synthesized by fibroblasts [38]. Accordingly, the reconstructed dermis of the

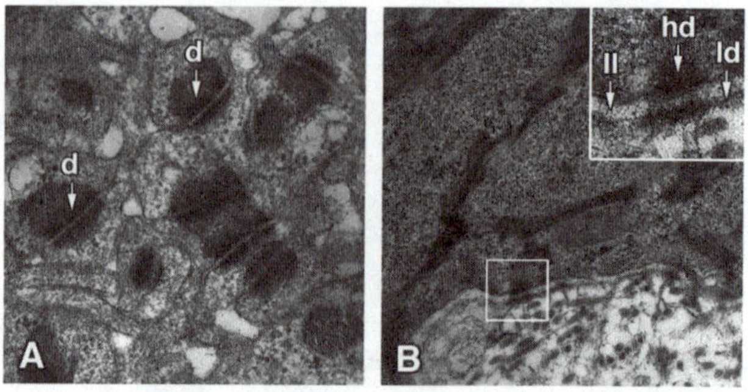

Figure 11.6 Ultrastructural analysis of the tissue-engineered skin substitute. Transmission electron microscopy of the tissue-engineered skin substitutes produced by the self-assembly approach showing (a) numerous desmosomes (indicated by the letter d) and (b) a well-structured basement membrane with hemidesmosomes (hd), lamina densa (ld), and lamina lucida (ll).

tissue-engineered skin substitute highly express type I [Figure 11.5(h)] and type III (not shown) collagens. During the culture, the integrity of the tissue-engineered dermis is preserved over time, thanks to the effect of ascorbic acid, which allows continuous collagen secretion and organization [39].

11.6 Discussion and Commentary

At the present time, animal models are commonly used in skin research and in the preclinical development of new drugs. The possibility to experimentally manipulate the mouse genome made the mouse a widely used model for advancing skin research. However, variations in histology and permeability exist between human and animal skin [11, 40]. For example, mouse dorsal skin presents two to three cell layers in the living portion of the epidermis, while human skin is rather constituted of about 5 to 15 living cell layers with less hair than mouse skin. In addition, hair follicle stem cells in mice and humans have a distinctive gene expression profile [41, 42]. Thus, the cellular and molecular characteristics of human epithelial cells could be different from those of rodents, supporting the use of human skin substitutes to take the forefront in research on human skin.

Skin substitutes provide an alternative to the animal models conventionally used [43]. As of September 11, 2004, a European Community Directive (ECD) strictly forbids the use of animals to test cosmetic products [4]. The successful application of tissue-engineered skin substitutes requires that the morphological and ultrastructural organization of the epidermis, dermis, and dermo-epidermal junction mimic the normal skin structure as closely as possible [4]. Skin substitutes produced with the self-assembly approach show an absence of exogenous material, a histologically normal and adequately differentiated tissue with the presence of a well-developed dermo-epidermal junction, and a complex collagen network [44]. This method can be used for both clinical and fundamental studies and it can be relatively adaptable to many in vitro applications: cutaneous physiology, skin disease, wound healing, stem cells, dermatopharmacology, toxicology, and angiogenesis studies.

It must be pointed out that despite the fact that the self-assembly approach has many advantages, it is a time-consuming method. It does not use any exogenous scaffold, but instead enough time must be allowed for the cells to proliferate and to produce their own extracellular matrix. On the other hand, the absence of a scaffold can be advantageous for mechanical study of the ECM. It should be noted that beside all similarities (see Table 11.3), there are some differences in the expression and localization of various specific differentiation markers. These differences include the presence of "hyperproliferative" keratins (K6, K16, and K17) as in other models of skin substitutes produced in vitro and sometimes of the precoce appearance of involucrin, transglutaminase, and filaggrin in the lower suprabasal layers of the reconstructed epidermis. K6, K16, and K17 are considered markers of keratinocyte activation, and their expression is coordinately increased in wounded skin and in hyperproliferative skin disorders such as psoriasis [45–48]. During the in vitro culture period, most skin substitutes exhibit a hyperproliferative phenotype similar to wounded skin and Smiley et al. have confirmed that engineered skin substitutes have a gene expression profile similar to those of wounded human skin [49]. A model that lacks this hyperproliferative phenotype still needs to be developed as an in vitro mimic of normal human skin [46].

The various methods used for the reconstruction of living skin give different epidermal-cell phenotypes depending on the culture conditions [2]. However, immunohistochemistry is not a quantitative technique. In this regard, Western Blot and FACS analysis allow more subtle comparisons between the expression level of a given protein and can be run in parallel on cells harvested from the reconstructed tissue. Also, it should be noted that for some markers, the stratum corneum can exhibit an important background [see Figure 11.5(c)] corresponding to the nonspecific binding of the primary or secondary antibody. This is true for tissue-engineered skin substitutes as well as for native skin.

Progress in tissue engineering and cell culture has led to the development of various organized tissues offering the possibility of creating a model for research applications. As examples, human tissue-engineered skin substitutes have been produced and applied to percutaneous absorption studies and to the characterization of epithelial stem cells [10, 11, 21, 50]. Wound-healing models composed of human epithelial and mesenchymal cells also permit a better understanding of the mechanisms of reepithelialization of the skin and cornea [51, 52]. The similarities observed between these models and the in vivo wound healing process support their use for investigating basic mechanisms. Using living cells, the reconstruction of organs such as the bladder, blood vessels, cornea, and adipose tissue have been developed and represent alternative tools to the use of animal in research [12, 15, 52, 53].

11.7 Application Notes

- When cultured in flasks immersed in the medium, keratinocytes do not reach the late stage of terminal differentiation. To lead keratinocyte differentiation study, the utilization of skin substitute cultured at the air-liquid interface is indicated.
- The expression and localization of differentiation markers in various skin models provide good parameters for the evaluation of the quality of skin substitutes.
- Epidermal differentiation is characterized by a series of coordinated morphological and biochemical changes over time, which allow for a step-by-step evaluation of the differentiation process.

Troubleshooting Table

Problem	Explanation	Potential Solutions
Tissue-engineered skin substitute		
The surface of the reconstructed tissue appears macroscopically nonhomogenous	Irregular distribution of keratinocytes	Be careful to seed keratinocytes slowly, drop by drop, on the entire surface delimited by the seeding ring.
Detachment or blistering between dermal sheets	Poor attachment between stromal sheets	During the superimposition of the stromal sheets, be sure to eliminate bubbles.
Immunofluorescence		
No labeling	The primary and the secondary antibodies must exhibit a good match	Use antibodies made from different species; the secondary antibody must be raised against the immunoglobulin type corresponding to that of the primary antibody.
	Photobleaching	Protect immunolabeled slides from light.
	Microscope equipped with the wrong filter	Microscope must be equipped with the filter corresponding to the dye used.
No or unexpected labeling	The choice of the antibody is crucial since there are many nonspecific ones on the market	Always run in parallel an appropriate positive control.
	Sources and characterization performed such as Western blot analyses and purification must be carefully checked before use	Prepare cell extract and use Western blotting to verify that the targeted protein is present.
High background level	Primary antibody is too concentrated	Run a titration curve to determine the optimal dilution.
	Secondary antibody too concentrated	Run a titration curve to determine the optimal dilution.
Basal keratinocyte marker expressed in first suprabasal cell layers [for example, see VM-2 labeling in Figure 11.5(c)]	Slope angle in the cross-section	Reorient the tissue block in order to cut in a vertical plane perpendicular to the skin surface.
Epidermal cyst within the dermal compartment	Slope angle in the cross-section	Orient the tissue block in order to cut in a vertical plane perpendicular to the skin surface.

- According to the proliferation-differentiation ratio (+/– proliferation or +/– differentiation), the appearance of the tissues will change: number of layers and thickness of tissues or delay/advance in the expression of epidermal differentiation markers.
- The number of proliferating cells could be modulated by the culture conditions.
- In vitro tissue-engineered skin models with genetically modified cells can be used as a tool to understanding mechanisms regulating keratinocyte growth and differentiation (as an example, see [54]).

11.8 Summary Points

1. The in vitro system described here combines the three critical skin components: epithelia, basement membrane, and stroma.
2. The tissue-engineered skin substitute produced by the self-assembly approach represents a promising tool both for basic and clinical applications or pharmacological research.

3. Different skin markers are available and represent useful tools to monitor the quality of a skin substitute or its differentiation status when the appropriate culture conditions are used.
4. The tissue-engineered skin substitute produced by the self-assembly approach is an interesting alternative to the utilization of animal models for skin research.
5. The self-assembly approach of tissue engineering can be adapted for the in vitro reconstruction of various organs.
6. Future studies will tell whether modifications of the culture conditions allow for the reconstruction of a skin substitute that fully mimics human skin.

Acknowledgments

The authors would like to thank current and former members of the LOEX laboratory who have contributed to develop the foregoing protocols. We especially thank the Saint-Sacrement Hospital Pathology facility and the Electron microscopy service of Laval University. We are grateful to Amélie Lavoie and Annie Beauparlant for their contributions to the figures and to Todd Galbraith for the revision of the manuscript.

References

[1] Boyce, S. T., "Cultured Skin Substitutes: A Review," *Tissue Engineering*, Vol. 2, No. 4, 1996, pp. 255–266.

[2] Basset-Seguin, N., et al., "Reconstituted Skin in Culture: A Simple Method with Optimal Differentiation," *Differentiation*, Vol. 44, No. 3, 1990, pp. 232–238.

[3] Ponec, M., "In Vitro Cultured Human Skin Cells as Alternatives to Animals for Skin Irritancy Screening," *Int. J. Cosmet. Sci.*, Vol. 14, No. 6, 1992, pp. 245–264.

[4] Black, A. F., et al., "Optimization and Characterization of an Engineered Human Skin Equivalent," *Tissue Eng.*, Vol. 11, No. 5-6, 2005, pp. 723–733.

[5] Asselineau, D., et al., "Human Epidermis Reconstructed by Culture: Is It 'Normal'?" *J. Invest. Dermatol.*, Vol. 86, No. 2, 1986, pp. 181–186.

[6] Blanpain, C., and E. Fuchs, "Epidermal Stem Cells of the Skin," *Annu. Rev. Cell Dev. Biol.*, Vol. 22, 2006, pp. 339–373.

[7] Radoja, N., et al., "Transcriptional Profiling of Epidermal Differentiation," *Physiol. Genomics*, Vol. 27, No. 1, 2006, pp. 65–78.

[8] Marionnet, C., et al., "Morphogenesis of Dermal-Epidermal Junction in a Model of Reconstructed Skin: Beneficial Effects of Vitamin C," *Exp. Dermatol.*, Vol. 15, No. 8, 2006, pp. 625–633.

[9] Fuchs, E., "Epidermal Differentiation: The Bare Essentials," *J. Cell Biol.*, Vol. 111, No. 6, Pt. 2, 1990, pp. 2807–2814.

[10] Pouliot, R., et al., "Reconstructed Human Skin Produced In Vitro and Grafted on Athymic Mice," *Transplantation*, Vol. 73, No. 11, 2002, pp. 1751–1757.

[11] Michel, M., et al., "Characterization of a New Tissue-Engineered Human Skin Equivalent with Hair," *In Vitro Cell Dev. Biol. Anim.*, Vol. 35, No. 6, 1999, pp. 318–326.

[12] L'Heureux, N., et al., "A Completely Biological Tissue-Engineered Human Blood Vessel," *Faseb. J.*, Vol. 12, No. 1, 1998, pp. 47–56.

[13] Auger, F.A., et al., "The Self-Assembly Approach for Organ Reconstruction by Tissue Engineering," *e-biomed: a Journal of Regenerative Medicine*, Vol. 1, 2000, pp. 75–86.

[14] Germain, L., et al., "Principles of Living Organs Reconstruction by Tissue Engineering," in M. J. Yaszemski et al., (eds.), *Tissue Engineering and Novel Delivery Systems*, New York: Marcel Dekker, 2004, pp. 197–228.

[15] Vermette, M., et al., "Production of a New Tissue-Engineered Adipose Substitute from Human Adipose-Derived Stromal Cells," *Biomaterials*, Vol. 28, No. 18, 2007, pp. 2850–2860.

[16] Jean, J., et al., "Development of an In Vitro Psoriatic Skin Model by Tissue Engineering," *J. Dermatol. Sci.*, Vol. 53, No. 1, 2009, pp. 19–25.

[17] Larouche, D., et al., "Regeneration of Skin and Cornea by Tissue Engineering," *Methods Mol. Biol.*, Vol. 482, 2009, pp. 233–256.

[18] Royal, I., et al., "Polyomavirus Middle T Selective Action on Cytokeratin 14 Gene Expression in Liver Nonparenchymal Epithelial Cells," *Exp. Cell Res.*, Vol. 220, No. 1, 1995, pp. 171–177.

[19] Douziech, M., et al., "Localization of the Mixed-Lineage Kinase DLK/MUK/ZPK to the Golgi Apparatus in NIH 3T3 Cells," *J. Histochem. Cytochem.*, Vol. 47, No. 10, 1999, pp. 1287–1296.

[20] Rousselle, P., et al., "Kalinin: An Epithelium-Specific Basement Membrane Adhesion Molecule That Is a Component of Anchoring Filaments," *J. Cell Biol.*, Vol. 114, No. 3, 1991, pp. 567–576.

[21] Ponec, M., et al., "Characterization of Reconstructed Skin Models," *Skin Pharmacol. Appl. Skin Physiol.*, Vol. 15, Suppl. 1, 2002, pp. 4–17.

[22] Fartasch, M. and M. Ponec, "Improved Barrier Structure Formation in Air-Exposed Human Keratinocyte Culture Systems," *J. Invest. Dermatol.*, Vol. 102, No. 3, 1994, pp. 366–374.

[23] Kannan, S., et al., "Alterations in Expression of Terminal Differentiation Markers of Keratinocytes During Oral Carcinogenesis," *Pathobiology*, Vol. 62, No. 3, 1994, pp. 127–133.

[24] Yuspa, S. H., et al., "Expression of Murine Epidermal Differentiation Markers Is Tightly Regulated by Restricted Extracellular Calcium Concentrations In Vitro," *J. Cell Biol.*, Vol. 109, No. 3, 1989, pp. 1207–1217.

[25] Kanitakis, J., et al., "Involucrin Expression In Keratinization Disorders of the Skin—A Preliminary Study," *Br. J. Dermatol.*, Vol. 117, No. 4, 1987, pp. 479–486.

[26] Watt, F. M., "Involucrin and Other Markers of Keratinocyte Terminal Differentiation," *J. Invest. Dermatol.*, Vol. 81, No. 1 Suppl, 1983, pp. 100s–103s.

[27] Robitaille, H., et al., "The Mitogen-Activated Protein Kinase Kinase Kinase Dual Leucine Zipper-Bearing Kinase (DLK) Acts as a Key Regulator of Keratinocyte Terminal Differentiation," *J. Biol. Chem.*, Vol. 280, No. 13, 2005, pp. 12732–12741.

[28] Robitaille, H., et al., "The Small Heat-Shock Protein Hsp27 Undergoes ERK-Dependent Phosphorylation and Redistribution to the Cytoskeleton in Response to Dual Leucine Zipper-Bearing Kinase Expression," *J. Invest. Dermatol.*, 2009.

[29] Germain, L., et al., "The Mixed Lineage Kinase Leucine-Zipper Protein Kinase Exhibits a Differentiation-Associated Localization in Normal Human Skin and Induces Keratinocyte Differentiation upon Overexpression," *J. Invest. Dermatol.*, Vol. 115, No. 5, 2000, pp. 860–867.

[30] Gerdes, J., et al., "Cell Cycle Analysis of a Cell Proliferation-Associated Human Nuclear Antigen Defined by the Monoclonal Antibody Ki-67," *J. Immunol.*, Vol. 133, No. 4, 1984, pp. 1710–1715.

[31] Barker, C. L., et al., "The Development and Characterization of an In Vitro Model of Psoriasis," *J. Invest. Dermatol.*, Vol. 123, No. 5, 2004, pp. 892–901.

[32] Morhenn, V. B., et al., "A Monoclonal Antibody Against Basal Cells of Human Epidermis. Potential Use in the Diagnosis of Cervical Neoplasia," *J. Clin. Invest.*, Vol. 76, No. 5, 1985, pp. 1978–1983.

[33] Russell, D., et al., "Mechanical Stress Induces Profound Remodelling of Keratin Filaments and Cell Junctions in Epidermolysis Bullosa Simplex Keratinocytes," *J. Cell Sci.*, Vol. 117, Pt. 22, 2004, pp. 5233–5243.

[34] Marinkovich, M. P., et al., "Cellular Origin of the Dermal-Epidermal Basement Membrane," *Dev. Dyn.*, Vol. 197, No. 4, 1993, pp. 255–267.

[35] Boyce, S. T., et al., "Vitamin C Regulates Keratinocyte Viability, Epidermal Barrier, and Basement Membrane In Vitro, and Reduces Wound Contraction After Grafting of Cultured Skin Substitutes," *J. Invest. Dermatol.*, Vol. 118, No. 4, 2002, pp. 565–572.

[36] Levy, L., et al., "Beta1 Integrins Regulate Keratinocyte Adhesion and Differentiation by Distinct Mechanisms," *Mol. Biol. Cell*, Vol. 11, No. 2, 2000, pp. 453–466.

[37] Jones, P. H., and F. M. Watt, "Separation of Human Epidermal Stem Cells from Transit Amplifying Cells on the Basis of Differences in Integrin Function and Expression," *Cell*, Vol. 73, No. 4, 1993, pp. 713–724.

[38] Berthod, F., et al., "Differential Expression of Collagens XII and XIV in Human Skin and in Reconstructed Skin," *J. Invest. Dermatol.*, Vol. 108, No. 5, 1997, pp. 737–742.

[39] Geesin, J. C., J. S. Gordon, and R. A. Berg, "Regulation of Collagen Synthesis in Human Dermal Fibroblasts by the Sodium and Magnesium Salts of Ascorbyl-2-Phosphate," *Skin Pharmacol.*, Vol. 6, No. 1, 1993, pp. 65–71.

[40] Bartek, M. J., J. A. LaBudde, and H. I. Maibach, "Skin Permeability In Vivo: Comparison in Rat, Rabbit, Pig and Man," *J. Invest. Dermatol.*, Vol. 58, No. 3, 1972, pp. 114–123.

[41] Tumbar, T., et al., "Defining the Epithelial Stem Cell Niche in Skin," *Science*, Vol. 303, No. 5656, 2004, pp. 359–363.

[42] Morris, R.J., et al., "Capturing and Profiling Adult Hair Follicle Stem Cells," *Nat. Biotechnol.*, Vol. 22, No. 4, 2004, pp. 411–417.

[43] Whalen, E., et al., "The Development of Three-Dimensional In Vitro Human Tissue Models," *Hum. Exp. Toxicol.*, Vol. 13, No. 12, 1994, pp. 853–859.

[44] Auger, F. A., et al., "Tissue-Engineered Skin Substitutes: From In Vitro Constructs to In Vivo Applications," *Biotechnol. Appl. Biochem.*, Vol. 39, Pt. 3, 2004, pp. 263–275.

[45] Smiley, A. K., et al., "Keratin Expression in Cultured Skin Substitutes Suggests That the Hyperproliferative Phenotype Observed In Vitro Is Normalized After Grafting," *Burns*, Vol. 32, No. 2, 2006, pp. 135–138.

[46] Ojeh, N. O., J. D. Frame, and H. A. Navsaria, "In Vitro Characterization of an Artificial Dermal Scaffold," *Tissue Eng.*, Vol. 7, No. 4, 2001, pp. 457–472.

[47] McGowan, K. M., and P. A. Coulombe, "Onset of Keratin 17 Expression Coincides with the Definition of Major Epithelial Lineages During Mouse Skin Development," *J. Cell Biol.*, Vol. 143, 1998, pp. 469–486.

[48] Paladini, R. D., et al., "Onset of Re-Epithelialization After Skin Injury Correlates with a Reorganization of Keratin Filaments in Wound Edge Keratinocytes: Defining a Potential Role for Keratin 16," *J. Cell Biol.*, Vol. 132, 1996, pp. 381–397.

[49] Smiley, A. K., et al., "Microarray Analysis of Gene Expression in Cultured Skin Substitutes Compared with Native Human Skin," *J. Invest. Dermatol.*, Vol. 125, No. 6, 2005, pp. 1286–1301.

[50] Schafer-Korting, M., et al., "The Use of Reconstructed Human Epidermis for Skin Absorption Testing: Results of the Validation Study," *Altern. Lab. Anim.*, Vol. 36, No. 2, 2008, pp. 161–187.

[51] Laplante, A.F., et al., "Mechanisms of Wound Reepithelialization: Hints from a Tissue-Engineered Reconstructed Skin to Long-Standing Questions," *Faseb. J.*, Vol. 15, No. 13, 2001, pp. 2377–2389.

[52] Carrier, P., et al., "Characterization of Wound Reepithelialization Using a New Human Tissue-Engineered Corneal Wound Healing Model," *Invest. Ophthalmol. Vis. Sci.*, Vol. 49, No. 4, 2008, pp. 1376–1385.

[53] Magnan, M., et al., "Tissue Engineering of a Genitourinary Tubular Tissue Graft Resistant to Suturing and High Internal Pressures," *Tissue Eng Part A*, Vol. 15, No. 1, 2009, pp. 197–202.

[54] Szabowski, A., et al., "c-Jun and JunB Antagonistically Control Cytokine-Regulated Mesenchymal-Epidermal Interaction in Skin," *Cell*, Vol. 103, No. 5, 2000, pp. 745–755.

CHAPTER

12

3D Culture of Primary Chondrocytes, Cartilage, and Bone/Cartilage Explants in Simulated Microgravity

Nathalie Steimberg, Jennifer Boniotti, and Giovanna Mazzoleni

General Pathology & Immunology Unit, Department of Biological Sciences and Biotechnologies, School of Medicine, University of Brescia, Brescia, Italy
Corresponding author: Giovanna Mazzoleni, MD, General Pathology & Immunology Unit, address: Department of Biological Sciences and Biotechnologies, School of Medicine, University of Brescia, viale Europa 11, I-25123 Brescia, Italy, e-mail: giovanna.mazzoleni@med.unibs.it,
phone: + 39-030-37.17.273/281, fax: + 39-030-37.01.157

Abstract

In vitro tissue models are becoming increasingly important for basic research purposes, as well as for pharmacotoxicology and clinical applications, but *their relevance* is often limited by the rapid loss of cell viability and differentiated functions that usually occur when traditional culture conditions are employed. In this chapter we describe various procedures that allow the three-dimensional (3D), long-term culture of articular chondrocytes, cartilage, and bone/cartilage tissue explants. The key point of our methods is the use of the Rotary Cell Culture System (RCCS) bioreactor that, by reproducing critical aspects of microgravity, generates a particular microenvironment where high mass transfer is attained with low shear stress, thus providing optimal conditions for cell survival and function within large-sized 3D constructs. The results obtained demonstrate that our culture methods can preserve cell viability and differentiated phenotype over several weeks of culture, confirming their value for a wide range of biomedical applications.

Key terms: bone, hyaline cartilage, microenvironment
primary chondrocytes, RCCS bioreactor
simulated microgravity, 3D models
tissue culture

12.1 Introduction

Although appearing to be relatively simple, articular (hyaline) cartilage is, in reality, a highly specialized, dynamic tissue, with unique mechanical properties, that provides a low-friction gliding surface, absorbs mechanical shocks, distributes the joint loads more evenly across the subchondral bone, and can vary its properties in response to differences in loading [1–3]. Hyaline cartilage is an avascular tissue, composed of one single cell type of mesenchymal origin, the chondrocytes (and their progenitor cells), encased in a complex extracellular matrix (ECM) that they produce, and that is structurally organized into four different zones (and three compartments), with different morphological features and mechanical properties. Mature, terminally differentiated articular chondrocytes are unable to proliferate and possess low residual synthetic capacity [4–6]. Specific ECM components include collagen (type II accounts for the 90–95% of total), proteoglycans (small molecules and large aggregating monomers, known as aggrecans), and noncollagenous proteins. This dynamic network of macromolecules controls the hydration level of the tissue, responsible for its mechanical and physical properties, and it is involved in the regulation of chondrocytes' homeostasis and behavior, as well as in the store of specific growth factors and cytokines, essential for tissue morphogenesis and function [1, 7–9].

The degenerative processes of joints' cartilage represent one of the major causes of disability that may affect patients of all ages. Due to its intrinsic biological features, articular hyaline cartilage has, in fact, little capacity for spontaneous self-healing in vivo. In spite of the great effort that has been made over the past four decades by basic scientists and surgeons for developing new strategies aimed at enhancing cartilage regeneration/repair (or its replacement), none of these approaches, at present, fulfills the needs of clinical applications [10, 11]. This is mainly due to the high complexity of this tissue, difficult to reproduce by the procedures currently employed in orthopedic tissue engineering, and to the fact that our knowledge on cartilage physiopathology and responses to environmental physical/chemical stimuli are still largely incomplete. Mechanistic studies are very complicated to be performed on this tissue in vivo, and, from the other side, the difficulty in maintaining isolated viable and fully differentiated native chondrocytes strongly limits the possibility of their study in vitro. It is well known, for example, how chondrocytes' isolation and traditional two-dimensional (2D) culture, prerequisites for cell amplification, entail rapid cell dedifferentiation and a loss of functional specificity [12]. The dedifferentiation process is accompanied by a shift towards a fibroblast-like phenotype, which is characterized by a stop in the expression of aggrecans and type II collagen (articular cartilage-specific markers), followed by an increased expression of type I collagen [13, 14]. However, we, in addition to other groups, clearly demonstrated that this process may be reverted by relocating cells into a more physiological, three-dimensional (3D) microenvironment [15–17].

Since it is well established that the 3D microenvironment plays a pivotal role in cell survival, differentiation, and activity [18], the rapid advances in cell biology and material sciences have greatly contributed to the development of a broad range of 3D in vitro models of cartilage tissue in the past 10 years. Examples include models based on the use of a number of different biomaterials that, by mimicking the mechanobiological properties of hyaline cartilage, were specifically designed to reproduce the physical microenvironment typical of this tissue. Among them, the most commonly employed mechanical supports comprise natural/synthetic nanostructured matrices (e.g., polymers and hydrogels, organized as meshes or sponges) and protein- or polysaccharide-based naturally derived/synthetic microcarriers or scaffolds [9, 19, 20]. The high metabolic requirements of 3D cell/tissue constructs, substantially different from those of flat cell monolayers grown in static environments in liquid media, also promoted the development of a wide variety of bioreactors, designed to generate dynamic culture conditions aimed at increasing mass transfer [21]. Ranging from the simplest spinner flasks to the more complex rotating clinostats, packed-bed and air-lift bioreactors, or their most sophisticated versions (that include load- and perfusion-controlled culture systems) [19, 22–25], all these devices provided significant progress in our knowledge of critical parameters that have to be considered for cartilage engineering.

Nevertheless, while demonstrating the feasibility of such culture approaches, none of these in vitro models is, at present, able to generate a reliable tissue analogous to native cartilage, with respect to quality and stability. Promising strategies to overcome the current limitations may now rely on innovative cell-based strategies, which, improved by emerging technologies, should better take into account the physiological requirements of this tissue and the specificity of its microenvironment. The recent progress in our understanding of regulators of tissue morphogenesis and function has, in fact, led in recent years to give an increasing importance to tissue-specific microenvironments that, in addition to dimensionality, also comprise factors such as the physical properties (stiffness) and the molecular composition of the ECM, the presence of specific soluble factors (e.g., chemical morphogens, growth factors, hormones and cytokines), and the reciprocal interactions that physiologically take place between cells or between cells and ECM [18].

Critical environmental factors that have to be considered in the strategies aimed to generate engineered hyaline cartilage have been discussed by Martin et al. [26] and more recently by Vinatier et al. [9], and comprise the appropriate choice of chondrogenic cell source, the accurate reproduction of the physicochemical characteristics of native cartilage microenvironment, and the use of specific bioreactors, which must fulfill the needs of functional cartilage constructs for sufficiently extended time periods.

In this chapter we describe the experimental procedure that we chose to generate different 3D in vitro models of hyaline cartilage tissue. While each model is based on the same attempt to recreate/preserve the original tissue-specific microenvironment, it presents a different degree of complexity with respect to the native tissue in vivo. In the most simplified model (primary culture of homotypic aggregates of isolated hyaline chondrocytes), the synthesis of neo-ECM is promoted by the culture conditions, while in the most complex models (cartilage and bone/cartilage tissue explants), the original ECMs' composition, cell ECM, and cell-cell interactions are preserved. The key technological point of our methods consists of the use of the Rotary Cell Culture System (RCCS) bioreactor. The RCCS technology was chosen for our purposes since, among all bioreactors actually available for tissue engineering, it represents the unique device able to generate a special microenvironment where high mass transfer is attained with low shear stress. This condition, known as simulated microgravity, has been already largely demonstrated, also by our group, to provide the best conditions to allow long-term in vitro preservation of viability and differentiated functions in cells and tissue explants from various origins [27], including embryonic and adult cartilage/bone cells/explants [28–31].

12.2 Experimental Design

12.2.1 Culture models

Three RCCS bioreactor-based 3D models of hyaline cartilage tissue are described that differ in their degree of complexity with respect to the native tissue in vivo (illustrated in Figure 12.1):

- Homotypic (preformed or spontaneously-induced) cell aggregates from isolated primary chondrocytes (complete loss of the native microenvironment);
- Articular cartilage explants (preservation of the native cartilage microenvironment);
- Bone/cartilage explants (preservation of both native bone and cartilage microenvironments and crosstalks).

12.2.2 The RCCS bioreactor and its operational conditions

The 3D cell and tissue culture was performed by the use of the Rotary Cell Culture System (RCCS) bioreactor (Synthecon Inc., Houston, Texas). A result of N.A.S.A. technolog-

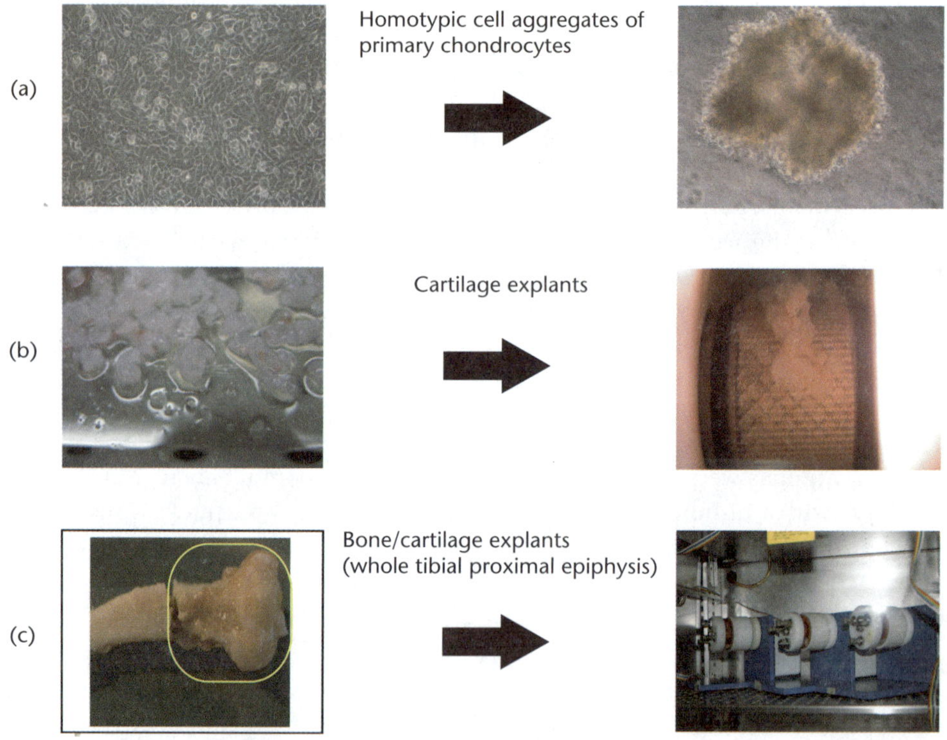

Figure 12.1 Culture models. The various RCCS-based 3D culture methods are considered according to the possibility to recreate/preserve the original tissue microenvironment. Reconstruction of a neo-matrix. (a) After cell amplification in monolayer, chondrocytes form multicellular aggregates in the RCCS bioreactor. Maintenance of the native cartilage microenvironment: the original ECM (composition and structure) is maintained by culturing (b) cartilage explants or (c) articular cartilage with their underlying bone tissue in the RCCS bioreactor. Concerning the culture of tibial proximal epiphyses, the bone/cartilage unit is preserved (the different components of each tissue may reciprocally influence each other).

ical research, and with no internal moving parts, the RCCS device is a transparent, horizontally rotating clinostat, consisting of a cylindrical [slow turning lateral vessel (STLV)] or disk-shaped [high aspect ratio vessel (HARV)] culture chamber; a motorized stand carries the culture vessel, while a controller unit permits to regulate its rotational speed [Figure 12.2(a)] [32]. Culture chambers are equipped with a silicon rubber gas exchange membrane, which allows an optimal diffusion of gas (oxygen) inside the vessel, without bubble generation [Figure 12.2(b)]. Within the culture chamber, no head space is then left between atmosphere and culture medium, so that shear forces and turbulence normally associated with impeller-driven stirred bioreactors are reduced to a minimum; sedimentation and inadequate gas/nutrients supply are also avoided, thus guaranteeing the most favorable conditions for cell/tissue culturing [33]; see also the Synthecon Web site at http://www.synthecon.com for the latest information on the different cell types/tissues that have been successfully cultured in RCCSs.

By adjusting the rotational speed of the culture chamber according to the specific experimental needs (e.g., sample number, density, or dimensions), it is possible to obtain a stable condition where the gravitational field is time-averaged to near zero over each revolution (vector-averaged gravity), thus effectively negating the influence of gravitational sedimentation (balanced with centrifugation and fluid drag) and reproducing some aspects of microgravity (simulated microgravity) [34].

All during the experimental procedures, RCCS operational conditions were constantly monitored and optimized for obtaining a laminar flow of the fluid medium inside the culture chamber, thus reducing to a minimum the mechanical stress (shear force) on the biological samples surface; the rotational speed of the chamber was also continuously controlled and adjusted, so that the samples (nonrotationally stabilized) remained in a constant orientation with respect to the chamber wall and moved in a near-solid body rotation with the fluid, thus fulfilling the requirements needed to successfully simulate microgravity condition [35].

12.2.3 Animals

Newborn New Zealand rabbits were used as cartilage donors. Young (7 to 8 weeks old) male Sprague-Dawley rats, from 200g to 225g of body weight, were used as cartilage/bone donors.

Animals were housed in a climate-controlled room (at about 22°C and 5% humidity), with a 12-hour light-dark cycle; they received tap water and the standard laboratory rat diet ad libitum. All during the experimental procedure, animals were handled in accordance with NIH guidelines for the care and use of laboratory animals. Sacrifice was performed by decapitation.

12.2.3.1 Technical notes

- Cells and tissues may also be obtained from any kind of experimental animal, of both sexes, at any age. Laboratory animals can be chosen in accord with the experimental objective. In the case of tissue explants, it will be necessary to adjust their shape and size, also taking into account key parameters of the RCCS operational conditions (e.g., rotational speed of the culture chamber), as described afterwards.

(a)

(b)

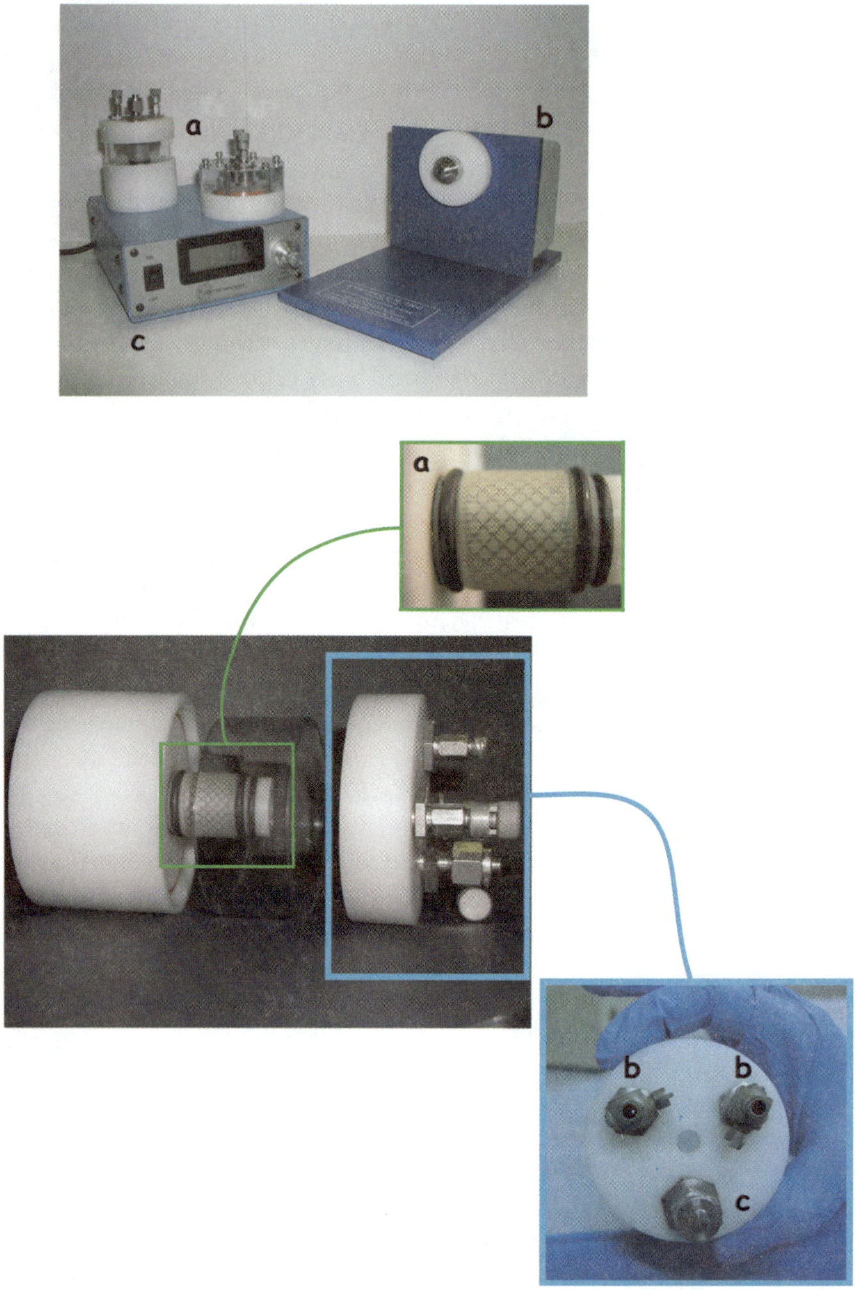

Figure 12.2 The RCCS bioreactor. (A) The RCCS device consists of: (a) a cylindrical (STLV) or disk-shaped (HARV) culture chamber that rotates around a horizontal axis; (b) a motorized stand that carries the culture vessel; and (c) a controller unit that allows for the adjustment of the rotational speed of the culture vessel. (B) The detailed structure of the STLV culture vessel. The culture chamber has a central core, covered by (a) an oxygenator membrane that permits gas exchanges; on the front-end cap, (b) two luer lock syringe ports and (c) a larger fill port (1/4-inch fill port) permit access to the chamber.

- Animal environment, housing, and management should follow local authorites' and institutional policies and responsibilities for the proper care, use, and humane treatment of animals used in research, testing, and education.

12.3 Materials

12.3.1 Equipment for cell/tissue culture and preparation of samples

- Standard incubator with humidified atmosphere at 37°C and 95% air 5% CO_2
- Sterile hood cabinet (class II biosafety cabinet should be more appropriate, especially if a possibly pathologic material is handled)
- High-speed rotary device, equipped with a diamond wheel (Dremel or other)
- Inverted optical microscope
- Rotary Cell Culture System (RCCS) bioreactor (Synthecon, United States)
- STLV and HARV vessels (Synthecon, United States)
- 20-ml and 50-ml sterile syringes
- Sterile 100-mm diameter Petri dishes, treated and not for cell culture (Falcon)
- Sterile scalpels n°21 and n°10
- Sterile small surgical scissors and anatomical fine forceps
- 15-ml and 50-ml centrifuge tubes (Falcon)
- 70-μm nylon cell strainer (Falcon)
- 0.22-μm pore filters (Millipore)
- Sterile Pasteur pipettes

12.3.2 Chemicals

12.3.2.1 Culture media and preparation of samples (sterile or aseptically prepared)

- Invitrogen, unless otherwise indicated
- Trypsin powder
- Collagenase powder (type II, Worthington)
- Ham's F12 nutrient mixture medium (Ham F12 medium)
- RPMI 1640 medium
- Fetal bovine serum (FBS)
- Antibiotic-antimycotic (100×) mix [10,000 units of penicillin G, 10,000 μg of streptomycin, and 25 μg of amphotericin B/ml], liquid (antibiotic mix)
- L-Glutamin 200 mM (100×), liquid
- Hanks' Balanced Saline Solution (HBSS), without calcium and magnesium
- Trypsin 0.05% EDTA solution
- 0.4% Trypan-blue solution (Sigma) (possibly carcinogen, health hazard)

12.3.2.2 Sample embedding and histochemistry

- OCT embedding medium: Tissue-Tek (Electron Microscopy Sciences)
- Liquid nitrogen
- Alcian Blue 8GX (Sigma)
- Dulbecco's Phosphate-Buffered Saline (PBS) solution (Sigma)

Technical and safety notes

1. Nearly all steps of culture procedures must be performed under sterile conditions; all solutions and implements coming into contact with the cells must be germ-free.
2. Always wear normal clean laboratory-work clothing, gloves, and, when necessary, glasses.
3. Refer to local and institutional safety rules for working with animal cells. Be careful when handling sharp instruments.
4. All materials used with cells should be regarded as potentially infected. All materials and solutions should be discarded after chemical disinfection and autoclaved/incinerated.
5. When working with liquid nitrogen, wear gloves, a face mask, and a closed lab coat; in order to avoid risk of asphyxiation from evaporated nitrogen, it is always recommended to work in a well-ventilated room, equipped with O_2 sensors.

12.4 Methods

12.4.1 Preparation of tissue explants

The described procedure refers to rabbit/rat samples, but it may be also adapted to samples of other origin.

1. Clean the skin of anterior and/or posterior limbs with an excess of alcohol (for dissecting tibio-femoral and/or scapulo-humeral articulations, respectively).
2. Carefully isolate the limbs from the body structures with sterile scissors and forceps.
3. Gently remove any hair and skin tissue.
4. Put excised limbs into a 50-ml sterile tube, containing Ham F12 medium supplemented with 2× antibiotic mix. You can preserve samples at 4°C for up to 12 hours.
5. From this step, all steps must absolutely be carried out under sterile conditions (hood cabinet).
6. Remove from the limbs any muscles and tendons and every other noncartilaginous tissue. Hold bones with sterile surgery gloves or with sterile pincers; sterile scissors or scalpel handles and blades are recommended for a quicker and cleaner work.
7. Put isolated bones into a 100-mm diameter Petri dish containing 10–15 ml of Ham F12 medium (supplemented with a 2× antibiotic mix solution). Change the scalpel blades frequently.
8. Eliminate any sticky connective tissue and the articular synovia (the use of sterile forceps and scalpels is recommended).
9. Keep the explanted bones in the 100-mm diameter Petri dish containing the Ham F12 medium (supplemented with a 2× antibiotic mix solution).

12.4.2 Isolation of chondrocytes

The procedure corresponds to that described by Green [36], and it is illustrated in Figure 12.3.

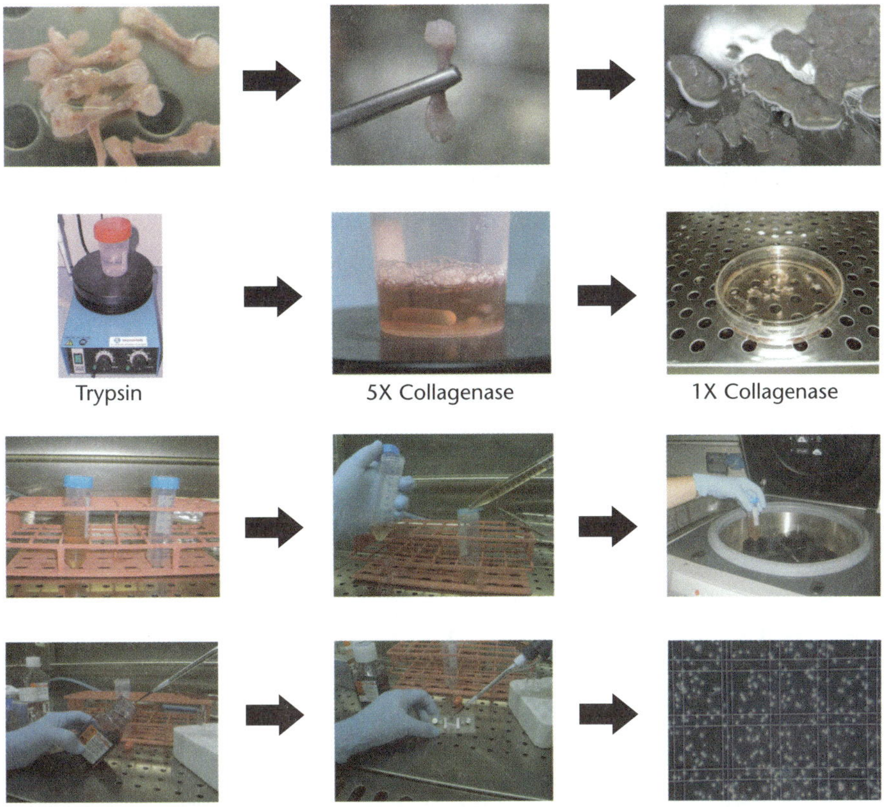

Figure 12.3 Procedure for the isolation of primary chondrocytes from hyaline cartilage. Articular cartilage is isolated from (a) long bones, (b) serial enzymatic digestions, (c) purification of chondrocytes, and (d) assessment of cell density and viability.

1. Hold bone specimens by their diaphyses, and gently isolate cartilage tissue; cut it into small slices, starting from the articular side (proximal part) to the growth plate (distal side). Put each slice into a new 100-mm diameter Petri dish, containing Ham F12 medium (supplemented with 2× antibiotic mix). You can stop the procedure at this step for 2–3 hours.
2. Proteolytic digestion of cartilage:
   - Preparation of Trypsin solution: for each animal, prepare 10 ml of trypsin solution (0.5 mg/ml), by dissolving 5 mg of trypsin into Ham F12 medium. Sterilize by filtration through 0.22-μm filter units.
   - Preparation of collagenase solution: Prepare 15 ml of a 3-mg/ml solution of collagenase, by dissolving the enzyme in Ham F12 (5× solution). Sterilize by filtration through 0.22-μm filter units. Prepare 1× collagenase solution, by adding 4 ml of 5× collagenase solution and 2 ml of FBS to 14 ml of Ham F12 medium (supplemented with 1× antibiotic mix).
3. Transfer cartilage slices into a sterile 150-ml plastic container, filled with 10 ml of trypsin solution. Incubate 25 minutes at room temperature under stirring (sterile magnetic stirring bar).
4. Eliminate trypsin after slices sedimentation.
5. Add 10 ml of 5× solution of collagenase (3 mg/ml) and let under stirrer agitation for 30 minutes at room temperature.

6. Let articular cartilage fragments decant, eliminate surnatant.
7. Add 10 ml of 1× collagenase solution into the 150-ml container, remove the magnetic stirring bar, and transfer all tissue fragments into a 100-mm diameter Petri dish (nontreated for cell culture) (1 dish for each animal); wash the 150-ml container with 10 ml of 1× collagenase solution and add them to the tissue fragments.
8. Close the dish and let the samples overnight at 37°C in the incubator.
9. Transfer all the content of the dish into a 50-ml tube; vortex two times.
10. Filtrate through a 70-μm disposable cell strainer unit (chondrocytes will pass through the filter, leaving fragments of matrix behind them on the filter surface).
11. Centrifugate at 2.300 rpm for 5 minutes at 4°C. Discard the surnatant and resuspend the cell pellets in the Ham F12 medium, completed with antibiotics mix and 10% FBS (complete the Ham F12 medium).
12. Determine the cell density with a hemocytometer (e.g., the Burker chamber).
13. Test the cells' viability by using the Trypan-Blue exclusion test. Typical cell yields range from 4×10^6 to 6×10^6 cells/tibio-femoral articulation.
14. According to the experimental procedure to be performed, cells can be cultured either in 2D or in 3D conditions.

12.4.3 2D culture of isolated chondrocytes (traditional monolayer in static fluid conditions)

1. Seed isolated cells onto 100-mm diameter Petri dishes at the density of 20,000 cells/cm^2.
2. Replace the complete culture Ham F12 medium twice a week.
3. When, usually after 1 week, cells reach confluence, discard the culture medium, rinse the dish with 4-ml HBSS (without Ca^{2+} and Mg^{2+}), and discard the HBBS solution.
4. Add a 500-μl trypsin-EDTA solution onto the cell monolayer, put the dish at 37°C, and verify, by optical microscopy, the time required for all cells to be detached from the plastic support.
    - Add to cell suspension 1 ml of the complete Ham F12 medium to stop the proteolytic activity of trypsin. Transfer the cell suspension into a sterile 15-ml tube.
    - Centrifugate at 2,300 rpm for 5 minutes at 4°C and discard the surnatant.
    - Add about 10-ml complete Ham F12 medium, assess the cell density, and then seed cells into 100-mm Petri dishes at a density of 2×10^4 cells/cm^2.
    - The primary culture of chondrocytes can be kept in a monolayer and propagated until the fourth passage.

12.4.4 3D culture of isolated chondrocytes (homotypic aggregates)

1. Before starting the culture, place the RCCS engine (rotator base) inside the CO_2 humidified incubator and connect it to its own controller unit and then to the electric power.

2. To form cell aggregates directly inside the culture chamber (STLV), we propose to start by loading 1.5×10^6 cells/ml of the culture medium. Alternatively, cell aggregates can be preformed before the RCCS culture (see Section 12.4.4.1).
3. Prepare the STLV culture chamber and verify that the center bolt is constricted (if it is loosened, gently tighten it with the Allen wrench).
4. Slowly add the cell suspension into the culture chamber from the 1/4-inch fill port [Figure 12.2, B(c)]. Use a sterile 10-ml syringe. Close the port.
5. Adjust the two syringes into the luer lock syringe ports on the front of the vessel [Figure 12.2, B(b)]. Fill up with the complete culture medium. Carefully eliminate any air bubble.
6. Close the two syringe ports and replace the end caps.
7. Place the culture chamber onto the motor axis. Turn on the power unit and regulate the rotational speed of the STLV chamber between 8 and 10 rpm.
8. For harvesting the cells, turn off the power and work under a sterile hood.
9. Open the two syringe ports and gently get the necessary volume of cell suspension, aggregates, or culture medium. Alternatively, if cell aggregates are very big, you can remove them from the 1/4-inch port with a 2-ml pipette.
10. Then follow the steps in Sections 12.4.5 to 12.4.7 steps.
11. Since, with time, the size of cell aggregates increases, it will be necessary to adjust the rotational speed of the culture chamber in order to guarantee the optimal operational conditions of the bioreactor (see Section 12.1.2).
12. Change the culture medium every 2 to 3 days.

12.4.4.1 Preformation (induction) of multicellular aggregates

This procedure is based on the so-called hanging-drops technique, usually employed to culture ES cells, and it is illustrated in Figure 12.4. You have to prepare:

1. Several 100-mm diameter Petri dishes (open them and upturn their lids);
2. A cell suspension at a density of 1×10^6 cells/ml.

On the inner side of the inverted lid of the dish, load 20-μl drops of cell suspension (use a multichannel pipette). Let at least 0.5 cm between each individual drop in order to avoid their mixing. When the lid surface is quite full, quickly replace the lid onto the bottom part of the Petri dish (previously filled with about 10 ml of sterile PBS, supplemented with 1× antibiotic mix), and let cells fall on the bottom of the drop spreads. Gently place the dishes inside the incubator, and let them between 24 to 48 hours at 37°C, 95% humidity and 5% CO_2. Wash the dish cover with the complete medium to retrieve all cell aggregates, and load them into the STLV chamber, as described for isolated cells.

Note that this method gives rise to the formation of aggregates uniform in size, but, since it is impossible to refresh the culture medium for the whole period of time required for their formation, it may be disadvantageous (some high-metabolically active cell types suffer from the limited diffusion of nutrients/gas in the deeper part of the aggregates, leading to central necrosis).

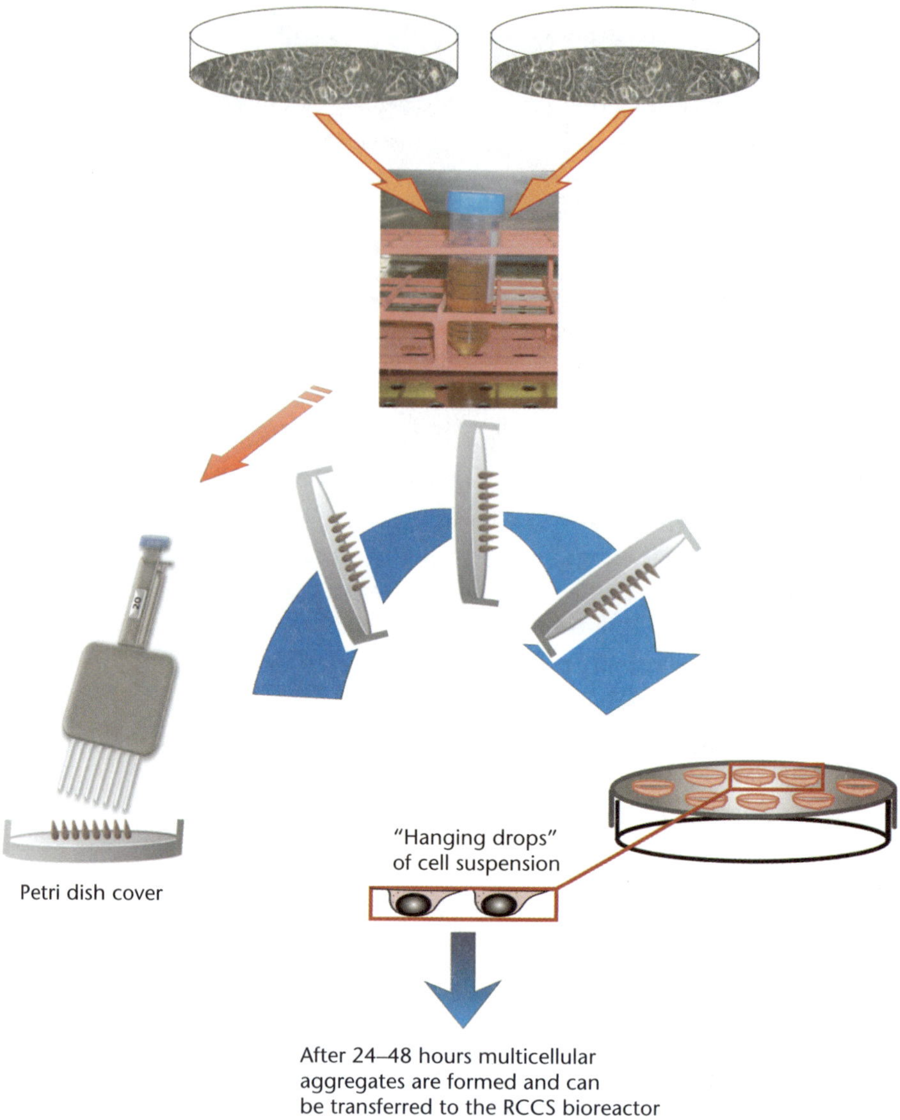

Figure 12.4 Preformation of cell aggregates: main steps of the procedure. Cells are amplified in monolayer for 1 week, and then resuspended by a trypsin treatment; 20 μl of cell suspension are dropped on a Petri dish cover and let to form aggregates for 24 to 48 hours.

12.4.5 3D culture of fragments of articular cartilage explants

1. Carefully hold the bones (previously prepared as described in Section 12.4.1) by their diaphysis and gently cut the articular cartilage tissue into 1 mm^3 pieces. Put the fragments into a 100-mm diameter Petri dish, containing Ham F12 (supplemented with 2× antibiotics mix).
2. When all the tissue fragments have been prepared, open the STLV culture vessel by unscrewing the center bolt located on its top and carefully take away the front-end cap. Do it under sterile conditions.
3. Load tissue fragments into the STLV vessel. Lock the vessel 1/4-inch fill port. Then follow the steps in Sections 12.4.5 to 12.4.7.

4. Use the 1/4-inch fill port to harvest the tissue samples; follow the steps in Sections 12.4.4.5 to 12.4.4.7.

12.4.6 3D culture of undissected, complete proximal tibial epiphyses

1. You can work with whole tibial epiphyses derived from small rodents. In the case of larger donors, you have to reduce the size of tissue samples (and adapt their shape), to meet the operational requirements of the RCCS bioreactor in order attain the microgravity condition.
2. Prepare the tissue samples by following steps 1–9 in Section 12.4.1.
3. Install the alcohol-sterilized diamond wheel on your high-speed rotary device (let alcohol air dry before using it and wear protective glasses).
4. Cut the tibia's bone as required for your experimental aim. Take care to wash the diamond wheel with the culture medium (without serum) between each incision, in order to avoid warming of the samples. Put the sectioned samples into a container filled with the culture medium to preserve the cell viability.
5. When the samples are ready, load them into the STLV vessel, as previously described for cartilage fragments (see Section 12.4.5).
6. Since bone explants are difficult to introduce into the culture chamber through the 1/4-inch fill port, you can open it completely.
7. The rotational speed must be regulated as a function of the size of the bone samples (usually higher than 50 rpm).
8. Change the RPMI complemented medium twice a week.
9. Harvest samples by completely and sterilely opening the culture vessel.

12.4.7 Histomorphological study of chondrocytes and cartilage tissue

Prepare samples for cryosectioning by embedding them in Tissue-Tek OCT and subsequently snap-freeze them in liquid nitrogen. Embedded samples should be kept at –20°C until use. Prepare 5–10-μm-thick serial slices by using a common cryostat and then keep them at –20°C for further investigations.

Hematoxylin-eosin staining

Let the frozen sections warm at room temperature and subsequently incubate them for about 10 minutes in Harris' Hematoxilin solution, wash them twice in tap water, and incubate them for 15 seconds in an eosin solution. After washing, dehydrate them with serial passages in 70° ethanol (1 minute), 95° ethanol (1 minute), pure ethanol (1 minute), and xylol (1 minute). Mount the sections with a DPX mounting medium (Electron Microscopy Sciences, United States).

Alcian-blue staining

This staining can be used to study cartilaginous ECM composition (proteoglycans) on frozen slices of homotypic cell aggregates and tissue explants or on freshly harvested cell monolayers (2D culture). Fix the samples in 2% glacial acetic acid in absolute ethanol (5 minutes), and rehydrate them with serial passages in ethanol. Stain samples overnight in 0.5% alcian-blue solution (0.1N HCl). Wash twice with 0.1N HCl to remove unbound dye and rinse with tap water. Mount the sections with EUKITT mounting medium (Electron Microscopy Sciences, United States).

Technical notes on the operational conditions of the RCCS bioreactor

- Regarding the sample size (and shape), remember that you have to reach an equilibrium between a sample mass and the rotational speed of the culture vessel (samples should not exceed certain dimensions).
- Since the main advantage of using RCCS technology consists of its capacity to generate a simulated microgravity environment inside the culture chamber, specific requirements must be satisfied concerning its operational conditions (treated in Section 12.1.2) (refer to [35] for a detailed discussion of these requirements in the specific context of STLV and HARV chambers). Note that the condition of free fall, where 3D constructs are maintained by being suspended in a relatively stable position, does not correspond to simulated microgravity, and may also imply undesired collisions between the samples themselves and between samples and the chamber walls (mechanical damage).

12.5 Anticipated Results

The most important characteristics of the described culture methods are illustrated in Figure 12.5:

- The procedure for isolating primary articular chondrocytes implies serial enzymatic treatments that cause significant cell stress; it is also a time-consuming step (about 24 hours versus 1–2 hours required for the culture of tissue explants).
- Even if it allows cell amplification, the 2D context induces a rapid cell dedifferentiation (second subculture passage) [Figure 12.6(a)], and a limited cell survival (fourth subculture passage), only short-term studies are allowed.
- RCCS-based 3D culture (spontaneously formed or preinduced cell aggregates and tissue explants) favors cell survival and the maintenance of differentiated phenotype expression (up to 4 weeks of culture); the preservation of the original microenvironment (tissue explants) reproduces more closely the in vivo situation [Figure 12.6(b)].

With our bioreactor-based culture method, isolated chondrocytes can also be seeded within various naturally derived or synthetic scaffolds and maintained successfully in a culture for several weeks (results are not shown). Various tissue-engineering approaches aimed at generating functional 3D substitutes of hyaline cartilage that fulfill the requirements of an appropriate cell source, mechanical support, and bioactive factors should then be tested for specific applications.

Moreover, one of the main advantages of all the described RCCS-based culture models consists in the possibility of applying to the 3D constructs advanced and innovative analysis procedures, in addiction to all the other histo-/immunomorphological, biochemical, and molecular methods currently employed for analyzing freshly explanted tissue samples. An example is shown in Figure 12.6(c), which displays a 3D reconstruction of high-resolution microcomputed tomography (micro-CT) scans of a rat tibial proximal epiphysis after 4 weeks of culture in the RCCS bioreactor. Micro-CTs were per-

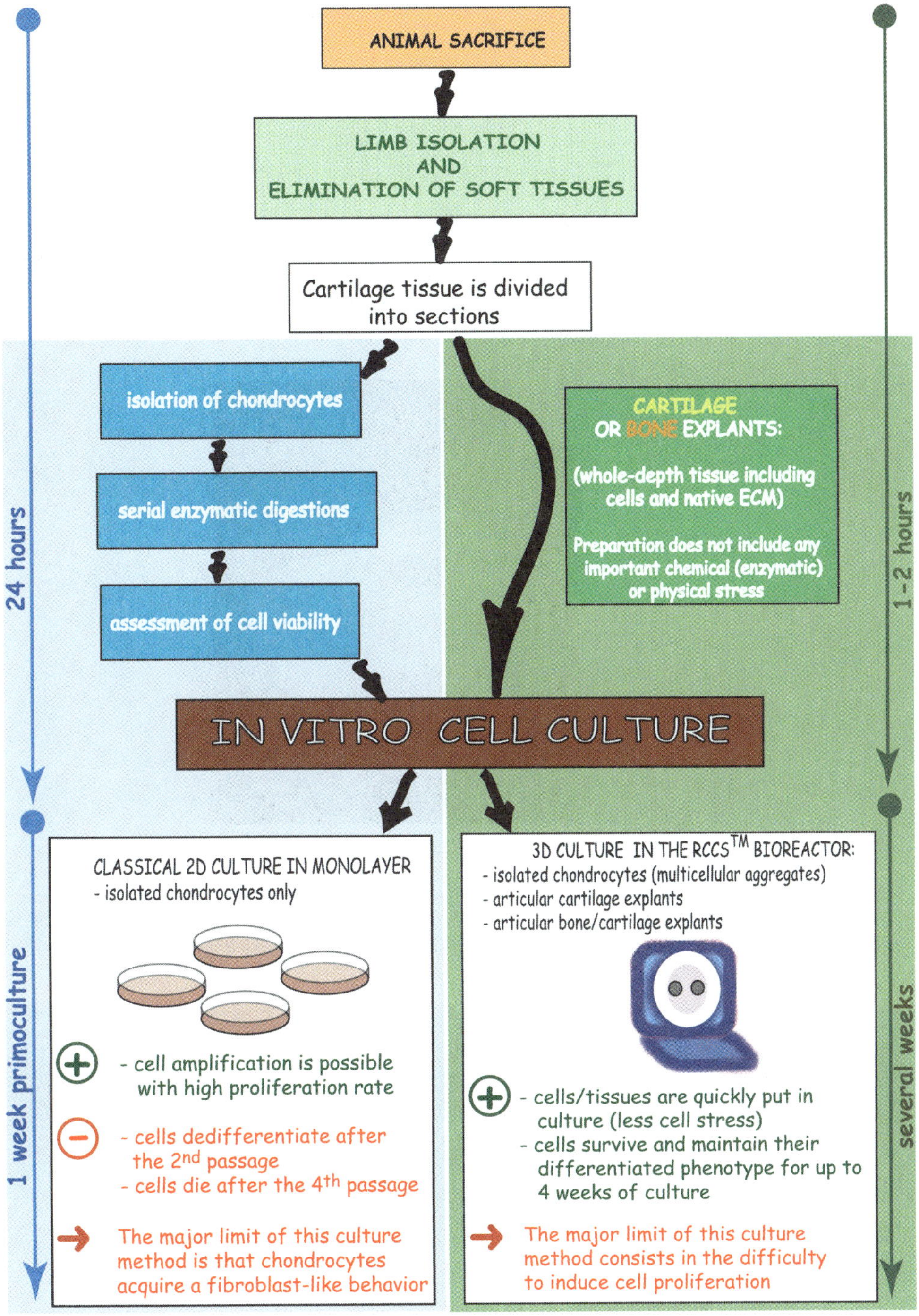

Figure 12.5 Summary of the culture procedures.

formed at the Elettra Synchrotron facility (Trieste, Italy). In the reported example, we have quantified, by a recently introduced numerical method, changes in the mechanical properties of the trabecular structure of bone specimens kept in simulated microgravity conditions for different time periods (3 days to 4 weeks) [30]. This was also the first

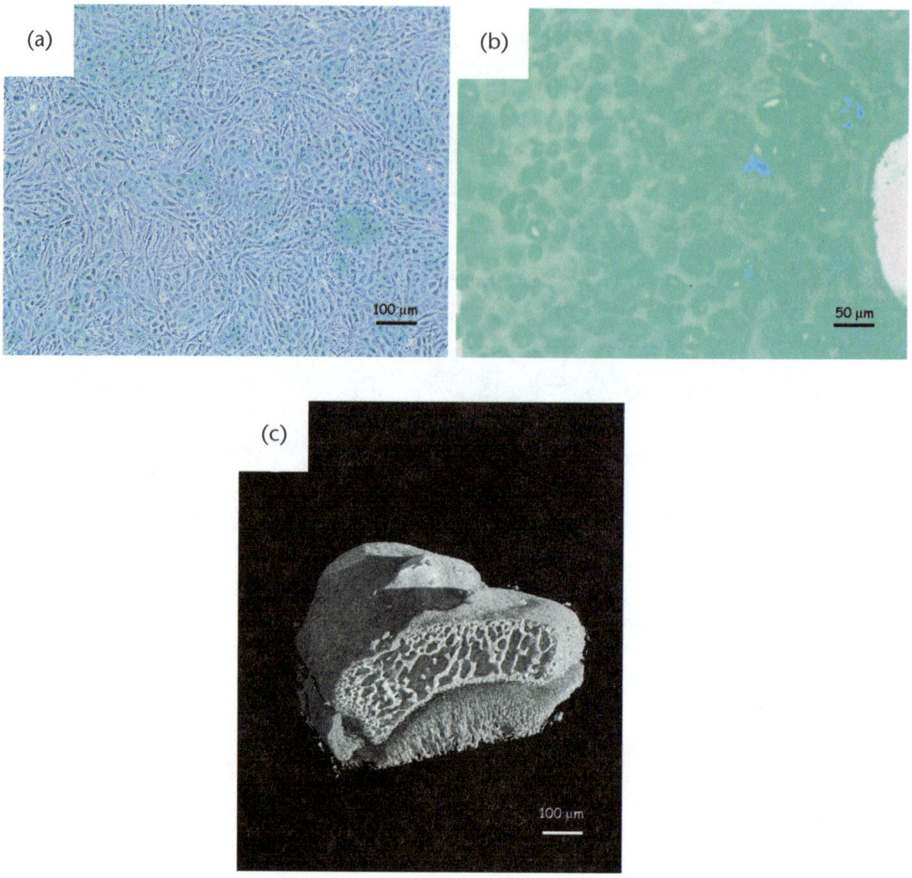

Figure 12.6 Morphological analysis of cell/tissue samples. Synthesis of ECM proteoglycans was estimated after Alcian blue staining: cells were cultured (a) in a monolayer for 1 week or (b) in the RCCS bioreactor (whole tissue explants) for 4 weeks. (c) A 3D reconstruction of the microtomographed structure of a whole tibial proximal epiphysis after 4 weeks of culture in the RCCS bioreactor in simulated microgravity conditions is proposed to illustrate the wide range of analytical methods that can be applied to our culture models.

attempt to try to establish, by the use of the RCCS bioreactor, a 3D in vitro model of whole bone tissue/cartilage (organ culture) from adult organisms.

12.6 Discussion

12.6.1 Discussion of pitfalls

Tissue digestion for primary cells isolation must be performed according to the indicated time schedule, in order to prevent low yield of isolated cells, or low cell viability.

The main trouble encountered while operating with the RCCS bioreactor is the formation of air bubbles inside the culture vessel that can compromise cell viability and function; the empirical definition of the optimal angular velocity represents another critical point of its use.

Culture methods should be carefully chosen (and adapted) in accordance with the specific experimental aims/needs.

RCCS-based 3D culture is expensive and requires a good-level experience in cell/tissue culturing to be successfully exploited.

12.6.2 General discussion and commentary

First developed to simplify in vivo models, in vitro cell culture contributes to physiopathological and pharmacotoxicological issues. Nevertheless, the most diffused 2D primary culture of isolated cells is progressively revealing its limits for numerous applications, due to the rapid loss of cell viability and differentiated functions it entails. In the case of cartilage specifically, primary chondrocytes usually dedifferentiate after the first subculture passage. This loss of cartilage-specific phenotype has also been observed in our experimental conditions, where, already after 1 week of monolayer culture, proteoglycans expression is drastically downregulated [Figure 12.6(a)].

Due to the low repair response of hyaline cartilage and to the consequent high clinical and socioeconomic impact of joint disease, a wide range of different in vitro tissue engineering approaches aimed at generating functional substitutes of cartilage tissue have been established and evaluated during the past years. Among all culture systems and bioreactors that have been used, RCCSs devices were proven to provide the best conditions to generate and grow, even for long time periods, 3D cartilagineous constructs that assemble and possess properties similar to those of native tissue [6, 27, 28; see also RCCS-based models of cartilage tissues at http://www.synthecon.com]. However, although promising, the structural and mechanical properties of these constructs still remained subnormal with respect to those of hyaline cartilage in vivo [6, 28].

Since one of the major aspects that has to be considered in generating functional 3D substitutes of native tissues consists in trying to reproduce the characteristics of native microenvironment where cells develop and grow, we propose here three 3D RCCS-based hyaline cartilage models that take into account the complexity of the original tissue-specific microenvironment. These models range from the simpler configuration of homotypic aggregates of isolated hyaline chondrocytes to the more complex configurations of tissue (articular cartilage explants) or organ (articular bone/cartilage explants) culture.

Even after several weeks of culture, all the 3D-designed models allow the preservation of cell viability and morphology (aggregates), tissue-specific histoarchitecture (tissue and organ culture), and differentiated properties, demonstrated by the expression of proteoglycans, one of the major hyaline cartilage components [31].

A summary of the characteristics and potential applications of each model proposed in this chapter (advantages and shortcomings included) is shown in Table 12.1.

For basic research purposes, for example, it is surely the preservation of the most complex tissue-specific microenvironment (organ/tissue culture), which could be of interest. Being closer to the in vivo situation, this condition may, in fact, allow the gain of new knowledge about the physiopathological processes that regulate cell fate and responses to various environmental stimuli (e.g., chemical/physical/biological agents). For testing aims, on the contrary, more easy-to-use and simplified models will be useful (providing that they still preserve the needed critical biological responses). In the case of clinical use, finally, culture models should satisfy, at a time, several needs. For example, they must be simple to handle, standardized, and safe; they must also permit the expansion of autologous chondrocytes in vitro (human articular biopsies are quite small and difficult to obtain), and, finally, they must provide engineered constructs complex enough to replace damaged tissue (implants) or sufficiently sophisticated to favor its repair (cell-based grafts). Subsequent use of 2D (cell amplification) and 3D RCCS-based models (generation of physiologically relevant 3D constructs that may also rely on all the possibilities offered by new biomaterials) seems a promising strategic approach to this challenging issue of medicine.

Lastly, in the vision and strategy of the 3Rs concept of alternatives to animal use for basic research and testing purposes [37], even if relying on experimental animals for primary cells/tissue supply, our 3D culture models fulfill all the three requirements.

1. *Refine*: Exclusively used as organ donors, animals can be easily handled in accordance with international guidelines and will be not subjected to painful procedures.
2. *Reduce:* Used for clarifying mechanisms of physiopathological processes and for identifying new significant toxicological endpoints, all models can reduce the need of experimental animals required for research and testing purposes. Moreover, single donor animals will provide multiple experimental samples and internal controls. With respect to the traditional in vivo testing procedure, for example, we have estimated that, even in its tissue culture configuration, our model system may permit the reduction of the number of required animals from 36 to 1.
3. *Replace:* In its cell-based configuration, especially if it will be able to integrate in its design new concepts and technologies, our culture model has the potential to provide more reliable and physiologically significant tissue substitutes that, particularly if they are based on human samples, may even supersede animal models.

The possibility to use either normal or pathological tissues of human origin, even obtained from poorly invasive surgical techniques (e.g., needle biopsies), the rapid development and optimization of specific culture protocols, and new analysis methods (with their subsequent standardization and validation) will also offer to these models concrete prospects of filling the gap between animal- and human-based studies, also opening new opportunities for their application in basic research, pharmacotoxicology, and clinical fields.

12.7 Application Notes

The 3D culture of isolated chondrocytes or articular-derived tissues in the RCCS bioreactor represents a potent tool that can be applied, as an in vitro alternative to animal use, in a wide range of studies of hyaline cartilage physiopathology, ranging from basic research to pharmacotoxicology and testing. This culture system could present also good perspectives for the development of engineered tissues for clinical applications in the orthopedic field.

The proposed RCCS-based culture models can be used either with isolated cells (as monotypic or heterotypic aggregates), provided or not with mechanical supports (e.g., scaffolds), or with whole tissue explants of human origin (normal/pathologic), to overcome the problems linked to cross-species extrapolation of results (research) or to provide reliable tissue-engineered substitutes (clinical applications). New concept (e.g.,

Troubleshooting Table

Problem	Step of the Procedure or Culture Conditions	Possible Explanation	Potential Solution
Low yield of isolated chondrocytes	Tissue digestion	Low enzyme specific activity.	Always choose the same lot number of collagenase after testing it.
Low yield of isolated chondrocytes or low cell viability	Tissue digestion	Old animals (articular cartilage presents low cellularity).	Use younger animals to reach your experimental aim.
		Young animals (ECM is less dense and cells are more susceptible to the proteolytic action of trypsin and/or collagenase).	Decrease the time of enzymatic treatment.
Irregular cell seeding	2D monolayer culture (Petri dish)	Plastic ware sometimes presents some irregularity that hinders cell adhesion.	Check the quality of glassware support.
Too quick loss of differentiated phenotype	2D monolayer culture in (Petri dish)	Chondrocytes in the monolayer culture usually lose collagen expression after the second passage; in our hands, a huge difference exists between properties of different lots of serum to this concern.	Test the batch of serum; always use the same batch number for a determinate sets of experiments.
High yield of dead cells	3D culture in RCCS (all models)	Culture medium acidification (usually observed at the beginning of the culture).	Take care to replace culture medium at least twice a week. Reduce cell density in the culture vessel.
	3D culture in RCCS (explants or aggregates)	Peripherical localization of dead cells: too high shear stress.	Eliminate air bubbles. Down-regulate rotational speed of the STLVI.
	3D culture in RCCS (explants or large aggregates)	Central necrosis, dead cells (inner part of 3D constructs/explants): low mass transfer (poor nutrients/gas diffusion, waste accumulation).	Increase the rotational speed of the culture vessel to improve waste/nutrient/gas exchanges.
	3D culture in RCCS (bone explants)	Superficial layers of cells are denaturated.	Bone fragments must be cut hurriedly and chilled at once in order to avoid burn up of cells.

systems biology) technological approaches and analytical methods may be integrated in the RCCS technology to increase the physiological relevance of the models.

12.8 Summary Points

1 In this chapter we describe three different 3D culture models of articular hyaline cartilage that are based on the use of the RCCS bioreactor, whose operational conditions allow the generation of a highly controlled, fluid dynamic microenvironment, where high mass transfer is attained with low shear stress. This condition is known as relative microgravity.

2. The proposed models present different degrees of complexity with respect to the native tissue (original microenvironment) in vivo. In the most simplified model (the primary culture of homotypic aggregates of isolated hyaline chondrocytes), the synthesis of neo-ECM is promoted by the culture conditions, while in the most

complex models (cartilage and bone/cartilage tissue explants), the original ECMs' composition and cell-ECM and cell-cell interactions are preserved.

3. These models, easy to settle up and to use, enable the long-term preservation of cell viability and cell differentiated functions, can be applied on samples of human origin, and allow all types of histo-/immunomorphological, biochemical, and molecular analyses currently employed to study tissue samples.
4. The highly controlled microenvironment generated inside the culture chamber permits the evaluation of long-term effect(s) of various chemical/physical/biological agents on cell viability and behavior, confirming their value for a wide range of biomedical applications in the fields of basic and applied research.
5. The possibility to improve the models with new concepts and technologies also offers good perspectives for the development of more physiologically relevant cartilage substitutes for clinical applications.

Acknowledgments

We thank Professor Ugo Pazzaglia and Dr. Guido Zarattini (Orthopedic Clinic, Brescia University) for the fruitful discussions and their significant contribution to the histomorphological analyses of cartilage explants, and we thank Mrs. Patrizia Barone for her skilled technical support during the sample preparation at Brescia University. This work has been partly supported by European Union grant LSHB-CT-2006-037168 (EXERA) and by funds of the University of Brescia to G.M.

References

[1] Wong, M., and D. R. Carter, "Articular Cartilage Functional Histomorphology and Mechanobiology: A Research Perspective," *Bone*, Vol. 33, No. 1, 2003, pp. 1–13.

[2] Alsalameh, S., et al., "Identification of Mesenchymal Progenitor Cells in Normal and Osteoarthritic Human Articular Cartilage," *Arthritis Rheumatism*, Vol. 50, No. 5, 2004, pp. 1522–1532.

[3] Bhosale, A., and J. B. Richarson, "Articular Cartilage: Structure, Injuries and Review of Management," *British Medical Bulletin*, Vol. 87, 2008, pp. 77–95.

[4] Buckwalter, J. A., and H. J. Mankin, "An Instructional Course Lecture, The American Academy of Orthopaedic Surgeons," *Journal of Bone Joint Surgery,* Vol. 79-A; 1997, pp. 600–611.

[5] Salter, D. M., "The Tissues We Deal with (II) Cartilage," *Current Orthopaedics,* Vol. 12, 1998, pp. 251–257.

[6] Temenoff, J. S., and A. G. Mikos, "Review: Tissue Engineering for Regeneration of Articular Cartilage," *Biomaterials*, Vol. 21, No. 5, 2000, pp. 431–440.

[7] Guilak, F., et al., "The Deformation Behavior and Mechanical Properties of Chondrocytes in Articular Cartilage," *Osteoarthritis and Cartilage*, Vol. 7, No. 1, 1999, pp. 59–70.

[8] Blain, E. J., "Involvement of the Cytoskeletal Elements in Articular Cartilage Homeostasis and Pathology," *International Journal of Experimental Pathology*, Vol. 90, No. 1, 2009, pp. 1–15.

[9] Vinatier, C., et al., "Cartilage Engineering: A Crucial Combination of Cells, Biomaterials and Biofactors," *Trends in Biotechnology*, Vol. 27, No. 5, 2009, pp. 307–314.

[10] Hunziker, E., "Articular Cartilage Repair: Basic Science and Clinical Progress. A Review of the Current Status and Prospects," *Osteoarthritis and Cartilage*, Vol. 10, 2001, pp. 432–463.

[11] Getgood, A., et al., "Articular Cartilage Tissue Engineering," *The Journal of Bone and Joint Surgery*, Vol. 91-B, 2009, pp. 565–576.

[12] Darling, E. M., and K. A. Athanasiou, "Rapid Phenotypic Changes in Passaged Articular Chondrocyte Subpopulations," *Journal of Orthopaedic Research*, Vol. 23, 2005, pp. 425–432.

[13] Benya, P. D., and J. D. Shaffer, "Dedifferentiated Chondrocytes Re-Express the Differentiated Collagen Phenotype When Cultured in Agarose Gels," *Cell*, Vol. 30, No. 1, 1982, pp. 215–224.

[14] Schnabel, M., et al., "Dedifferentiation-Associated Changes in Morphology and Gene Expression in Primary Human Articular Chondrocytes in Cell Culture," *Osteoarthritis Cartilage*, Vol. 10, 2002, pp. 62–70.

[15] Lemare, F., et al., "Dedifferentiated Chondrocytes Cultured in Alginate Beads: Restoration of the Differentiated Phenotype and of the Metabolic Responses to Interleukin-1 beta," *Journal of Cellular Physiology*, Vol. 176, No. 2, 1998, pp. 303–313.

[16] Domm, C., et al., "Redifferentiation of Dedifferentiated Bovine Articular Chondrocytes in Alginate Culture under Low Oxygen Tension," *Osteoarthritis Cartilage*, Vol. 10, 2002, pp. 13–22.

[17] Malda, J., et al., "Expansion of Bovine Chondrocytes on Microcarriers Enhances Redifferentiation," *Tissue Engineering*, Vol. 9, 2003, pp. 939–948.

[18] Mazzoleni, G., D. Di Lorenzo, and N. Steimberg, "Modelling Tissues in 3D: The Next Future of Pharmaco-Toxicology and Food Research?" *Genes & Nutrition*, Vol. 4, No. 1, 2009, pp. 13–22.

[19] Schulz, R. M., and Bader A., "Cartilage Tissue Engineering and Bioreactor Systems for the Cultivation and Stimulation of Chondrocytes," *European Biophysics Journal*, Vol. 36, 2007, pp. 539–568.

[20] Chung, H. J., et al., "Highly Open Porous Biodegradable Microcarriers: In Vitro Cultivation of Chondrocytes for Injectable Delivery," *Tissue Engineering Part A*, Vol. 14, No. 5, 2008, pp. 607–615.

[21] Martin, Y., and P. Vermette, "Bioreactors for Tissue Mass Culture: Design, Characterization, and Recent Advances," *Biomaterials*, Vol. 26, 2005, pp. 7481–7503.

[22] Pei, M., et al., "Bioreactors Mediate the Effectiveness of Tissue Engineering Scaffolds," *The FASEB Journal*, Vol. 16, 2002, pp. 1691–1694.

[23] Hussein, M. A., et al., "On the Lattice Boltzmann Method Simulation of a Two-Phase Flow Bioreactor for Artificially Grown Cartilage Cells," *Journal of Biomechanics*, Vol. 41, 2008, pp. 3455–3461.

[24] Schulz, R. M., et al., "Development and Validation of a Novel Bioreactor System for Load- and Perfusion-Controlled Tissue Engineering of Chondrocyte-Constructs," *Biotechnology and Bioengineering*, Vol. 101, No. 4, 2008, pp. 714–728.

[25] Gigout, A., M. D. Buschmann, and M. Jolicoeur, "Chondrocytes Cultured in Stirred Suspension with Serum-Free Medium Containing Pluronic-68 Aggregate and Proliferate While Maintaining Their Differentiated Phenotype," *Tissue Engineering, Part A*, Vol. 15, No. 8, 2009, pp. 2237–2248.

[26] Martin, Y., O. Démarteau, and A. Braccini, "Recent Advances in Cartilage Tissue Engineering: From the Choice of Cell Sources to the Use of Bioreactors," *JSME International Journal-Series C*, Vol. 45, No. 4, 2002, pp. 851–861.

[27] Unsworth, B. R., and P. I. Lelkes, "Growing Tissues in Microgravity," *Nature Medicine*, Vol. 4, 1998, pp. 901–907.

[28] Vunjak-Novakovic, G., et al., "Microgravity Studies of Cells and Tissues," *Annual New York Academy of Sciences*, Vol. 974, 2002, pp. 504–517.

[29] Klement, B. J., et al., "Skeletal Tissue Growth, Differentiation and Mineralisation in the NASA Rotating Wall Vessel," *Bone*, Vol. 34, 2004, pp. 487–498.

[30] Cosmi, F., et al., "Structural Analysis of Rat Bone Explants Kept In Vitro in Simulated Microgravity Conditions," *Journal of the Mechanical Behavior of Biomedical Materials*, Vol. 2, 2009, pp. 164–172.

[31] Steimberg, N., et al., "3D Culture of Articular Cartilage Explants: New Perspectives for Future Possible Clinical Applications," *Journal of Bone and Joint Surgery—British Volume*, Vol. 91-B, Supp. II, 2009, p. 269.

[32] Hammond, T. G., and J. M. Hammond, "Optimized Suspension Culture: the Rotating-Wall Vessel," *American Journal of Physiology. Renal Physiology*, Vol. 281, 2001, pp. 12–25.

[33] Schwarz, R. P., T. J. Goodwin, and D. A. Wolf, "Cell Culture for Three-Dimensional Modeling in Rotating-Wall Vessels: An Application of Simulated Microgravity," *Journal of Tissue Culture Methods*, Vol. 14, 1992, pp. 51–57.

[34] Klaus, D. M., "Clinostats and Bioreactors," *Gravitational Space Biology Bulletin*, Vol. 14, 2001, pp. 55–64.

[35] Ayyaswamy, P. S., and K. Mukundakrishnan, "Optimal Conditions for Simulating Microgravity Employing NASA Designed Rotating Wall Vessels," *Acta Astronautica*, Vol. 60, 2007, pp. 397–405.

[36] Green, W. T., "Behavior of Articular Chondrocytes in Culture," *Clinical Orthopaedics and Related Research*, Vol. 75, 1971, pp. 248–260.

[37] Russell, W., and R. Burch, *The Principles of Humane Experimental Technique*, London, U.K.: Methuen & Co., 1959.

CHAPTER

13

Alternatives for Absorption Testing

Monika Schäfer-Korting and Sarah Küchler

Institut für Pharmazie, Freie Universität Berlin, Berlin, Germany,
Corresponding author: Monika Schäfer-Korting,
address: Institut für Pharmazie Königin-Luise-Str. 2-4, D-14195 Berlin, Germany
e-mail: msk@zedat.fu-berlin.de, phone: (+49) 030-838 53284, fax: (+49) 030-83854399

Abstract

This chapter deals with validated and standardized methods to determine the skin absorption of substances avoiding animal testing. Hereby it is distinguished between skin penetration and permeation. There is a variety of published test methods, and we focus on the most common procedures. The current gold standards—excised human skin, pig skin, and commercially available 3D skin models—are considered. For data evaluation the most important procedures—high-performance liquid chromatography, scintillation counting, and fluorescence microscopy—are described.

Key terms
excised skin
Franz diffusion cell
reconstructed human skin
skin absorption

13.1 Introduction

Knowledge about the amount of skin absorption is important as side effects like sensitization or systemic effects can occur when topically applied substances make contact with the skin. In fact, next to i.v. administration, topical drug application is the exposure with the most risks with respect to allergic side effects excluding, for example, the topical use of penicillin. Moreover, the majority of pesticide intoxications are due to dermal uptake. The additional major influence of the formulation besides the intrinsic absorbability of a compound adds to the need of testing. Therefore, the assessment of dermal penetration and the permeation of xenobiotics including drugs are crucial alternatives for testing and are needed as trials with humans and animals raise ethical questions. Thus, several methodologies for in vitro experiments are established. Three different aspects have to be considered: the in vitro test systems, the test matrix, and detection methods. For absorption testing the most common in vitro methods use Franz diffusion cells [1], either static-typed or flow-through cells, whereby both approaches lead to identical results [2]. For the determination of skin absorption in vitro, the use of excised human or animal skin is widely accepted. Hereby, viable full-thickness skin, nonviable full thickness skin, dermatomed skin, isolated epidermis, and, more rarely, separated stratum corneum are used.

Problems can arise from the limited quantities of human skin obtained from surgery or postmortem. This will be most relevant in the near future due to the major increase in toxicological screening in the EU because of REACH (registration, evaluation, authorisation and restriction of chemicals), which asks for data documentation of all compounds produced or imported in amounts of 1 ton per year or more. Therefore, animal skin of mainly mice and pigs is used instead and a test guideline was approved [3]. Yet the OECD also adopted a guideline for in vitro testing of skin absorption and skin penetration [4, 5].

Domestic pig skin is most similar to human skin on morphological base. In contrast, in furry animals like mice, skin is more permeable than human skin owing to the higher amount of skin appendages, but also in the differences in stratum corneum lipids and skin hydration [5–7].

It is important to notice that the animal skin is not obtained from animals that were killed just because of the skin. Collaborations with facilities of experimental animal sciences or local abattoirs can allow for fresh skin. Thus, the 3Rs principle (refine, reduce, and replace) is applied by reducing the number of animals and circumventing the potential harm of tested agents in living animals.

For further improvement of in vitro testing in the last decade, great efforts have been made to develop human-based reconstructed skin models. Today, reconstructed human epidermis (RHE) (e.g., EPISKIN, EpiDerm, SkinEthic) is commercially available and validated for skin absorption testing [5]. The epidermal architecture is highly similar to native skin. All major epidermal layers are present and the intra- and interbatch variability is low [8]. In comparison to human and pig skin, RHEs are clearly more permeable and the lag time (the delay of drug permeation after application) for RHEs is low or absent [5, 9]. These effects are due to its insufficiently developed stratum corneum [8]. The content of important barrier lipids like free fatty acids is lower as in native skin, providing that differences in lipid lamellar organization and the profiles of ceramides are incomplete. Dysfunctions in the differentiation process also contribute to the impaired barrier properties. Yet there is a continuous approach for an improvement.

Perfusion models were hoped to be the missing link of in vivo and in vitro methods [10]. Since there is not formal validation, much work remains to be done before the perfused skin model will be used regular for skin absorption testing. Alternatively, mathematical models have been described. These models are based on relationships between physicochemical properties of a compound and quantities such as permeability. Mathematic models often suffer from being too complex to be practically useful or too simplified to sufficiently depict the complex processes of skin absorption in human skin [11].

Depending on the aim of the skin absorption studies, either to study skin penetration or permeation, various detection methods are established. Depending on the tested agent and the approach, HPLC and scintillation counting being most relevant [5], fluorescence reading is often used for marker agents [12].

To obtain deeper insight into the field of alternatives for absorption testing, a comprehensive overview is given in the following.

13.2 Materials

13.2.1 Franz diffusion cell

- Franz diffusion cell, for example, Model 4G-01-00-15-12 (PermeGear, Bethlehem, Pennsylvania)
- Thermostat
- Magnetic stirrer
- Connecting tubes

13.2.2 Consumables

- Human, pig, or reconstructed skin
- Disposable syringes
- Forceps
- Scalpels
- Knives
- Kleenex tissues
- Parafilm or Nescofilm
- Franz cell inserts (if needed when using small piece of RHE)
- Tissue freezing medium
- Peel Aways

13.2.3 Chemicals and solutions

- Phosphate buffered saline pH 7.4 (PBS): KCl 0.2g, NaCl: 8.0g, KH_2PO_4: 0.2g, $Na_2HPO_4 \times 2H_2O$: 1.44g, or Na_2HPO_4: 1.1486g.
- Dissolve in aqua bidest to 1l.
- Store at 4°C.
- Include additives to increase solubility in PBS receptor medium (e.g., bovine albumine, ethanol, PEG).

13.2.4 Technical equipment

- Dermatome (e.g., Aesculap GA 630)
- Fluorescence microscope
- Freeze microtome
- Microplate reader

13.3 Methods

13.3.1 Skin preparation

The described methods refer to the preparation of excised human skin from patients who have undergone plastic surgery (please note: an allowance by an ethical committee may be needed) or of excised pig skin. The preparations of commercially available reconstructed human skin models have to be performed according the manufacturer information.

1. Take the fresh skin and clean the skin's surface with PBS moist cotton. Avoid contact between the subcutaneous fat tissue and the skin surface (Figure 13.1).
2. Remove the subcutaneous fatty tissue using a scalpel and forceps. The skin surface is cleaned again (Figure 13.2).
3. For storage the skin surface is wrapped in aluminium foil and put into impermeable polyethylene bags.
4. The package can be stored up to 6 months at –20°C until use.

13.3.2 Determination of skin penetration using the Franz cell setup

1. Franz cells are filled with receptor fluid, for example, phosphate buffered saline (PBS) pH 7.4 and preheated with the water bath to 33.5°C ensuring a skin surface temperature of about 32°C. Receptor fluid is stirred at 500 r/m.

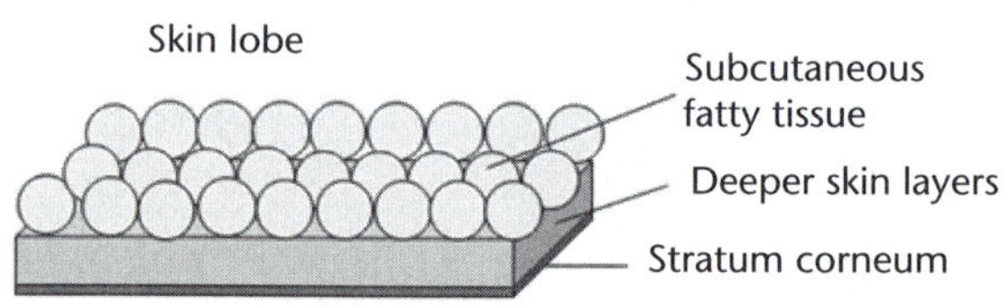

Figure 13.1 Skin preparation.

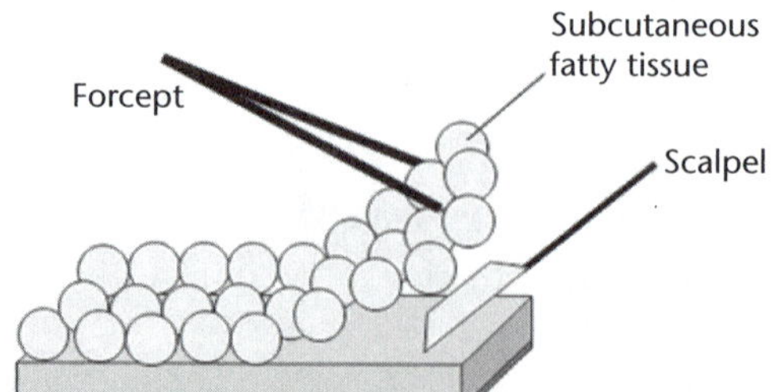

Figure 13.2 Example depiction of a Franz diffusion cell, static type.

2. Pig or human skin is thawed and 1,000 μm ± 100 μm skin sheets are prepared using a Dermatome (e.g., Aesculap, Tuttlingen, Germany). Alternatively, the epidermis of human skin can be separated from the dermis by heating to 60°C in a water bath for 90–120 seconds, then the epidermis can be removed carefully. Skin is nourished in phosphate buffered saline until further processing to avoid swelling.
3. Skin is punched to 2-cm diameter disks.
4. Skin disks are mounted to Franz cells with the horny layer facing the air and the dermis having contact with the receptor fluid and left for equilibration for 30 minutes.
5. Aiming a finite-dose approach, about 1–2 mg/cm^2 of the test formulations are applied onto the skin surface. In the case of the infinite-dose approach, about 500 μl for 1.76 cm^2 can be applied. The finite-dose approach is more relevant concerning the exposure of toxic compounds as well as a clinical practice and the use of cosmetics.
6. The donor compartment is sealed with, for example, Nescofilm to avoid evaporation.
7. After residue time the skin is removed from the Franz cells and surplus formulations are gently removed by carefully washing the skin using PBS.
8. Treated skin areas are isolated, embedded in tissue freezing medium (e.g., Jung, Nussloch, Germany), and stored in Peel-Aways (e.g., Polyscience, Eppelheim, Germany) at a temperature of –80°C.

13.3.3 Determination of skin permeation using the Franz cell setup

1. Follow the instructions of Section 13.3.2 until point 6.
2. After 30 minutes, 60 minutes, 90 minutes, 120 minutes, 6 hours, 24 hours, and 48 hours, eventually more than 300 μl of the receptor fluid are removed using a disposable syringe.
3. Add 300 μl of the fresh receptor fluid to replace the removed volume, taking care to avoid air.
4. Freeze the obtained samples for further processing.

13.3.4 Skin absorption studies with commercially available 3D skin models

Skin absorption studies can be performed analogous to the procedures described for human and pig skin. Then shorter test periods are adequate because of the increased permeability.

Due to the increasing number of reconstructed skin models on the market, no universal handling instructions can be given. Thus, it is important to handle and store the skin models strictly according to the manufacturer's instructions. Some skin models are too small for being placed directly onto the Franz cells (e.g., EPISKIN). In this case special Teflon inserts can and have to be used [13].

13.3.5 Quality control

To overcome the widely variable permeability of human skin, the OECD recommends parallel testing of standard substances such as caffeine (hydrophilic) and/or testosterone (lipophilic) [4]. This is also recommended when using RHE because of an enhanced barrier function over time due to improved culture conditions by the manufacturers [5].

13.3.6 Data evaluation

Various methods are possible to determine the skin absorption of substances. The following section will focus on the most important approaches.

13.3.6.1 High-performance liquid chromatography (HPLC)—skin penetration

1. The frozen skin disks are cut horizontally into 5-μm slices using a freeze microtome. Alternatively, tape stripping can be performed. Hereby parts of the skin layers are removed that may also be analyzed.
2. For a complete extraction of skin slices, thawing and freezing cycles are recommended. The slices are incubated in fluid nitrogen for 1 minute and afterwards put in a water bath at 37°C for 5 minutes. This procedure has to be repeated seven times. Then the slice is extracted twice with the extraction medium. For the extraction of the tapes, they are incubated with an appropriate extraction medium (e.g., ethyl acetate, methanol, and so forth).
3. The extract is filtrated using filter membranes with a pore size of 0.22 μm, and the filtrate is injected into the HPLC. Note that the thickness of human skin layers is about 10–20 μm for the stratum corneum, 30–80 μm for the viable epidermis, and 100–500 μm for the dermis. The layer thickness of animal skin and reconstructed skin models differs. Thus, it is recommended to previously determine the average thickness of the skin layers.

13.3.6.2 High-performance liquid chromatography (HPLC)—skin permeation

1. The samples of the receptor fluid are injected into the HPLC.
2. Be aware that you have to calculate the dilution of your receptor fluid after removing the single samples. The following equation can be used for calculation:

$$C_{cor} = C_{measured} + (\text{sample volume}/\text{receptor volume}) * (C_1 + C_2 + \ldots + C_{measured-1})$$

13.3.6.3 Scintillation counting

Specimens sampled for analysis of the substances are stored at –80°C until analyzed. Substance concentrations in the receptor medium are quanti?ed by scintillation counting (e.g., Microbeta Plus,Wallac, Turku, Finland). To overcome the quenching effect of BSA or media components, calibration curves are obtained dissolving the spiked substance in identical solvents.

Standard operating procedures for skin absorption testing using human and pig skin as well as RHE have been set up in the German Validation Study and are freely available by the Society of Dermopharmacy (Gesellschaft für Dermopharmazie; http://www.gd-online.de/german/fgruppen.htm).

13.3.6.4 Fluorescence—method establishment and validation

1. Excitation wavelength and camera integration time have to be chosen according the applied dye. Image treatment software (for example, Axiovision 3.1.3.1, ZEISS) is required.
2. The fluorescence intensity of the dye has to be studied for linearity in relation to the dye concentration in solution. For instance, 500 μl of dye solutions (various concentrations, for example, 0.0020 to 8.0 μg/ml) are subjected to fluorescence microscopy. Four samples for each dye concentration are tested, and measurements are repeated on different days.
3. Calculate in which concentration range a linear relation between dye concentrations and fluorescence intensities exists. Data evaluation is possible (e.g., with Prism).

13.3.6.5 Fluorescence—skin penetration of dyes or fluorescent-labeled substances

1. The frozen skin disks are cut into vertical slices of 20-μm thickness using a freeze microtome. Slices are stored at +4°C and analyzed within 24 hours.
2. Skin slices are subjected to normal light and fluorescence light using an inverted fluorescence microscope equipped with a monochrome camera (AxioCam HR, version 5.05.10, Zeiss) in the microscope. Adequate magnification is chosen.
3. Arbitrary pixel brightness values (ABU) give the relative dye content within the areas of interest (e.g., stratum corneum, epidermis, and dermis).
4. Pixel intensities should be corrected by subtracting background fluorescence intensity units obtained from the measurement of the native fluorescence of the respective skin layers of untreated skin. Corrected ABUs are obtained.

13.3.6.6 Fluorescence—skin permeation of dyes or fluorescent-labeled substances

1. Approximately 150 μl of the removed receptor fluid are filled in per well of a 24-well plate. The samples per time point are measured in duplicate.
2. The fluorescence intensity of the samples is quantified using a microplate reader (e.g., FLUOstar Optima, BMG Labtech), choosing the adequate excitation and emission wavelength depending on the dye.

13.3.7 Biostatistics

Several analysis options are possible. For skin penetration experiments, corrected ABU values (in the case of fluorescence measurements) or drug concentrations determined by HPLC are obtained from substance uptake following an application of different pharmaceutical formulations. These values can be related to uptake data from a reference. By the means of the resulting penetration enhancing effect (PEE), for example, the impact of the pharmaceutical formulations on the skin penetration can be determined. The PEE of the reference is 1 by definition.

For data evaluation of skin permeation experiments, the most prevalent analysis option is to plot the drug concentrations over the time. The results should be presented as average values ± standard deviation (SD) or standard error of the mean (SEM). At least three independent experiments, meaning skin from three donors, have to be performed. Suitable tools for statistical analysis of skin absorption are the H-test of Kruskal and

Wallis followed by a post hoc test using an X^2-approach or the Wilcoxon matched pair test.

13.4 Results and Discussion

Two-compartment in vitro test systems like Franz diffusion cells have been used for decades to examine the skin absorption of a compound. The determination of skin penetration is of great interest when new systems for topical drug delivery or drugs for topical treatment of skin diseases are being developed, but also for the safety assessment of chemical compounds, in particular, those of pesticides. To obtain information about possible systemic (side) effects of topically applied substances, knowledge about the magnitude of their skin permeation is essential.

The percutaneous or cutaneous absorption is a diffusion process whose important parameters and dependencies are described by the Fick's first law [14]:

$$J = [(K * D)/h] * \Delta c$$

where J is the diffusion velocity (μg/cm^2*h), K is the distribution coefficient (cm/h), D is the diffusion coefficient of the membrane (cm^2/h), h is the thickness of the membrane (cm), and Δc is the concentration differences (μg/cm^3).

Parameters such as the dimension of treated skin area, the residence time of the pharmaceutical formulation, and the amount and solubility of the drug in the formulation and the skin determine the velocity and the quantity of skin penetration [14]. The preparation of split skin (e.g., 1,000-μm thickness) is recommended, as according to Fick's first law, the thickness of the "membrane" significantly determines the diffusion velocity.

Franz diffusion cell experiments are an extensively reviewed and validated experimental set up to conduct the skin absorption of substances [4, 5, 15]. They allow studying skin penetration and permeation according to the principles of Fick's diffusion law. Several papers describe the essential physicochemical characterization of the test drug, for example, with respect to solubility in the receptor fluid [16] and a program for automated data analysis in case of the infinite dose approach [17].

Furthermore, skin absorption is dependent on the anatomical region of application; drug absorption is best from the forehead and eyelids, worse from the arm and belly, and worst from the horny skin of the palms and soles [18, 19]. Thus, for in vitro evaluation the site of the skin is crucial. From a pig, the skin of the axillaries region or skin of the ears are preferable. If human skin from plastic surgery is used, skin obtained from breast corrections or abdominal wall tightening are preferred. Facial skin should be avoided, as those parts are often intensively pretreated with cosmetics.

Due to the limited availability of human skin, alternatives were needed. Unfortunately no single animal species exists where skin morphology and physiology matches perfectly with human skin, not even the skin of the genetically closest primates. Rodents, rabbits, and guinea pigs are often used in preclinical skin absorption and dermatotoxicity studies [4, 20] as they are also used predominantly in routine toxicology testing or for immunological considerations [21]. Due to anatomical and physiological differences (e.g., higher amount of hair follicles, accelerated desquamation, thinner stratum corneum) in comparison to human skin such experiments often lead to an overestimation of the transdermal absorption. With respect to the barrier properties, pig skin is comparable to human skin and already well established for in vitro testing [5, 15]. Pig skin is similar in morphology considering the low number of hair follicles and the properties of the epidermis, but differs in the vascularization of the skin, which is higher in humans, and in the types of sebaceous glands [22] and epidermal lipids [23]. However, the drug penetration across pig skin may be lower in comparison to human skin, possibly due to a thicker stratum corneum in porcine skin [5, 9]. A similar dosing area, surface characteristics, and a comparable skin surface area–to–body weight ratio allow for extrapolations to human skin. This holds true for in vitro and in vivo models. Aiming to study skin absorption of diseased or injured skin, an impairment of the barrier properties by tape-stripping or treatment with irritant substances like acetone or sodiumdodecylsulfate are possible.

The permeability properties of skin are preserved after excising, since penetration is a passive process and is controlled by the nonviable stratum corneum. No active transport was observed; therefore, an appropriate storage in the freezer up to 6 months is possible, unless metabolic activity is required [2, 4].

Reconstructed skin models are the first step in completely avoiding the use of animal skin for dermal absorption testing. Several reconstructed human epidermis models (RHEs) are already validated for the testing of dermal absorption (e.g., Episkin and Epiderm), but the skin models are still clearly more permeable compared to pig and human skins [5, 9]. This holds also true with respect to full-thickness skin models, which are built of an epidermis and dermis and have been introduced only recently into the market [24, 25]. Intralaboratory and interlaboratory variabilities of in-house skin models have been tested by respective studies to compare data to those of validated RHEs.

In addition to skin absorption studies, in vitro skin models are suitable tools to examine dermatotoxicological aspects such as skin corrosion, irritation, sensitization, and photo-aging. Respective test guidelines have been adopted by the OECD and EU. Full-thickness skin models were developed enabling in vitro studies of aspects that are not strictly related solely to the epidermis [26], such as glucocorticoid-induced skin thinning [27] or wound healing [28, 29].

Alternatively, for the reconstructed skin models, the development of stratum corneum substitutes (SCS), which may function as predictive and standardized models for cutaneous absorption, also gained interest. Essential for adequate SCSs are uniform lipid distributions, the compactness of the lipid layer, and the orientation of lipid lamellae. To achieve this, various techniques such as airbrushing are performed for the production of the lipid layers [30, 31]. A lipid mixture that mimics the properties of native human stratum corneum very well consists of 50% epidermal ceramides, 28% cholesterol, 17% free fatty acids, and 5% cholesteryl sulphate [7]. The first penetration studies revealed comparable barrier properties of those lipid matrices and proved the possibility even to observe formulation effects on the penetration [32]. Thus, these approaches may be used in the future following a respective validation study.

Ultimately, by altering the lipid composition of those systems, diseased and dry skin can be mimicked and important insights concerning altered absorption of drugs due to the disease pattern may be provided. Although it is most challenging to mimic the lipid organization and orientation of the human stratum corneum, problems such as high intervariability related to excised skin may be circumvented by such artificial skin barriers, but there is still a great necessity for further research in this field.

HPLC, LSC, and fluorescence measurements are the most common methods to determine skin absorption. Several other techniques can also be applied, depending on the purpose of the study. For example, electron spin resonance (ESR) measurements were shown to give useful information on the properties of different regions of the skin and their changes by the addition of various drugs, cosmetics, and absorption enhancers [33, 34]. ESR imaging allows a spatial resolution of paramagnetic centers in different tissues, and combined with modulated field gradients, this method provides the possibility for the differentiation of ESR spectra in selected compartments.

13.5 Discussion of Pitfalls and Troubleshooting

1. Strictly avoid contact between the fat tissue of the excised skin and the skin surface. In this case the skin surface may become even more hydrophobic, resulting in an erroneously lower skin absorption.
2. Avoid a swelling and soaking of the excised skin during its preparation for the experiment. Excessive hydration weakens the connectivity of the corneocytes, leading to a misleading higher skin absorption.
3. In the case of a need for dermatomed, nonviable skin, it is recommended to dermatome the skin when it is still half-frozen.
4. Do an integrity check by a visual inspection. There should be no fluid appearance at the skin/RHE surface after the equilibration period.
5. To allow for even skin absorption, the applied formulation should be spread homogeneously over the skin. In the case of reconstructed skin models, avoid sharp items and spread carefully; otherwise, the tissue may be harmed.
6. Avoid air bubbles in the water jacket around the Franz cell and ensure direct contact between the skin and the acceptor fluid. Keep in mind that fluorescence light emission can change with the surrounding interface and media supplements (e.g., with Nile red).
7. For fluorescence read-out, remove carefully and completely the on the skin applied formulation using a cotton tissue. Otherwise, an overall staining will preclude exact data evaluation with the microscope.

13.6 Summary

To date, the Franz diffusion cell is the most important alternative for animal testing to examine the skin absorption of drugs, additives of cosmetics, and chemical compounds. The technique is easily manageable and cheap and results are reproducible. The gold standard in the testing today is the use of excised animal or human skin whereby the application in living animals is avoided. Furthermore, the development of reconstructed skin models to replace excised skin completely and to increase the reproducibility further is a good step in the right direction, although to date still the skin absorption is overestimated with reconstructed skin models due to differences in the formation of the respective skin layers. Parallel tests with standard agents allow for a thorough monitoring of all the experiments. In the future the application of stratum corneum substitutes, which can mimic the penetration barrier may be an option, but here again there is still a lot of work to do. In conclusion, in the future further research efforts are required to establish new alternative or to improve current approaches.

References

[1] Franz, T. J., "Percutaneous Absorption on the Relevance of In Vitro Data," *J. Invest. Dermatol.*, Vol. 64, No. 3, 1975, pp. 190–195.

[2] Schreiber, S., et al., "Reconstructed Epidermis Versus Human and Animal Skin in Skin Absorption Studies," *Toxicol. In Vitro*, Vol. 19, No. 6, 2005, pp. 813–822.

[3] OECD, "Guidance Document No. 28 for the Conduct of Skin Absorption Studies," *35th Joint Meeting August 2003*, 2003.

[4] OECD, "Test Guideline 428: Skin Absorption: In Vitro Method," adopted April 13, 2004.

[5] Schäfer-Korting, M., et al., "The Use of Reconstructed Human Epidermis for Skin Absorption Testing: Results of the Validation Study," *Altern. Lab. Anim.*, Vol. 36, No. 2, 2008, pp. 161–187.

[6] Brain, K. R., K. A. Walters, and A. C. Watkinson, "Methods for Studying Percutaneous Absorption," in K. A. Walters, (ed.), *Dermatological and Transdermal Fomulations*, New York: Marcel Dekker, 2002.

[7] Cevc, G., "Transfersomes, Liposomes and Other Lipid Suspensions on the Skin: Permeation Enhancement, Vesicle Penetration, and Transdermal Drug Delivery," *Crit. Rev. Ther. Drug Carrier Syst.*, Vol. 13, 1996, pp. 257–388.

[8] Ponec, M., et al., "Characterization of Reconstructed Skin Models," *Skin Pharmacol. Appl. Skin Physiol.*, Vol. 15, Suppl. 1, 2002, pp. 4–17.

[9] Schäfer-Korting, M., et al., "Reconstructed Human Epidermis for Skin Absorption Testing: Results of the German Prevalidation Study," *Altern. Lab. Anim.*, Vol. 34, No. 3, 2006, pp. 283–294.

[10] Riviere, J. E., K. F. Bowman, and N. A. Monteiro-Riviere, "On the Definition of Viability in Isolated Perfused Skin Preparations," *Br. J. Dermatol.*, Vol. 116, No. 5, 1987, pp. 739–741.

[11] Anissimov, Y. G., *Mathematical Models for Different Exposure Conditions*, New York: Informa Healthcare, 2008.

[12] Lombardi Borgia, S., et al., "Lipid Nanoparticles for Skin Penetration Enhancement-Correlation to Drug Localization Within the Particle Matrix as Determined by Fluorescence and Parelectric Spectroscopy," *J. Control Release*, Vol. 110, No. 1, 2005, pp. 151–163.

[13] Netzlaff, F., et al., "TEWL Measurements as a Routine Method for Evaluating the Integrity of Epidermis Sheets in Static Franz Type Diffusion Cells In Vitro. Limitations Shown by Transport Data Testing," *Eur. J. Pharm. Biopharm.*, Vol. 63, No. 1, 2006, pp. 44–50.

[14] Halkier-Sorensen, L., "Dermatopharmacology of Topical Preparations: A Product Development-Oriented Approach," in B. Gabard et al., (eds.), *The Absolute Fundamentals of Transdermal Permeation*, New York: Springer, 2000.

[15] Diembeck, W., et al., "Test Guidelines for In Vitro Assessment of Dermal Absorption and Percutaneous Penetration of Cosmetic Ingredients," *European Cosmetic, Toiletry and Perfumery Association, Food Chem. Toxicol.*, Vol. 37, No. 2-3, 1999, pp. 191–205.

[16] Kaca, M., et al., "The Physicochemical Parameters of Marker Compounds and Vehicles for Use in In Vitro Percutaneous Absorption Studies," *Altern. Lab. Anim.*, Vol. 36, No. 2, 2008, pp. 189–200.

[17] Niedorf, F., E. Schmidt, and M. Kietzmann, "The Automated, Accurate and Reproducible Determination of Steady-State Permeation Parameters from Percutaneous Permeation Data," *Altern. Lab. Anim.*, Vol. 36, No. 2, 2008, pp. 201–213.

[18] Feldmann, R. J., and H. I. Maibach, "Percutaneous Penetration of Steroids in Man," *J. Invest. Dermatol.*, Vol. 52, No. 1, 1969, pp. 89–94.

[19] Rougier, A., C. Lotte, and H. I. Maibach, "In Vivo Percutaneous Penetration of Some Organic Compounds Related to Anatomic Site in Humans: Predictive Assessment by the Stripping Method," *J. Pharm. Sci.*, Vol. 76, No. 6, 1987, pp. 451–454.

[20] OECD, "Test Guideline 431: In Vitro Skin Corrosion: Human Skin Model Test," adopted April 13, 2004.

[21] Roberts, M. S., and K. A. Walters, *Dermal Absorption and Toxicity Assessment*, 2nd ed., New York: Informa Healthcare, 2008.

[22] Simon, G. A., and H. I. Maibach, "The Pig as an Experimental Animal Model of Percutaneous Permeation in Man: Qualitative and Quantitative Observations—An Overview," *Skin Pharmacol. Appl. Skin Physiol.*, Vol. 13, No. 5, 2000, pp. 229–234.

[23] Stahl, J., F. Niedorf, and M. Kietzmann, "Characterisation of Epidermal Lipid Composition and Skin Morphology of Animal Skin Ex Vivo," *Eur. J. Pharm. Biopharm.*, Vol. 72, No. 2, 2009, pp. 310–316.

[24] Ackermann, K., et al., "The Phenion®-Full Thickness Skin Model for Percutaneous Absorption Testing," *Skin Pharm. Phys.*, in press.

[25] Schäfer-Korting, M., et al., "Reconstructed Epidermis and Full-Thickness Skin for Absorption Testing: Influence of the Vehicles Used on Steroid Permeation," *Altern. Lab. Anim.*, Vol. 36, No. 4, 2008, pp. 441–452.

[26] Hayden, P. J., et al., "Epiderm™ Full Thickness (Epiderm FT™). A Dermal/Epidermal Skin Equivalent with a Fully Developed Basement Membrane," *J. Invest. Dermatol.*, Vol. 121, 2003, p. Abstract #543.

[27] Zoller, N. N., et al., "Evaluation of Beneficial and Adverse Effects of Glucocorticoids on a Newly Developed Full-Thickness Skin Model," *Toxicol. In Vitro*, Vol. 22, No. 3, 2008, pp. 747–759.

[28] Kandarova, H., et al., "An In Vitro Skin Irritation Test (SIT) Using the EpiDerm Reconstructed Human Epidermal (RHE) Model," *J. Vis. Exp.*, Vol. 29, 2009.

[29] Wolf, N. B., et al., "Influences of Opioids and Nanoparticles on In Vitro Wound Healing Models," *Eur. J. Pharm. Biopharm.*, Vol. 73, No. 1, 2009, pp. 34–42.

[30] de Jager, M., et al., "Preparation and Characterization of a Stratum Corneum Substitute for In Vitro Percutaneous Penetration Studies," *Biochim. Biophys. Acta.*, Vol. 1758, No. 5, 2006, pp. 636–644.

[31] de Jager, M. W., et al., "Lipid Mixtures Prepared with Well-Defined Synthetic Ceramides Closely Mimic the Unique Stratum Corneum Lipid Phase Behavior," *J. Lipid Res.*, Vol. 46, No. 12, 2005, pp. 2649–2656.

[32] Jaeckle, E., U. F. Schaefer, and H. Loth, "Comparison of Effects of Different Ointment Bases on the Penetration of Ketoprofen Through Heat-Separated Human Epidermis and Artificial Lipid Barriers," *J. Pharm. Sci.*, Vol. 92, No. 7, 2003, pp. 1396–1406.

[33] Kitagawa, S., M. Kasamaki, and A. Ikarashi, "Effects of N-Alkyltrimethylammonium on Skin Permeation of Benzoic Acid Through Excised Guinea Pig Dorsal Skin," *Chem. Pharm. Bull. (Tokyo)*, Vol. 48, No. 11, 2000, pp. 1698–1701.

[34] Moll, K. P., et al., "Changes of the Properties in the Upper Layers of Human Skin on Treatment with Models of Different Pharmaceutical Formulations—An Ex Vivo ESR Imaging Study," *Chem. Med. Chem.*, Vol. 3, No. 4, 2008, pp. 653–659.

CHAPTER

14

A 3D Model of the Human Epithelial Airway Barrier

Andrea Lehmann,[1] Christina Brandenberger,[1] Fabian Blank,[2] Peter Gehr,[1] and Barbara Rothen-Rutishauser[1]

[1]Institute of Anatomy, Division of Histology, University of Bern, Bern, Switzerland
[2]Department of Clinical Research, University of Bern, Bern, Switzerland
Corresponding author: Barbara Rothen-Rutishauser, PhD., address: Institute of Anatomy, Division of Histology, University of Bern, Baltzerstrasse 2, CH-3000 Bern 9, Switzerland,
e-mail: barbara.rothen@ana.unibe.ch, phone: ++41-31-631-8441, fax: ++41-31-631-3807

Abstract

A triple cell coculture system that simulates the human epithelial airway barrier of the respiratory system was established and is described here in detail. The model consists of lung epithelial cells, monocyte-derived macrophages, and monocyte-derived dendritic cells. This system allows investigation of the interaction of xenobiotics, for example, particulate material, which are deposited either in suspension or at the air-liquid interface to lung cells. In addition, the interplay between the different cells, such as epithelial cells and the cells of the defense system, and the cellular responses can be studied following the xenobiotic exposure. In this chapter, the experimental procedure is outlined on how to grow, evaluate and experimentally use the triple-cell coculture system.

Key terms dendritic cells
epithelial airway/alveolar barrier
epithelial cells
macrophages

14.1 Introduction

With every breath millions of particles and antigens are inhaled and enter the respiratory system. A series of structural and functional barriers protect the respiratory system against harmful and innocuous particulate material. This is important as the internal surface area of the lungs is vast (the alveoli and airways are approximately 150 m^2) [1] and therefore facilitates broad access of inhaled substances to the lung tissue. The thin tissue barrier of the gas exchange area in the alveoli is only a few hundred nanometers in thickness and separates the blood from the air space. Efficient protecting and clearing mechanisms are therefore essential.

Most of the clearing structures are part of the barrier components of the lungs (airways) such as the surfactant film [2–4], the aqueous surface lining layer including the mucociliary escalator [5], a population of macrophages in the airways and in the alveoli (professional phagocytes) [6, 7], the epithelial cellular layer endowed with tight junctions and adherens junctions between the cells [8, 9] and a network of dendritic cells inside and underneath the epithelium [10, 11]. In addition to the lung epithelium, the basement membrane [12, 13], the connective tissue [14], and the capillary endothelium [15, 16] serve as structural barriers against inhaled particulate material.

Nonetheless, a number of epidemiological studies have shown that ambient particulate matter causes adverse health effects associated with increased pulmonary and cardiovascular morbidity and mortality [17–19], suggesting that the particles can overcome the barrier components and cause systemic effects. Moreover, some studies have indicated a specific toxicological effect from inhaled combustion derived nanoparticles [20, 21]. These small particles, with a diameter of less than 100 nm in at least one dimension, not only can cross the blood-air barrier, but also have the ability to cross cellular membranes [22–24]. Hence, they are able to enter different compartments of the cell, dependent on particle size and particle surface coatings. Once inside the cells, nanosized particles may cause several biological responses including the generation of reactive oxygen species [25, 26], the enhanced expression of proinflammatory cytokines [27], and DNA strand breaks [28, 29].

There is a strong need for in vitro test systems to assess the toxicity of particulate matter and especially nanoparticles [30, 31]. Many lung cell culture models exist, providing an alternative to animal exposures for analyzing the effects of different types of particles. It has been suggested that the 3D models represent a more physiologically relevant model [32]. It is not only the 3D structure that is important but also the cocultures of different cell types that have been shown to have an influence on the outcome of the results. For instance, in the airway mucosa epithelial cells, macrophages and dendritic cells continuously communicate in vivo through intercellular signaling to maintain homeostasis and to coordinate immune responses [33]. In vitro models of mucosal surfaces are now in use, particularly to characterize the mechanism of particle sampling by intraepithelial dendritic cells [34–36]. Recently the authors developed an in vitro triple cell coculture model of the human airway barrier to study the cellular interplay and the cellular response of epithelial cells (EC), human blood monocytes derived macrophages, and dendritic cells to particles [34, 37–39]. In this chapter the development of the triple cell coculture model is described, with its evaluation by measurement of the transepithelial electrical resistance and the detection of certain protein by immunofluorescent proteins, as well as its experimental use.

14.2 Experimental Design

In the triple cell coculture model of the human epithelial barrier, monolayers of two different epithelial cell lines, A549, an alveolar type II–like cell line [40], and 16HBE14o-, a bronchiolar epithelial–like cell line [41, 42], were grown on a microporous membrane in a two-chamber system. In parallel, human blood–derived monocytes were isolated and differentiated into monocyte-derived macrophages (MDM) and monocyte-derived dendritic cells (MDDC) and then added at the apical side and the basal side of the epithelium, respectively. After a thorough evaluation [34, 39], this model was exposed to particles (either airborne or suspended in medium) of different materials (polystyrene, titanium dioxide, diesel exhaust particles, single-walled carbon nanotubes) and of different sizes (= 1 μm) [34, 38, 43, 44].

14.3 Materials

14.3.1 General materials

- Sterile bench
- Sterile general glassware and plastics
- Incubator (37°C, 5%CO_2)
- Water bath (37°C)
- Centrifuge
- Pipettes and pipette tips
- Parafilm (Alcan Packaging)

14.3.2 Epithelial cell cultures—thawing

- Fetal calf serum (FCS) (Gibco)
- L-Glutamine (5 ml at 200 mM) (Gibco)
- Penicillin (5 ml at 100U.ml^{-1})/Streptomycin (5 ml at 0.1C g.ml^{-1}) (Gibco)
- MEM (Gibco) for 16HBE14o- cells
- RPMI 1640 (Gibco) for A549 cells
- 25 cm^2 cell culture flask (Fisher Scientific)
- General glassware and plastics
- Ice
- Nitrogen tank
- Pipettes (Socorex, Gilson)
- Polystyrene box
- Safety goggles

14.3.3 Epithelial cell cultures—culturing

- FCS (Gibco)
- L-Glutamine (5 ml at 200 mM) (Gibco)

- Penicillin (5 ml at 100U.ml^{-1})/Streptomycin (5 ml at 0.1Cg.ml^{-1})(Gibco)
- MEM (Gibco) for 16HBE14o- cells
- RPMI 1640 (Gibco) for A549 cells
- Trypan Blue (Sigma)
- Trypsin/EDTA (Gibco)
- Human fibronectin (BD Biosciences) for coating of flask for 16HBE14o- cell culturing
- Bovine serum albumin (BSA) (Sigma)
- Bovine collagen, Type I (BD Biosciences)
- LHC Basal Medium (Lucerna Chemie)
- 25 cm^2 and 75 cm^2 cell culture flasks (Fischer Scientific)
- Neubauer chamber (hemocytometer)

14.3.4 Isolation of monocyte-derived macrophages (MDM) and dendritic cells (MDDC)

- Granulocyte macrophage colony-stimulating factor (GM-CSF; stock solution 1 μg/ml) (Sigma)
- Interleukin-4 (IL-4; stock solution 1 µg/ml) (R & D Systems)
- L-Glutamine (5 ml at 200 mM) (Gibco)
- Penicillin (5 ml at 100U.ml^{-1})/Streptomycin (5 ml at 0.1Cg.ml^{-1}) (Gibco)
- Phosphate buffered saline (PBS) (pH 7.4)
- RPMI 1640 (Sigma Aldrich)
- 6-well plates (Falcon)
- Clamp
- Stand
- Scissors or scalpel

14.3.5 Triple cell coculture

- Large glass Petri dishes (Sarsted)
- Cell scrapers
- Medium (RPMI 1640 + 5% Human Serum + 1% L-Glutamine + 1% Penicillin/Strepyomycin)
- 6-well plates (Falcon)
- Cell culture inserts with a surface area of 4.2 cm^2 for 6-well plates, 3.0-μm pores, transparent PET membrane (Falcon)

14.3.6 Transepithelial electrical resistance (TEER) measurements

- RPMI 1640 (Gibco)
- 70% ethanol (Merck)
- Sterile high purified water

- General glassware and plastics
- TEER sensor (Millipore AG)
- Millicell-ERS system MERS 000 01 (Millipore AG)

14.3.7 Staining for laser scanning microscopy (LSM)

- 3% Paraformaldehyde (PFA) (diluted in PBS)
- 0.1M Glycine (diluted in PBS)
- 0.2% TritonX-100 (diluted in PBS)
- Mowiol (PBS:glycerol (2:1) containing 170 mg/ml Mowiol 4–8 (Calbiochem, VWR International AG))
- Cover slips

14.3.8 Embedding for transmission electron microscopy (TEM)

- Glutaraldehyde (GA) (in potassium phosphate buffer)
- 70% ethanol (Merck)
- Sterile high purified water
- General glassware and plastics

14.4 Methods

14.4.1 Epithelial cells

14.4.1.1 Characterization and description of the currently used cell lines

16HBE14o- cell line

The immortalized and by the SV40 large T-antigen transformed human bronchial epithelial cell line, 16HBE14o-, is a normal human airway epithelial cell line and was kindly donated from Dieter Gruenert (Cardiovascular Research Institute, University of California, San Francisco). This cell line has already been used for the triple cell coculture model. The advantage of these cells is that they form polarized monolayers with extensive tight junction belts. Therefore, this cell line is a useful tool to study tight junction properties and permeability functions of human bronchiolar epithelium and may be a suitable candidate for an in vitro model for mechanistic studies of drug transport processes involved in the smaller airways [42, 45, 46]. For cell culturing all flask and plastics for experiments, including wells, inserts, and chamber slides, have to be treated with fibronectin coating solution containing bovine serum albumin (0.1 mg/ml) (Sigma, Fluka Chemie GmbH, Buchs, Switzerland), 1% bovine collagen, Type I (BD Biosciences, Basel, Switzerland), and 1% human fibronectin (BD Biosciences) in LHC basal medium (Lucerna Chemie AG). Without this coating the cells do not grow to confluence.

A549 cell line

The alveolar type II like cell line A549, which originates from human lung carcinoma [40], is part of the highly characterized and most widely used in vitro models [47]. It is available from the American Type Culture Collection and it has been shown that the A549 cells have many important biological properties of alveolar epithelial type II cells, such as membrane-bound inclusions, which resemble lamellar bodies of type II cells [48], and other ultrastructural characteristics common to type II cells, for example, distinct polarization, tight junctions, and extensive cytoplasmic extensions, which have been described [49]. Contradictory results about the capability of this cell line to express tight junctions and to generate transepithelial electrical resistance have been published. However, under the culture conditions currently used, the authors found that the A549 cells express tight junction proteins and built a transepithelial electric resistance [34, 39, 50].

14.4.1.2 General protocol for cell thawing, using the 16HBE14o- cells as an example

1. Write the cell line, passage number, and date on a 25 cm^2 cell culture flask.
2. Fill the flask with 10 ml of the cell culture medium (MEM), containing 1% L-Glutamine, 1% Penicillin/Streptomycin and 10% FCS, and place it in an incubator at 37°C for at least 1 hour prior to use.
3. Place a 15-ml tube containing 9 ml of the culture medium in a water bath.
4. Remove the required frozen cell sample from the nitrogen tank and place it immediately on ice. Wear safety goggles until step 5 is finished.
5. Gently thaw the cell sample by holding the sample tube in a warm water bath. As soon as the ice has melted, remove the tube and clean the outside of the tube with 70% ethanol.
6. While working on a sterile bench, pipette the cells into the 15-ml tube containing a 9-ml culture medium.
7. Centrifuge the tube, containing the cell suspension, for 5 minutes at 40 rcf.
8. Remove the supernatant and resuspend the cell pellet with 1 ml of the cell culture medium.
9. Place the cell suspension in the prepared 25 cm^2 cell culture flask and place it in an incubator.
10. Use the cell lines between 20 to 25 passages afterwards thaw new cells.

14.4.1.3 Epithelial cell line cultures

When using the 16HBE14o- cells, all flasks and wells have to be coated with fibronectin collagen. Ensure that enough flasks and wells are prepared with the coating.

The fibronectin coating solution contains: 0.1 mg/ml bovine serum albumin (Sigma, Fluka Chemie GmbH, Buchs, Switzerland), 1% bovine collagen, Type I (BD Biosciences, Basel, Switzerland), and 1% human fibronectin (BD Biosciences) in LHC Basal Medium (Lucerna Chemie AG). Cover the bottom of the growing area of the culture flasks and wells with the fibronectin coating solution for at least 2 hours in an incubator. Following incubation, remove the fibronectin coating solution and use the flasks and wells as soon as they are dry. It is possible to store the prepared coated cell culture material in the fridge (but for no longer than 1 month).

Examine the cells under the microscope prior to the beginning of their culture. If qualitative assessment deems the cells viable, perform the technique. If the cell culture is contaminated in any form, dispose the cells.

1. Warm the cell culture medium in a water bath up to 37°C. For 16HBE14o- cells, use MEM supplemented with 1% L-Glutamine, 1% Penicillin/Streptomycin, and 10%

FCS. For A549 cells use RPMI 1640 supplemented with 1% L-Glutamine, 1% Penicillin/Streptomycin, and 10% FCS

2. For the 16HBE14o- cells, wash the cells with prewarmed Typsin/ethylenediaminetetraacetic acid (EDTA) (1–2 ml/25 cm^2 flask; 3 ml/75 cm^2 flask). Shake the cells with Trypsin/EDTA for about 10 seconds and then remove the excess. For the A549 cells, wash the cells with prewarmed RPMI 1640 without additives (1–2 ml/25 cm^2 flask; 3 ml/75 cm^2 flask). Shake the cells with RMPI 1640 for about 10 seconds and then remove the excess.
3. Treat the cells with Trypsin/EDTA (1–2 ml/25 cm^2 flask; 3 ml/75 cm^2 flask). Shake the cells with Trypsin/EDTA for about 10 seconds and then remove the excess.
4. Place the flask in an incubator at 37°C, 5% $CO_{2.}$ The 16HBE14o- cells must be incubated for 8 minutes and the A549 cells for 4 minutes until they detach from the flask. Observe the cells under the light microscope; it may be that they need a longer time for detachment.
5. Add the fresh cell culture medium to the cells (2 ml per 25 cm^2 flask; 6 ml per 75 cm^2 flask) and subsequently resuspend them.
6. Count the cells with a Neubauer chamber (Haemocytometer) using a 1:5 dilution of cell suspension in a cell culture medium. (200-μl Trypan Blue, 200-μl cell culture medium, and 100-μl cell suspension). The cell density is calculated as followed: Total cell number refers to the mean number of cell counted $\times 5 \times 10^4 \times$ amount of the medium (ml). Determine cell viability by counting cells that are not stained with Trypan Blue.
7. Place the cell suspension in a fresh cell culture flask and seed the cells using the following concentrations: 0.5×10^6 cells per 25 cm^2 flask or 1×10^6 cells per 75 cm^2 flask.
8. Put the flask in an incubator at 37°C, 5% CO_2 until required.
9. Cell lines require a culture twice a week.

14.4.1.4 Seeding of the cells

Seed the epithelial cells 7 days before using them for the triple cell coculture model and within the regular time point of cell splitting.

1. Warm the cell culture medium in a water bath to 37°C. For 16HBE14o- cells use MEM supplemented with 1% L-Glutamine, 1% Penicillin/Streptomycin, and 10% FCS. For A549 cells use RPMI 1640 supplemented with 1% L-Glutamine, 1% Penicillin/Streptomycin, and 10% FCS as well as RMPI 1640.
2. Write the cell line, passage number, and date on a new 6-well plate and add 3 ml of the prewarmed culture medium to each well.
3. For experimental cultures seed the cells on transparent cell culture inserts. Cell culture inserts for 16HBE14o- cells must be pretreated with fibronectin coating solution. Take care that the inserts do not touch the edge of the respective well.
4. Count the cells with a Neubauer chamber (Hemocytometer) using a 1:5 dilution of cell suspension. (200 μl Trypan Blue, 200 μl cell culture medium and 100 μl cell suspension). The cell density is calculated as followed. Total cell number refers to the mean number of cells counted $\times 5 \times 10^4 \times$ amount of the medium (ml). Determine the cell viability by counting cells that are not stained with Trypan Blue.

5. Add 2-ml cell suspension to each well with the following concentration: 0.5×10^6 cells per ml (which equates to 1 million cells per insert).
6. Maintain the cell culture for 7 days at 37°C and 5% CO_2 in humidified atmosphere and culture the cells twice a week every 3 or 4 days.

14.4.2 Isolation of monocyte-derived macrophages (MDM) and dendritic cells (MDDC)

IMPORTANT: Before commencing this isolation protocol, be aware that the material you work with is human derived and not tested for disease or infections. An immunization of hepatitis B is recommended and gloves should be worn during the whole procedure. All waste occurring during the procedure is considered to be hazardous and must be autoclaved before disposed of into normal waste.

The following protocol is written for an isolation of one Buffy coat bag with 50 ml of content, which can be obtained from a blood donation center.

1. Use a fresh, uncooled bag of Buffy coat. Buffy coat can be stored at 4°C for 1 day.
2. Shake the bag and clean it with 70% ethanol. Hang it up at a stand with a tube clamp attached below [Figure 14.1(a)].

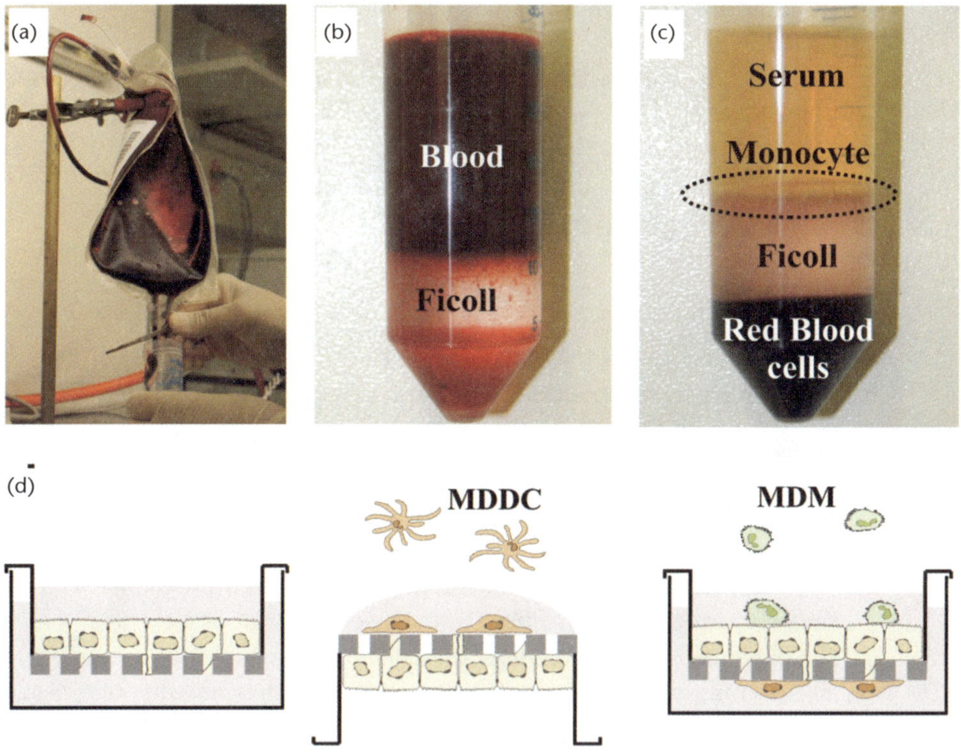

Figure 14.1 Isolation of monocytes and diagram of the triple cell coculture design. (a) A bag of Buffy coat is hanged up with a tube clamp. (b) The Ficoll is released under the blood in the tube to allow the Ficoll gradient to form. (c) After centrifugation four layers can be observed: the serum on top, followed by a thin layer containing the monocytes, followed by a layer containing the Ficoll and other white blood cells, and the pellet containing red blood cells at the bottom. (d) Design of the triple cell coculture. (*From:* [39]. © 2005 Reproduced with permission from *American Journal of Respiratory Cell and Molecular Biology*.)

3. Open the bag below the tube clamp and place two 50-ml tubes under the opening. Slowly open the clamp and distribute the content of the bag into the two 50-ml tubes (25 ml in each).
4. Fill up the two tubes to 50 ml with PBS (which must be at room temperature), close the tubes, and gently mix the suspension.
5. Take four new tubes and add 20 ml of the blood/PBS mixture into each tube (20 ml will remain and can be disposed of).
6. Carefully apply, with a 10-ml pipette, 13 ml of cooled Ficoll (4°C), and slowly release 12-ml Ficoll under the blood in the tube in order to allow the Ficoll gradient to form [Figure 14.1(b)].
7. Centrifuge the tubes for 20 minutes at 520 rcf, with weak acceleration and no braking at 25°C. This process produces four layers in the tube. The serum will be on top, followed by a thin layer containing the monocytes, followed by a layer containing the Ficoll and other white blood cells, and the pellet containing red blood cells at the bottom [Figure 14.1(c)]. This process lasts about 30 minutes in which the medium could be prepared.

Medium preparation

- Adherence medium for monocytes: 35-ml RPMI 1640 containing 10% human serum (4 ml), 1% L-Glutamine (400 μl), 1% Penicillin/Streptomycin (400 μl).
- Differentiation medium for MDM: 23-ml RMPI containing 5% human serum (1.25 ml), 1% L-Glutamine (250 μl), 1% Penicillin/Streptomycin.
- Differentiation medium for MDDC: 21-ml RMPI containing 5% human serum (1.25 ml), 1% L-Glutamine (250 μl), 1% Penicillin/Streptomycin, 1.25 ml of GM-GSF (end concentration of 50 ng/ml), and 850 μl of IL-4 (end concentration of 34 ng/ml).

Place the prepared medium into the water bath until they are required.

- Remove the tubes from the centrifuge and carefully apply with a 10-ml pipette the middle monocyte layer. Put the monocyte solution into a new 50-ml tube. Pool the four origin tubes into two tubes.
- Fill up the two tubes containing the monocyte suspension with PBS and mix gently.
- Centrifuge the solution for10 minutes at 270 rcf with a medium acceleration and brake with medium power.
- Resuspend the pellet with 5 ml of cell culture medium for the cell adherence (RPMI 1640 containing 1% L-Glutamine, 1% Penicillin/Streptomycin, 10% human serum). Then add the resuspended cell suspension to the rest of the adherence medium.
- Place a total of 3.3 ml of cell suspension into each well of two 6-well plates and allow the cells to adhere for 90 minutes.
- Remove the adherence culture medium and replace it with the respective medium. For differentiation into macrophages, give 3.7-ml new cell culture medium (RPMI 1640 containing L-Glutamine, 1% Penicillin/Streptomycin, 5% human serum, without growth factor) to the adhered cells in each well. For differentiation into dendritic cells, give a 3.7-ml new cell culture medium (RPMI 1640 containing

L-Glutamine, 1% Penicillin/Streptomycin, 5% human serum) containing growth factor, GM-CSF, and interleukin-4 (IL-4)) to the adhered cells in each well.

- Keep the cells for 7 days in the incubator without changing the medium, until the cells are required for experimentation.

Note that the cell medium may look muddy in the first 3–4 days. However, this cloudiness will disappear and the medium will become clear. As long as there are no bacterial contaminations, there is no need to dispose of the cultures.

14.4.2.1 Triple cell coculture design

The epithelial cells must be cultured for 7–8 days prior to the triple cell coculture protocol and the MDM and MDDC must be isolated 7 days before. A549 or 16HBE14o- cells should be maintained on transparent cell culture inserts in case of the 16HBE14o- coated with the coating solution. Please also refer to Figure 14.1(d), which shows some technical details.

1. Gently remove the adherent MDDC from the bottom of the 6-well plate, using a cell scraper.
2. Remove the medium from the upper side of the insert with the epithelial cells.
3. Remove the inserts from the plate and carefully place them inverted into a large plastic Petri dish.
4. Optional: Depending on the epithelial cell type used, it may be that in long-term cultures epithelial cells grown in monolayers may traverse the membrane and grow on the bottom of the membrane; therefore, the epithelial cells at the bottom can be abraded carefully with a cell scraper.
5. Remove the excess medium from the inverted insets with a one-use plastic pipette.
6. Place 300 μl of MDDC suspension on the top of the inverted insert. Be careful that no cell suspension drops down at the edge of the inserts.
7. Place the Petri dish into the incubator for 90 minutes.
8. Change the medium in the bottom of the 6-well plate and add 3-ml fresh medium to the wells. Put the plate back into the incubator until the inserts are placed back into the 6-well plate.
9. Following the 90-minute incubation period, remove the excess MDDC suspension and then transfer the inserts with the MDDC on the bottom back into the 6-well-plate with the fresh medium.
10. Gently remove the adherent MDM using a cell scraper.
11. Add 500 μl of the MDM suspension into each insert and place the Petri dish into the incubator for 90 minutes.
12. After 90 minutes remove the excess MDM suspension and add 2 ml of fresh medium into each upper chamber of the inserts.

Note that the triple cell coculture is now completed. After 4 hours the cocultures are ready to use; however, the best condition for experiments is 24 hours after the setup of the triple cell cocultures.

14.4.2.2 Air-liquid cultures

In many in vitro cell culture models, designed to study particle-cell interactions, the cells are immersed in medium and particles are added, suspended in liquid. This does not reflect the physiological condition of lung epithelial cells (airway barriers), which are exposed to air [50], separated by only a thin liquid lining layer (hypophase) with a surfactant film at the air-liquid interface [2–4]. As an alternative, particles can be nebulized (as a suspension or as a dry powder) over the air-exposed system using a spraying device (e.g., a Microsprayer) [34, 50] or an exposure chamber with an integrated aerosol generator [51]. Another experimental approach working with an exposure device based on a cell exposure system is the CULTEX [52].

The cultivation of epithelial cells on permeable supports allows the culture medium to remain separated on either side of the cultured epithelium, which leads to an increased differentiation of the cultured cells [53]. Furthermore, the medium can be removed from the upper side to expose the cells to air on the apical side and to allow them feed from the medium in the chamber underneath [54]. Air-exposed cell cultures allow investigation of the interaction of particles with cells in an environment that more closely mimics the in vivo situation. Of particular importance is that the cells are covered by a very thin liquid lining layer with a molecular surfactant-like film at the air-liquid interface. Surfactant plays an important role in particle displacement and retention [55].

It is important to mention that the air-cultured cells need time to produce a surfactant-like structure into the liquid layer, reducing the surface tension and protecting the cells. It has been shown that monolayers of A549 cells can be exposed to air for 2 days without losing their monolayer character. However, as of day 3, the cultures began to build multilayers [50]. A similar morphological change had already been observed by other researchers with A549 cells exposed to air for 4 weeks [56] and with the bronchial epithelial cell line 16HBE14o- cells exposed to air for 17 days [57]. A possible explanation for this behavior could be that cells grow on the surface of the cell layer under air-exposure conditions, since the cells are not removed during feeding (i.e., no medium is changed in the upper well when cells are exposed to air). Compared to submersed cultures, the air-exposed A549 cultures showed no difference in TEER values [50].

14.4.3 TEER measurements

TEER values give an indication about the cell layer integrity and confluence. Regularly check the TEER of cells cultured on inserts, if using them for particle and transport experiments where an appropriate tight cell layer is necessary.

1. Calibrate the sensor. Place the sterile, dry TEER sensor in a 15-ml falcon tube containing a 5-ml cell culture medium without any additions (use the correct medium for the respective cell line) for a minimum of 24 hours prior to calibration.
2. Place the sensor in 5-ml 70% ethanol for 15 minutes and subsequently allow to air-dry for 30 seconds.
3. Place the sensor in at least a 5-ml culture medium without any additions in a 15-ml tube for a further 15 minutes.
4. During this time, remove the samples from the incubator and place them in the laminar flow in order to equilibrate the cells to room temperature.

5. Perform five to six test measurements with the sensor until the values are stable. Do not touch the sides or base of the wells and insert and make sure that the sensor is totally covered with medium. Then start with the experiments.
6. For a sample on a 6-well membrane insert, take four recordings at different areas of the insert.
7. Between each sample, immerse the sensor into sterile high purified water.

14.4.4 Staining for LSM

Dependent on which cell types and cell structures are wanted to be stained, the combination of the different stains has to be considered.

1. Fix the cell culture by removing the medium and then adding 3% PFA diluted in PBS, to the upper and lower chambers of the wells for 15 minutes at room temperature. Make sure that all cells are completely covered with PFA.
2. Replace the 3% PFA with 0.1M Glycine (diluted in PBS) for at least 5 minutes at room temperature. Either continue with the staining procedure or store the cells in Glycine for a maximum period of 1 month in the fridge at 4°C.
3. If required for staining, permeablize the cells with 0.2% TritonX-100 (diluted in PBS) for 15 minutes at room temperature.
4. Remove the supernatant from either step 2 or 3 and wash the cells with PBS.
5. Incubate the cells with the first antibody for 60 minutes at room temperature in the dark.
6. Wash the cells three times with PBS.
7. Incubate the cells with the second antibody for 60 minutes at room temperature in the dark.
8. Wash the cells again three times with PBS.
9. Mount the preparations in PBS:glycerol (2:1) containing 170 mg/ml Mowiol 4–8 (Calbiochem, VWR International AG). Store the samples in the fridge at 4°C in the dark until further examinations. To be sure that the samples are well embedded, wait at least 24 hours until further examinations with the microscope.

Note that, dependent on which cell type is used, which cell structures have to be stained, and which laser wavelength are available, different staining combinations have to be chosen (Table 14.1). There are a couple of staining examples that are already tested and well established.

All antibodies were diluted in PBS as summarized in Table 14.1.

The combination of the fluorochromes depends on the filters that are available at the fluorescence microscopes. As a control, the specificity of the antibodies and the labeling procedure were tested with the secondary antibodies only.

14.4.4.1 LSM and image restoration

The cell cultures can be examined in a live or fixed manner. The appropriate cell culture material must be chosen dependent upon the experimental design. For live cell imaging, plastics or slides must be used that are specially made for live cell imaging, such as chambered cover glass slides (Nunc, Fisher Scientific AG, Wohlen, Switzlerand).

Table 14.1 Summary of Antibodies and Fluorescent Dyes Needed for the Characterization of the Cell Cultures

Primary Antibodies	Marked Structure/Protein	Dilution	Order Number/Company
Mouse antihuman CD14	CD14 on the surface of MDM	1:20	Clone UCHM-1, C 7673, Sigma
Mouse antihuman CD86	CD86on the surface of MDDC	1:20	Clone HB15e, 36931A, PharMingen, BD Biosciences
Rabbit antioccludin	Tight junction protein in epithelial cells	1:50	Clone Z-T22, 71-1500, Invitrogen, Basel Switzerland
Rabbit antihuman claudin-2	Tight junction protein in epithelial cells	1:50	51–6,100; Zymed
Rabbit antihuman ZO-3	Tight junction protein in epithelial cells	1:100	36–4,100; Zymed
Mouse antihuman E-cadherin	Adherens junction protein in epithelial cells	1:50	Clone HECD-1, 13–1700; Zymed
Secondary antibodies			
Goat antimouse cyanine 5	Detection of primary mouse antibodies	1:50	AP124S, Chemicon, VWR International AG, Life Sciences, Lucerne, Switzerland
Goat antirabbit cyanine 5	Detection of primary rabbit antibodies	1:50	AP132S, Chemicon, VWR International AG, Life Sciences, Lucerne, Switzerland
Fluorescent dyes			
Phalloidin Alexa 488	F-Actin	1:100	A12379, Molecular Probes, Invitrogen AG, Basel, Switzerland
Phalloidin Rhodamine	F-Actin	1:100	R415, Molecular Probes, Invitrogen AG, Basel, Switzerland
4‘,6-Diamidino-2-phenylindol (DAPI)	Cell nuclei	1:1000	D3571, Molecular Probes, Invitrogen AG, Basel, Switzerland

A Zeiss LSM 510 Meta with an inverted Zeiss microscope (Axiovert 200M, Lasers: HeNe 633 nm, HeNe 543 nm, Ar 488 nm, and diode laser 405 nm) is used as an example. For the detection of signals of the examined stained structure, a negative sample containing only the secondary antibodies should be scanned first and the detector gain should be adjusted so that no fluorescent signal of the secondary antibody (such as background) can be detected. The scans of the labeled samples are then acquired using the same detector settings. Image processing and visualization is done using IMARIS, a 3D multichannel image processing software for confocal microscopic images (Bitplane AG, Zurich, Switzerland).

14.4.5 Fixation and embedding of cells for transmission electron microscopy (TEM)

The observation of the triple cell coculture model at the ultrastructural level can be performed by TEM. The quality of the cells and the 3D arrangement can be analyzed. The procedure is not required for the characterization of the triple cell coculture model, but dependent on the question of research, an analysis of the cellular ultrastructure and morphology might be needed and therefore a description of the fixation and embedding protocol is given here in detail.

Note that the protocol should follow the Control of Substances Hazardous to Health (COSHH) standards, and general cell culture health and safety precautions apply. It must be noted that Glutaraldehyde, OsO_4, sodium-cacodyl acid, Epon, and its catalyst are highly toxic. Additionally, uranylacetat is radioactive and therefore all these substances must be handled with extreme caution. Gloves, protecting glasses, lab coats, and face masks must be worn at all times. The chemicals must be disposed properly in a special fridge and all glassware and plasticware must be washed or disposed of appropriately afterwards.

Solution preparation

- 0.164M potassium phosphate buffer: To prepare the solution, first prepare solution A. Add 31.82g $K_2HPO_4 \cdot 3\ H_2O$ in a measuring cylinder, fill it up to 850 ml with distilled water, and stir. Then prepare solution B. Add 5.58g KH_2PO_4 to 250 ml in a measuring cylinder and stir. Mix 800 ml of solution A and 200 ml of solution B to produce 1,000 ml 0.164M potassium phosphate buffer with a final pH of 7.4 (adjusted with solution A or B).
- 2.5% Glutaraldehyde (GA) in 0.03M potassium phosphate buffer – pH 7.4, 350mOsm: Dilute a 0.164M potassium phosphate buffer to produce a 0.03M potassium phosphate buffer by mixing a 183-ml 0.164M potassium phosphate buffer with 817-ml distilled water. Then mix 900-ml 0.03M potassium phopshate buffer with 100-ml GA 25% and adjust the pH to 7.4 using a 0.164M potassium phosphate buffer or potassium hydroxide. The osmolarity should be adjusted to 350 ± 10–20 with distilled water or saccharose.
- 2% Osmium tetroxide (OsO_4) solution: Work in a hood to prepare the solution. Break one glass vial, a unit of OsO_4 containing 1g OsO4, into a brown flask containing 50 ml of distilled water. Close the flask immediately with parafilm. Either sonify the solution straightaway for no longer than two times for 10 minutes, or store it for 24 hours at room temperature before sonification. Usually the osmium is prepared 24 hours ahead, since sonifying for only 20 minutes might not dissolve it all.
- 0.2M Na-cacodylat solution: Weigh 1.4g of sodium-cacodyl acid ($C_2H_6AsNaO_2 \cdot 3\ H_2O$) and place it in a measuring cylinder. Fill it up with distilled water to 500 ml. Stir and add 3.74g of sodium chloride (NaCl) to increase the osmolarity to 620–700.
- 1% OsO_4 in 0.1M sodium (Na)-cacodylat-hydrochloric (HCl) acid: Mix 50 ml of 0.2M Na-cacodylat buffered solution with 50-ml 2% OsO_4. Then adjust the pH with 1M HCl to 7.4 and the osmolarity to 350 ± 10^{-20} with NaCl or distilled water.
- 0.05M maleat-sodium hydroxide (NaOH) buffer – pH 5.0: Weigh 5.8g of maleic acid and place it in a measuring cylinder. Fill it up with distilled water to 1,000 ml. Adjust the pH to 5.0 with NaOH.
- 0.5% Uranylacetat in 0.05M maleat-NaOH buffer – pH 5.0: Weigh 0.5g uranylacetat and place it in a measuring cylinder. Fill it up with100-ml 0.05M maleat-NaOH buffer. Seal it with parafilm and stir.

Fixation and embedding protocol

1. Fix the cells with 2.5% GA in 0.03M potassium phosphate buffer and incubate the cells for at least for 24 hours at 4°C. The whole membrane with the cells should be completely covered with the GA buffer. Cover the 6-well plate, including the fixed cells, with parafilm, so that no vapors escape. Storage up to 1 month is possible at 4°C.

2. The whole insert membrane (with the cells attached) must be removed from the insert holder with a scalpel and incubated with 15 ml of 0.164M potassium phosphate. Buffer twice for 5 minutes and overnight at 4°C in a third step.
3. Perform a secondary fixing and staining step by incubating the cells on the membrane with 4 ml of 1% OsO_4 in 0.1M sodium-cacodylat-HCl buffer for at least 1 hour at 4°C. At this point, Epon should be removed from the freezer.
4. Wash the membranes three times with 15 ml of 0.05M Maleat-NaOH Buffer (which is stored at 4°C) for at least 5 minutes per washing step.
5. Place the membranes in 4 ml of 0.5% uranylacetat in a 0.05M maleat NaOH buffer and incubate them for at least 1 hour at 4°C. Then wash the membranes again three times for at least 5 minutes each time, with 15 ml of a 0.05M maleat NaOH buffer (4°C).
6. Four steps to dehydrate the membranes have to be performed. First, incubate the membranes at room temperature for 15 minutes with 15 ml of 70% ethanol. Second, incubate for 15 minutes with 15 ml of 80% ethanol at room temperature, then 15 minutes at room temperature with 15 ml of 96% ethanol. At this stage, mix the Epon (11 ml per membrane) with a catalyst (N-Benzyldimethylamin, 1.5%), using a measuring cylinder and a glass stirrer (nonmagnetic). Finally, for the fourth dehydration step, incubate the membranes three times for at least 10 minutes at room temperature, twice with 15 ml of 100% ethanol and once with 15 ml of 100% ethanol p.a.
7. Remove the required amount of Epon for the embedding (6 ml per membrane) from the measuring cylinder, place in a beaker, seal them with parafilm and store in a freezer at –20°C until required. Mix the remainder (5 ml per membrane) 1:1 with acetone and stir it for at least 15 minutes (nonmagnetic).
8. Use acetone as an intermedium and incubate the membranes twice at room temperature for 10 minutes with 15 ml of acetone.
9. Incubate the membranes with the Epon and acetone mix (10 ml) overnight at room temperature.
10. The following day, embed the membranes with cells into 6-ml Epon by placing them into a silicon mold of appropriate size (diameter of 45 mm) and cover them with Epon. Let the Epon polymerize at 60°C for 5 days.
11. Release the embedded samples from the mold. The samples can now be cut into ultrathin vertical sections with a microtome and placed on cooper grids for TEM observation.

14.5 Anticipated Results

In order to study the interaction of particles that have been deposited at the airway wall with the cells of the human airway barrier, an in vitro model simulating the cellular airway barrier was generated and characterized. The triple cell coculture model system composed of cuboidal epithelial cells, and MDM and MDDC were established and evaluated in terms of their functional relevance to the in vivo tissue. For this purpose the human (alveolar) epithelial cell line A549, which originated from human lung carcinoma [40], or the human immortalized bronchiolar epithelial cell line 16HBE14o- (kindly received from Dieter Gruenert, Cardiovascular Research Institute, University of California, San Francisco) was chosen. In the study MDM and MDDC derived from human blood monocytes [58] were combined with the epithelial cells. The triple cell cocultures were characterized in terms of their typical features (e.g., morphology of the cells), integrity of the epithelial layer, and expression of specific cell surface markers.

In order to clearly identify MDM and MDDC in the triple cell coculture system, the expression of specific markers was investigated with confocal LSM. For the visualization of the cells, their actin-containing cytoskeleton was stained with phalloidin rhodamine. After 24 hours in coculture, the cells were fixed and stained for the specific surface markers CD14 and CD86 for the labelling of MDM and MDDC, respectively. The MDM were localized on top of the epithelium, whereas MDDC could be observed on the bottom side (Figure 14.2) [39, 44]. The model was evaluated carefully and compared with in vivo data, and it was found that the morphological [37] as well as the quantitative occurrence of MDM and MDDC very closely resembled the in vivo situation [34]. In addition, TEM pictures of the cultures showed in more detail the 3D arrangement of the different cells in the model (Figure 14.3).

The tightness of the epithelial layers was assessed by measurement of TEER and the staining of the tight junction proteins by immune fluorescence methods. TEER values of around 500 Ωcm^2 were measured in 16HBE14o- cocultures and in A549 coculture of around 200 Ωcm^2. In addition, it was shown that A549 and 16HBE14o- cells express the tight junction proteins occludin, claudin-2, and zonula occludens-3 (Figure 14.4) and

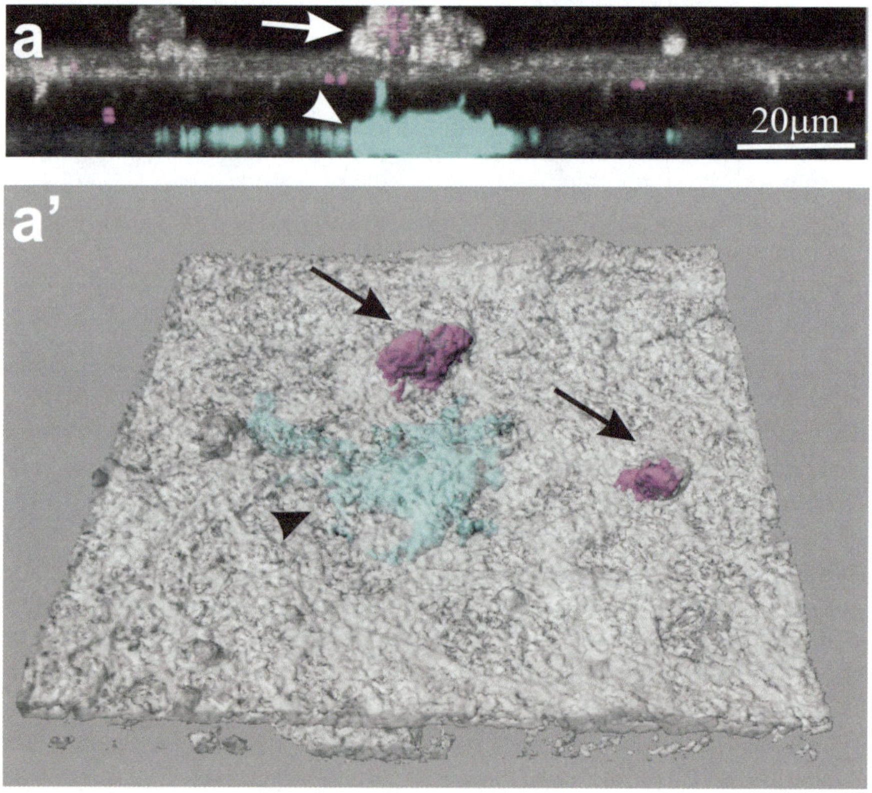

Figure 14.2 LSM image of the triple cell coculture model. EC (white), MDM on top (purple, black arrows), and MDDC underneath (light blue, white arrow) the epithelium are shown. The same data set is shown in (a) and (b). (a) *xz*-projection, and (b) 3D surface rendering of all cells, with EC made transparent.

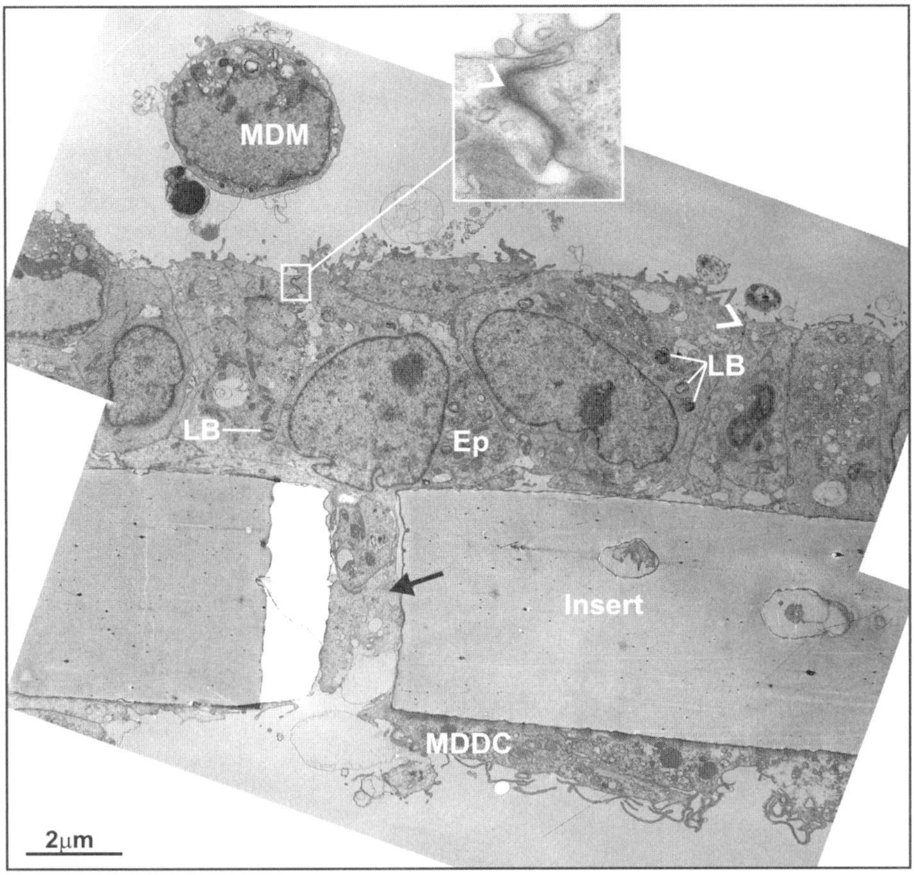

Figure 14.3 TEM images of the triple cell cocultures with EC, MDM, and MDDC grown on a filter insert. The A549 cells (Ep) containing lamellar bodies (LB) formed a monolayer of polar cells with tight junctions (inset, arrowhead), and with protrusions through the pores (black arrow). MDM were found at the apical side of the epithelial cells, whereas MDDC were localized at the bottom side of the insert. (*From:* [39]. © 2005 Reproduced and adapted with permission from *American Journal of Respiratory Cell and Molecular Biology*.)

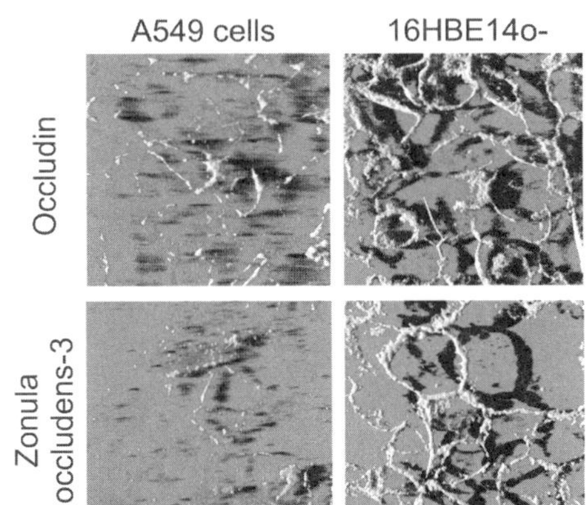

Figure 14.4 Expression of tight junction proteins in A549 and 16HBE14o- cells. LSM images of occludin and Zonula occludens-3 in both cell lines. The tight junction proteins are expressed and localized at the apical cell border in both cell types; however, in 16HBE14o- cells the tight junction arrangement is more regular. The images represent shadow projections.

Troubleshooting Table

Problem	Explanation	Potential Solutions
Cell death during cell thawing.	All cells are dead after the cell thawing procedure.	Also do a freezing control. In this way it can be determined whether the cells died during this procedure. Work fast and consider the temperature of the water bath.
There are not enough epithelial cells to seed out.	The cells did not grow as expected and there are not enough epithelial cells to seed out.	Check the growing conditions (for example, the medium, temperature, or gas mixture). Reduce the amount of the planned experiment and seed the cells in the normal concentration because it is important to have a confluent epithelial cell layer for the triple cell coculture model.
Epithelial cells do not detach.	During the splitting or seeding procedure, the adherent epithelial cells do not detach well from the flask bottom.	Enhance the incubation time in the incubator after the Trypsine/EDTA step and/or knock the flask for several times, with caution, on a hard surface.
A small amount of MDM and MDDC are cultured.	There are only a few MDM and MDDC harvested during the monocyte isolation procedure.	Perform an accumulation, by concentrating the cell suspensions either by centrifugation of the cell suspension and resuspension in a smaller amount of medium or by simply removing a part of the cell culture medium of the adhered cells in the 6-well plates If there are only a few monocytes harvested after the Ficoll separation, try to obtain all the cells in the respective middle band; otherwise, a lot of cells will be lost. Furthermore, it depends on the blood donor's state of health. If the donor has previously had any inflammation, the amount of white blood cells is enhanced. Therefore, the rate of yield will differ slightly from isolation to isolation.
No separation of cells.	During the cell separation using the Ficoll gradient, there is no clear band of red blood cells at the bottom, an overlay of serum, and a middle band of the white blood cells.	Take care that the Ficoll is applied to the tube carefully and that the Ficoll is released very slowly. If it is not performed carefully, then there will not be a clear separation.
The MDDC solution is dripping down to the inverted inserts.	By adding the MDDC on the bottom of the inverted inserts, the MDDC solution is dripping down into the Petri dish.	Put the MDDC solution on the bottom of the inverted insert very carefully. Remove any medium, with a small pipette, before adding the MDDC solution.
The MDDC solutions dry out on the inverted insert.	During the 90-minute incubation period, the MDDC solution dries out. The cells will be dead.	Check after 45 minutes if the insert membrane is still wet. If not, carefully add single droplets of medium onto the inverted insert membrane, but not so much that the medium drops down.
Unstable values of TEER measurements.	During the TEER measurement procedure, the values are unstable or differ from the habitual values.	Clean the sensor with the special abrasive paper. Do a new calibration with 70% ethanol followed by the medium as described in the manufacturer's protocol. Check by using light microscopy if the epithelial cell layer is confluent. If not, do not use the system for any transcytosis or particle uptake studies.

the adherens junction protein E-cadherin [34, 39, 50]. Corresponding with the higher TEER values in 16HBE14o- cells, the tight junction arrangement of both proteins is more regular. The analysis investigating if other tight junction proteins are expressed is ongoing work in our group.

14.6 Discussion and Commentary

The described triple cell coculture model may help to elucidate mechanisms of particle-cell interaction in the lung and airways which are assumed to induce toxic reactions. Despite a number of limitations, lung cell cultures (by using cell lines) offer a valuable tool to study effects of inhalable substances. Guidelines for good cell culture practice are required, including the control of the starting material (the cultured cells), the culture medium, and the culture substratum [59, 60]. An essential disadvantage is that cell culture models often do not exhibit all the differentiated and functional characteristics of the corresponding native epithelium or the entire organ. Therefore, a model for a certain development problem or scientific question should be selected very carefully, with considerations being given to its limitations, the experimental design and the interpretation of results. However, the presented 3D model of the epithelial airway barrier, which includes several cell types, is a step forward since it offers the possibility to study not only the reaction of individual cell types but also the interaction of the different cell types with each other.

14.7 Application Notes

The presented 3D model of the human epithelial airway barrier offers the possibility to study different basic research questions with lung cells in vitro. This approach offers the advantage of a 3D structure and also the communication of different cells and both aspects represent a more physiologically relevant situation. Another advantage of our model is the reproducibility of the system in regards to all tested parameters.

14.8 Summary Points

1. The authors have developed a 3D cell culture model of the human epithelial airway model. The system is composed of epithelial cells (cell lines of different origin) combined with MDDC and MDM.
2. A series of evaluation procedures have been developed for the accurate characterization of the triple cell coculture model.
3. The model can be used to study the effects of inhaled xenibiotics/particulate matter at the human epithelial airway barrier.

Acknowledgments

We thank Kirsten Dobson for proofreading of the chapter manuscript. This work was supported by the Swiss National Science Foundation (3100A0_118420), the German research foundation (SPP1313), the Swiss Agency for the Environment, the Doerenkamp-Zbinden Foundation, the Johanna Duermueller-Bol Foundation and the Animal Free Research.

References

[1] Gehr, P., M. Bachofen, and E. R. Weibel, "The Normal Human Lung: Ultrastructure and Morphometric Estimation of Diffusion Capacity," *Respir. Physiol.,* Vol. 32, 1978, pp. 121–140.

[2] Gehr, P., et al., "Particle Retention in Airways by Surfactant," *Jounal of Aerosol Medicine* 1990, 3, pp. 27–43.

[3] Gil, J., and E. R. Weibel, "Extracellular Lining of Bronchioles After Perfusion-Fixation of Rat Lungs for Electron Microscopy," *Anat. Rec.,* Vol. 169, 1971, pp. 185–199.

[4] Schürch, S., et al., "Surfactant Displaces Particles Toward the Epithelium in Airways and Alveoli," *Respiration Physiology* Vol. 80, 1990, pp. 17–32.

[5] Kilburn, K. H., "A Hypothesis for Pulmonary Clearance and Its Implications," *Am. Rev. Respir. Dis.,* Vol. 98, 1968, pp. 449–463.

[6] Brain, J. D., "Lung Macrophages: How Many Kinds Are There? What Do They Do?" *Am Rev Respir Dis* 1988, 137, pp. 507–509.

[7] Lehnert, B. E., "Pulmonary and Thoracic Macrophage Subpopulations and Clearance of Particles from the Lung," *Environ. Health Perspect.,* Vol. 97, 1992, pp. 17–46.

[8] Godfrey, R. W., "Human Airway Epithelial Tight Junctions," *Microsc. Res. Tech.,* Vol. 38, 1997, pp. 488–499.

[9] Schneeberger, E. E., and R. D. Lynch, "Tight Junctions. Their Structure, Composition, and Function," *Circ. Res.,* Vol. 55, 1984, pp. 723–733.

[10] Holt, P. G., and M. A. Schon-Hegrad, "Localization of T Cells, Macrophages and Dendritic Cells in Rat Respiratory Tract Tissue: Implications for Immune Function Studies," *Immunology,* Vol. 62, 1987, pp. 349–356.

[11] McWilliam, A. S., P. G. Holt, and P. Gehr, "Dendritic Cells as Sentinels of Immune Surveillance in the Airways," in P. Gehr and J. Heyder, (eds.), *Particle-Lung Interactions*, New York: Basel/Marcel Dekker, 2000, pp. 473–489.

[12] Maina, J. N., and J. B. West, "Thin and Strong! The Bioengineering Dilemma in the Structural and Functional Design of the Blood-Gas Barrier," *Physiol. Rev.,* Vol. 85, 2005, pp. 811–844.

[13] Yurchenco, P. D., et al., "Models for the Self-Assembly of Basement Membrane," *J Histochem. Cytochem.,* Vol. 34, 1984, pp. 93–102.

[14] Dunsmore, S. E., and D. E. Rannels, "Extracellular Matrix Biology in the Lung," *Am. J. Physiol.,* Vol. 270, 1996, pp. L3–27.

[15] Dudek, S. M., and J. G. Garcia, "Cytoskeletal Regulation of Pulmonary Vascular Permeability," *J. Appl. Physiol.,* Vol. 91, 2001, pp. 1487–1500.

[16] Schneeberger, E. E., "Ultrastructure of Intercellular Junctions in the Freeze Fractured Alveolar-Capillary Membrane of Mouse Lung," *Chest,* Vol. 71, 1977, pp. 299–300.

[17] Peters, A., et al., "Respiratory Effects Are Associated with the Number of Ultrafine Particles," *American Journal of Respiratory and Critical Care Medicine,* Vol. 155, 1997, pp. 1376–1383.

[18] Pope, I. C. A., D. W. Dockery, and J. Schwartz, "Review of Epidemiological Evidence of Health Effects of Particulate Air Pollution," Vol. 7, 1995, pp. 1–18.

[19] Schulz, H., et al., "Cardiovascular Effects of Fine and Ultrafine Particles," *Journal of Aerosol Medicine-Deposition Clearance and Effects in the Lung*, Vol. 18, 2005, pp. 1–22.

[20] Araujo, J. A., et al., "Ambient Particulate Pollutants in the Ultrafine Range Promote Early Atherosclerosis and Systemic Oxidative Stress," *Circ. Res.*, Vol. 102, 2008, pp. 589–596.

[21] Borm, P. J. A., and W. Kreyling, "Toxicological Hazards of Inhaled Nanoparticles—Potential Implications for Drug Delivery," *Journal of Nanoscience and Nanotechnology*, Vol. 4, 2004, pp. 521–531.

[22] Geiser, M., et al., "Ultrafine Particles Cross Cellular Membranes by Nonphagocytic Mechanisms in Lungs and in Cultured Cells," *Environ. Health Perspect.*, Vol. 113, 2005, pp. 1555–1560.

[23] Rothen-Rutishauser, B., "SSGP: Interaction of Particles with Membranes," in K. Donaldson and P. Borm, (eds.), *The Toxicology of Particles*, 2007, pp. 139–160.

[24] Rothen-Rutishauser, B. M., et al., "Interaction of Fine Particles and Nanoparticles with Red Blood Cells Visualized with Advanced Microscopic Techniques," *Environ. Sci. Technol.*, Vol. 40, 2006, pp. 4353–4359.

[25] Gonzalez-Flecha, B., "Oxidant Mechanisms in Response to Ambient Air Particles," *Mol. Aspects Med.*, Vol. 25, 2004, pp. 169–182.

[26] Li, N., T. Xia, and A. E. Nel, "The Role of Oxidative Stress in Ambient Particulate Matter-Induced Lung Diseases and Its Implications in the Toxicity of Engineered Nanoparticles," *Free Radic. Biol. Med.*, Vol. 44, 2008, pp. 1689–1699.

[27] Muller, J., et al., "Respiratory Toxicity of Multi-Wall Carbon Nanotubes," *Toxicol. Appl. Pharmacol.*, Vol. 207, 2005, pp. 221–231.

[28] Schins, R. P., and A. M. Knaapen, "Genotoxicity of Poorly Soluble Particles," *Inhal. Toxicol.*, Vol. 19, Suppl. 1, 2007, pp. 189–198.

[29] Vinzents, P. S., et al., "Personal Exposure to Ultrafine Particles and Oxidative DNA Damage," *Environ. Health Perspect.*, Vol. 113, 2005, pp. 1485–1490.

[30] Ayres, J. G., et al., "Evaluating the Toxicity of Airborne Particulate Matter and Nanoparticles by Measuring Oxidative Stress Potential—A Workshop Report and Consensus Statement," *Inhal. Toxicol.*, Vol. 20, 2008, pp. 75–99.

[31] Borm, P., et al., "Research Strategies for Safety Evaluation of Nanomaterials, Part V: Role of Dissolution in Biological Fate and Effects of Nanoscale Particles," *Toxicological Sciences*, Vol. 90, 2006, pp. 23–32.

[32] Carterson, A. J., et al., "A549 Lung Epithelial Cells Grown as Three-Dimensional Aggregates: Alternative Tissue Culture Model for Pseudomonas Aeruginosa Pathogenesis," *Infect. Immun.*, Vol. 73, 2005, pp. 1129–1140.

[33] Roggen, E. L., N. K. Soni, and G. R. Verheyen, "Respiratory Immunotoxicity: An In Vitro Assessment," *Toxicol. In Vitro*, Vol. 20, 2006, pp. 1249–1264.

[34] Blank, F., B. Rothen-Rutishauser, and P. Gehr, "Dendritic Cells and Macrophages Form a Transepithelial Network Against Foreign Particulate Antigens," *Am. J. Respir. Cell Mol. Biol.*, Vol. 36, 2007, pp. 669–677.

[35] Rescigno, M., et al., "Dendritic Cells Express Tight Junction Proteins and Penetrate Gut Epithelial Monolayers to Sample Bacteria," *Nat. Immunol.*, Vol. 2, 2001, pp. 361–367.

[36] Rescigno, M., et al., "Dendritic Cells Shuttle Microbes Across Gut Epithelial Monolayers," *Immunobiology*, Vol. 204, 2001, pp. 572–581.

[37] Rothen-Rutishauser, B., F. Blank, and C. G. P. Mühlfeld, "In Vitro Models of the Human Epithelial Airway Barrier to Study the Toxic Potential of Particulate Matter," *Exp. Opinion Drug Metabolism Toxicol.*, 2008, in press.

[38] Rothen-Rutishauser, B., et al., "Translocation of Particles and Inflammatory Responses After Exposure to Fine Particles and Nanoparticles in an Epithelial Airway Model," *Part Fibre Toxicol.*, Vol. 4, 2007, p. 9.

[39] Rothen-Rutishauser, B. M., S. G. Kiama, and P. Gehr, "A Three-Dimensional Cellular Model of the Human Respiratory Tract to Study the Interaction with Particles," *Am. J. Respir. Cell Mol. Biol.*, Vol. 32, 2005, pp. 281–289.

[40] Lieber, M., et al., "A Continuous Tumor-Cell Line from a Human Lung Carcinoma with Properties of Type II Alveolar Epithelial Cells," *Int. J. Cancer*, Vol. 17, 1976, pp. 62–70.

[41] Forbes, B., and C. Ehrhardt, "Human Respiratory Epithelial Cell Culture for Drug Delivery Applications," *Eur. J. Pharm. Biopharm.*, Vol. 60, 2005, pp. 193–205.

[42] Forbes, I. I., "Human Airway Epithelial Cell Lines for In Vitro Drug Transport and Metabolism Studies," Vol. 3, 2000, pp. 18–27.

[43] Muller, L., et al., "Oxidative Stress and Inflammation Response After Nanoparticle Exposure: Differences Between Human Lung Cell Monocultures and an Advanced Three-Dimensional Model of the Human Epithelial Airways," *J. R. Soc. Interface*, 2009.

[44] Rothen-Rutishauser, B., et al., "A Newly Developed In Vitro Model of the Human Epithelial Airway Barrier to Study the Toxic Potential of Nanoparticles," *ALTEX*, Vol. 25, No. 3, 2008, pp. 191–196.

[45] Ehrhardt, C., et al., "Drug Absorption by the Respiratory Mucosa: Cell Culture Models and Particulate Drug Carriers," *J. Aerosol. Med.*, Vol. 15, 2002, pp. 131–139.

[46] Wan, H., et al., "Tight Junction Properties of the Immortalized Human Bronchial Epithelial Cell Lines Calu-3 and 16HBE14o-," *Eur. Respir. J.*, Vol. 15, 2000, pp. 1058–1068.

[47] Foster, K. A., et al., "Characterization of the A549 cell line as a type II pulmonary epithelial cell model for drug metabolism," *Exp. Cell Res.*, Vol. 243, 1998, pp. 359–366.

[48] Shapiro, D. L., et al., "Phospholipid Biosynthesis and Secretion by a Cell Line (A549) Which Resembles Type II Aleveolar Epithelial Cells," *Biochim Biophys Acta*, Vol. 530, 1978, pp. 197–207.

[49] Stearns, R. C., J. D. Paulauskis, and J. J. Godleski, "Endocytosis of Ultrafine Particles by A549 Cells," *Am. J. Respir. Cell Mol. Biol.*, Vol. 24, 2001, pp. 108–115.

[50] Blank, F., et al., "An Optimized In Vitro Model of the Respiratory Tract Wall to Study Particle Cell Interactions," *J. Aerosol. Med.*, Vol. 19, 2006, pp. 392–405.

[51] Tippe, A., U. Heinzmann, and C. Roth, "Deposition of Fine and Ultrafine Aerosol Particles During Exposure at the Air/Cell Interface," *Aerosol Science*, Vol. 33, 2002, pp. 207–218.

[52] Aufderheide, M., and U. Mohr, "CULTEX—An Alternative Technique for Cultivation and Exposure of Cells of The Respiratory Tract to Airborne Pollutants at the Air/Liquid Interface," *Exp. Toxicol. Pathol.*, Vol. 52, 2000, pp. 265–270.

[53] Handler, J. S., N. Green, and R. E. Steele, "Cultures as Epithelial Models: Porous-Bottom Culture Dishes for Studying Transport and Differentiation," *Methods Enzymol.*, Vol. 171, 1989, pp. 736–744.

[54] Voisin, C., et al., "Effects of Nitrogen Dioxide on Alveolar Macrophages Surviving in the Gas Phase. A New Experimental Model for the Study of In Vitro Cytotoxicity of Toxic Gases (author's translation)," *Bull. Eur. Physiopathol. Respir.*, Vol. 13, 1977, pp. 137–144.

[55] Gehr, P., et al., "Airway Surfactant, a Primary Defense Barrier: Mechanical and Immunological Aspects," *J. Aerosol. Med.*, Vol. 9, 1996, pp. 163–181.

[56] Radyuk, S. N., et al., "In Vitro-Generated Respiratory Mucosa: A New Tool to Study Inhalational Anthrax," *Biochem. Biophys. Res. Commun.*, Vol. 305, 2003, pp. 624–632.

[57] Ehrhardt, C., et al., "Influence of Apical Fluid Volume on the Development of Functional Intercellular Junctions in the Human Epithelial Cell Line 16HBE14o-: Implications for the Use of This Cell Line as an In Vitro Model for Bronchial Drug Absorption Studies," *Cell Tissue Res.*, Vol. 308, 2002, pp. 391–400.

[58] Sallusto, F., et al., "Dendritic Cells Use Macropinocytosis and the Mannose Receptor to Concentrate Macromolecules in the Major Histocompatibility Complex Class II Compartment: Downregulation by Cytokines and Bacterial Products," *J. Exp. Med.*, Vol. 182, 1995, pp. 389–400.

[59] Gruber, F. P., and T. Hartung, "Alternatives to Animal Experimentation in Basic Research," *ALTEX*, Vol. 21, Suppl. 1, 2004, pp. 3–31.

[60] Gstraunthaler, G., and T. Hartung, "Good Cell Culture Practice: Good Laboratory Practice in the Cell Culture Laboratory for the Standardization and Quality Assurance of In Vitro Studies," in C. M. Lehr, (ed.), *Cell Culture Models of Biological Barriers: In Vitro Test Systems for Drug Absorption and Delivery*, London, U.K.: Taylor and Francis, 2002, pp. 112–120.

CHAPTER

15

Experimental Wear Assessment of Tibial Inserts for Total Knee Replacement

Michele Spinelli[1,2] and Saverio Affatato[1]

[1]Medical Technology Laboratory, Rizzoli Orthopaedic Institute, Bologna, Italy
[2]Industrial Bioengineering Laboratory, Department of Mechanical and Industrial Technologies, University of Florence, Florence, Italy
Corresponding author: Saverio Affatato, address: Medical Technology Laboratory, Rizzoli Orthopaedic Institute, Bologna, Italy, Via di Barbiano 1/10, 40136, Bologna, Italy
e-mail: affatato@tecno.ior.it, phone: +39-051-6366564, fax: +39-051-6366863

Abstract

We present an innovative procedure to assess the wear of tibial inserts of total knee replacement. By means of a knee joint wear simulator, we reproduced working conditions as close as possible to the in vivo ones by fixing the femoral component of the prosthesis to the distal part of a synthetic femur. The gravimetric wear of the tibial inserts was assessed at regular intervals on the basis of ISO guidelines. The wear patterns on tibial inserts were characterized through a standardized protocol based on digital image analysis. The wear tracks observed on tibial inserts and other inserts retrieved after some period of in vivo function support the efficacy of the new in vitro method presented.

Key terms
Gravimetric assessment
knee joint simulator
knee prosthesis
synthetic bone
wear test

15.1 Introduction

Joint prostheses are implanted in large numbers today. In the 30 member countries of the Organisation for Economic Co-operation and Development (OECD), the rate is around 50 to 140 lower limb joint replacements per year per each 100,000 inhabitants [1]. Out of the whole population, total knee replacement (TKR) (Figure 15.1) reached an incidence of about 500,000 prostheses implanted every year in the world [2]. Among these a large and increasing number of TKRs are performed to satisfy younger and more active patients with respect to better long-term expectations [3].

The formation and development of wear is still widely accepted as one of the major concerns in the long-term survivorship of contemporary TKR in vivo [3], despite the good results obtained with in vitro wear tests [4–6]. This suggests that there are one or more cofactors, which are not accounted in standard ISO 14243-1/3 wear tests that might induce severe wear in vivo.

Cadaver studies hint that changes on the orientation of the prosthetic components with respect to the loading directions as small as three degrees can change dramatically the contact pressure, one of the prime determinants of wear [7]. Thus, it might be possible that one of those cofactors driving the clinical results far from the ones observed in vitro might be related to the fact that, for in vitro wear tests, we usually assume the prosthetic components rigidly fixed with respect to the loading directions. Current laboratory methods make it possible to evaluate with good accuracy the changes of position and orientation of the prosthetic components with respect to the host bone, and the effects that these changes may induce in wear patterns of tibial inserts [8].

As the wear of TKR is very sensitive to working conditions [9–13], the way components are fixed to the simulator might be a crucial point in wear simulation. In TKR the load transfer between the bone and the prosthesis plays a relevant role in the observed wear and this might be attributable to the implant-bone fixation, alignment, and overall

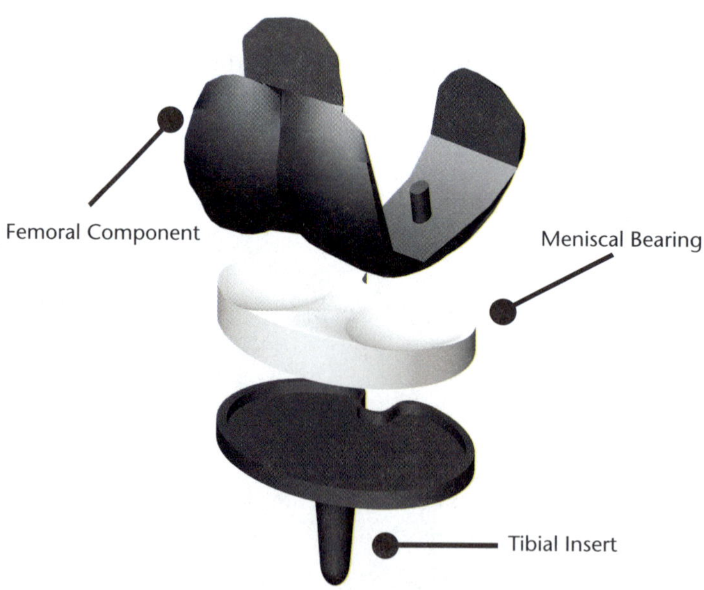

Figure 15.1 A schematic illustrating the state of the art for a total knee prosthesis. The femoral component is usually metallic (CoCrMo alloy) and articulated on a plastic insert (UHMWPE) that is supported by a metallic meniscal bearing CoCrMo alloy.

stiffness. One particular interest in our laboratory is to evaluate the wear of TKR, which includes not only a detailed reproduction of the kinematics and dynamic conditions but also realistic biomechanical conditions.

Hereby we present a new method for the experimental assessment of tibial inserts' wear under conditions that are as close as possible to an in vivo function. For this purpose, we used a "three-plus-one-station" knee joint wear simulator (three stations for the test specimens plus one for the soak control specimen) following the international standard (ISO 14243-1,2,3), but with the added feature of fixing the femoral component of the TKR to the distal part of a synthetic femur model. A comparison of the generated wear patterns on the tested inserts with those observed on available retrieved inserts showed the efficacy of this innovative experimental method for the wear assessment of tibial inserts.

15.2 Experimental Design

The femoral components were implanted in composite femur models, made of a glass fiber-reinforced epoxy shell surrounding polyethylene foam. It is well known that synthetic bone models allow a significant reduction in variability compared with human bones, and their use is standardized for assessing the long-term stability of hip stems [10, 14]. The distal end of the femur was resected and machined using a CNC milling machine to achieve the bone-cutting geometry indicated by the manufacturer of the knee components with the highest repeatability and to receive the metal femoral components. Special holders were custom-made to connect the composite femurs on the flexion-extension axis of the knee joint simulator (Figure 15.2). The femoral component was fixed to the distal end of the synthetic femur by a qualified orthopedic surgeon using an approved surgical procedure. The prosthesis was mounted in the simulator so that its multiradiused profile could be adapted with the single flexion/extension axis of the machine (Figure 15.3).

15.3 Materials

- UHMWPE tibial inserts (size 3, Multigen-Plus Knee, Lima-Lto, Udine, Italy) (at least one of the test specimens is used as a control to account for mass changes not directly related to the applied consistent load cycle)
- CoCrMo alloy femoral components (size 3, Multigen-Plus Knee, Lima-Lto, Udine, Italy)
- Meniscal bearings (size 3, Multigen-Plus Knee, Lima-Lto, Udine, Italy)
- Three-plus-one-station knee joint wear simulator (Shore Western Mfg., Los Angeles, California)
- Sterile calf serum (SIGMA Chemical, St. Louis, Missouri)
- Deionized water (SALF, Bergamo, Italy)
- Acetone (SIGMA Chemical, St. Louis, Missouri)
- Sodium azide (Merck, Darmstadt, Germany)
- Sodium hydroxide (SIGMA Chemical, St. Louis, Missouri)

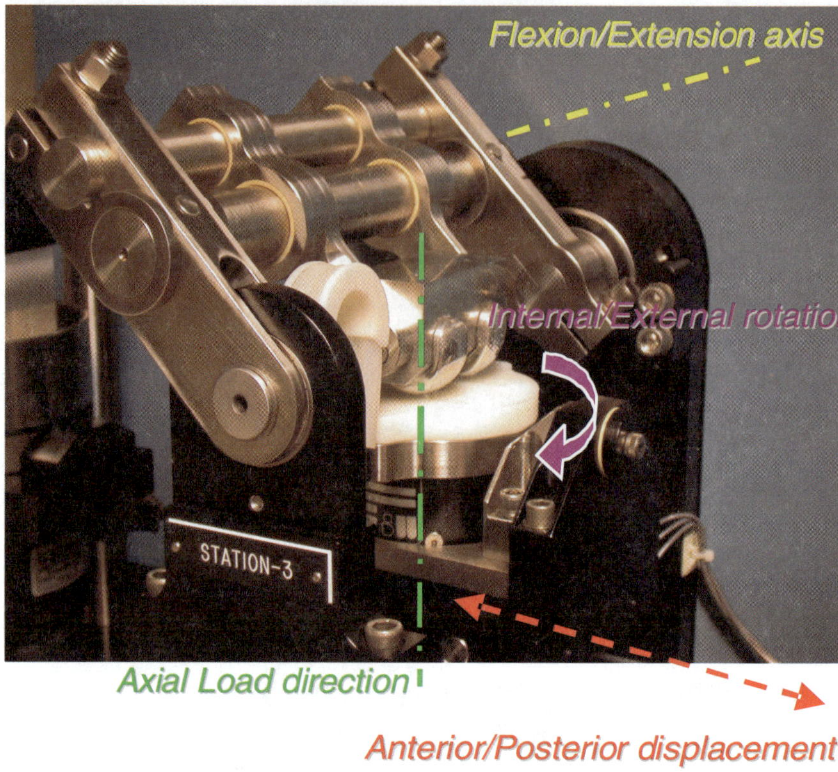

Figure 15.2 A graphical arrangement of the mechanical degrees of freedom of a knee simulator.

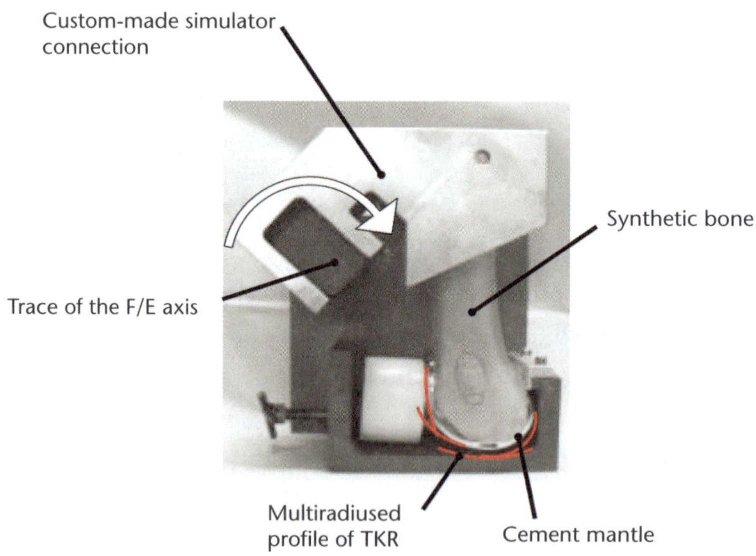

Figure 15.3 Custom-made holder to allow the connection of the composite femurs to the flexion/extension axis of the knee joint wear simulator. The femoral component is connected to the simulator through a synthetic femur in order to reproduce more realistic biomechanical conditions. The prosthesis was attached to the synthetic femur through a thin cement mantle by a qualified orthopedic surgeon.

- EDTA (SIGMA Chemical, St. Louis, Missouri)
- Level plane with scribing block and vertical ruler
- Analytical balance, precision of ±0.1 mg (Sartorius 210g, Alessandrini, Modena, Italy)
- Ultrasound washer (Transsonic 780/H, Alessandrini, Modena, Italy)
- Detergent: Clean 65 (Elma, Milano, Italy)
- Filter system, 0.22-μm porosity (Nalge Company, New York)
- Filtered inert gas (e.g., nitrogen)
- Vacuum pump (D.V.P., Vacuum Technology, Bologna, Italy)
- Bone cement (Palacos-R40, Schering-Plough, Brussels, Belgium)
- Digital camera (Mod. Coolpix 995, Nikon)
- Optical binocular reflection microscope (Mod. SMZ-2T, Nikon, Inv. N.: 18275/746F) with 10× eyepiece (with reference grid), 2× objective, and 2.5× lens for photo camera

15.4 Methods

15.4.1 Management of the specimens

15.4.1.1 Presoaking procedure for the tibial inserts

1. Remove the specimens from the package. Clean with a neutral detergent.
2. Iteratively plug and unplug tibial inserts from meniscal bearings (10 times in a row).
3. Soak the tibial inserts in deionized water for 4 weeks [15].
4. Weigh each tibial insert three times in rotation after 90 minutes of completion of the cleaning protocol.
5. Repeat the entire procedure at intervals (each couple of days) until the incremental mass change of the specimens over the interval is less than 10% of the previous cumulative mass change. These are to be considered the weights at time zero.

15.4.1.2 Cleaning procedure of the tibial inserts

1. Remove the specimens from the soaking bath.
2. Clean in deionized water and neutral detergent.
3. Rinse in deionized water.
4. Vibrate for 15 minutes in a mixture of detergent (Clean 65, 5% v/v) and deionized water at 50°C.
5. Rinse in deionized water.
6. Vibrate for 15 minutes in deionized water at 50°C.
7. Dry with a jet of inert gas.
8. Dry in a vacuum chamber (1.0 ± 0.1 bar) for at least 40 minutes.
9. Wait at least 90 minutes for the specimens to stabilize. Then weigh the specimens.

15.4.1.3 Gravimetric measurement procedure of the tibial inserts (ISO 14243-2)

This procedure applies to the initial management of the specimens and to each weight stop during the wear test.

1. Prepare the analytical balance on a rigid plane, isolated from vibrations.
2. Make sure that the specimens stabilize in a room with constant temperature and stable mechanical/ventilation isolation.
3. Put on clean plastic gloves.
4. Weigh each specimen three times in rotation.
5. Calculate the gravimetric wear as follows:

   $$W_n = W_{an} + S_n$$

   where W_n is the net mass loss after n cycles of loading, W_{an} is the average uncorrected mass loss, and S_n is the average variation in mass of the control specimen over the same period.
6. Calculate the average wear rate a_G using the following equation for the least-squares linear fit relationship between W_n and the number of loading cycles n:

   $$W_n = a_G * n + b$$

   where W_n is the net loss mass in mass after n cycles of loading and b is a constant. The zero time point shall not be used in the calculation.

15.4.1.4 Mounting procedure for the femoral components

1. Clean the femoral components with acetone.
2. Achieve the bone-cutting geometry indicated by the manufacturer of the implants through machining under a CNC milling machine.
3. Mix prechilled (4°C) bone cement.
4. With the help of an experienced surgeon, implant the femoral component on the prepared bone by interposing the bone cement.
5. Make sure that the specimens stabilize in a temperature- and humidity-controlled room (seasonal environmental conditions) for at least 24 hours.
6. Fix the implanted synthetic bone to the simulator frame in order to align the axis of flexion/extension of the prosthesis with the one of the machine.

15.4.2 Wear test procedure

15.4.2.1 Serum formulation

The entire formulation procedure should be executed under a laminar flow hood. A new formulation must be prepared after each weight stop.

1. Thaw the stocked calf serum at an ambient temperature 24 hours before proceeding to Step 2.
2. Filter the concentrated calf serum with the filter system (0.22-μm porosity).

3. Dilute with deionized water the filtered calf serum in a clean glass container at 25% vol/vol.
4. Add 0.2% sodium azide.
5. Add 20 mMol/dm^3 EDTA.
6. Add sodium hydroxide to balance the pH to 7.0.

15.4.2.2 Preparation of the simulator

1. Install one complete knee bearing (femoral and tibial components) for each one of the three testing stations.
2. Install one identical configuration in the control station (a soak control specimen is necessary in order to estimate the total change in mass of the tested specimens due to lubricant absorption as recommended by ISO 14243-2).
3. Calibrate all the degrees of freedom of the simulator in accordance with ISO 14243-3 (Figure 15.4).
4. Run the test for 5 million cycles.
5. Stop the simulator every 500,000 cycles for gravimetric measurements.

15.4.3 Examination of worn tibial inserts surfaces

1. Apply the cleaning procedure to the tibial inserts before examination.

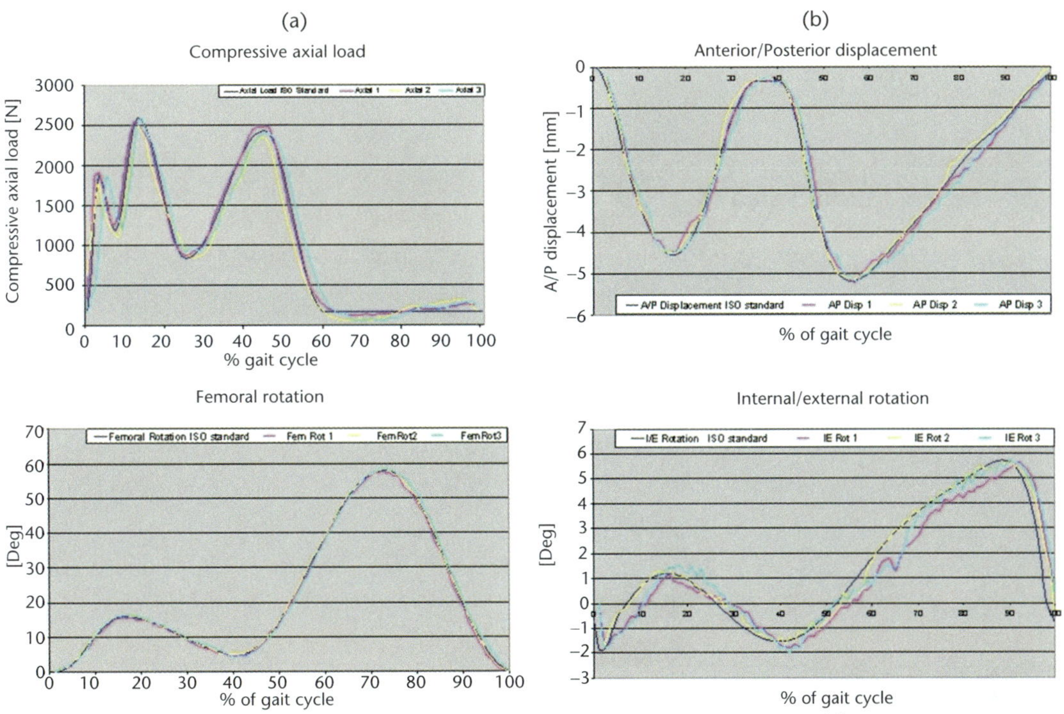

Figure 15.4 Knee simulator input profiles for: (a) A/P displacement, (b) axial force, (c) femoral rotation, and (d) I/E rotation. The picture depicts the recommended ISO standard profile (ISO14243-3, gray line) and the effective input to which the TKR is subjected (black line). The knee joint wear simulator is able to reproduce the four standardized range of motions with a maximum error of 5%. (Adopted from [16].)

2. Take a picture of the top view of each tibial insert on a scaled photographic scene.
3. Perform geometrical and optical calibrations of all digital images in order to equalize information content on the same perspective view and the same information content. A semiautomatic routine can be developed through the Macro-Editor of Image-Pro Plus for Windows (Media Cybernetics, Inc., Bethesda, Maryland).
4. Define the total contact area and the worn area (Figure 15.5).
5. Contour the worn area through the image analyzer software (Image-Pro Plus).
6. Assess the previous steps with three different observers to assure appropriate accuracy.

15.5 Anticipated Results

The wear behavior of the tibial inserts can be summarized through the mean weight loss ± SD (Figure 15.6). A careful visual examination of the all the tibial inserts, after 5 million cycles, revealed intraspecimen similarity. On tibial inserts two different worn areas can be observed for the medial and lateral compartments. The percentage of the damaged area usually ranges from 30% to 50% of the overall contact area [16]. The predominance of adhesive wear is suggested by a small amount of burnishing [4]. Usually some scratches are visible on the inserts along the anterior/posterior (A/P) direction. These scratches can be produced by third-body particles shaped in turn by the polishing phenomena and entrapment between the bearing surfaces [17]. A comparison between in

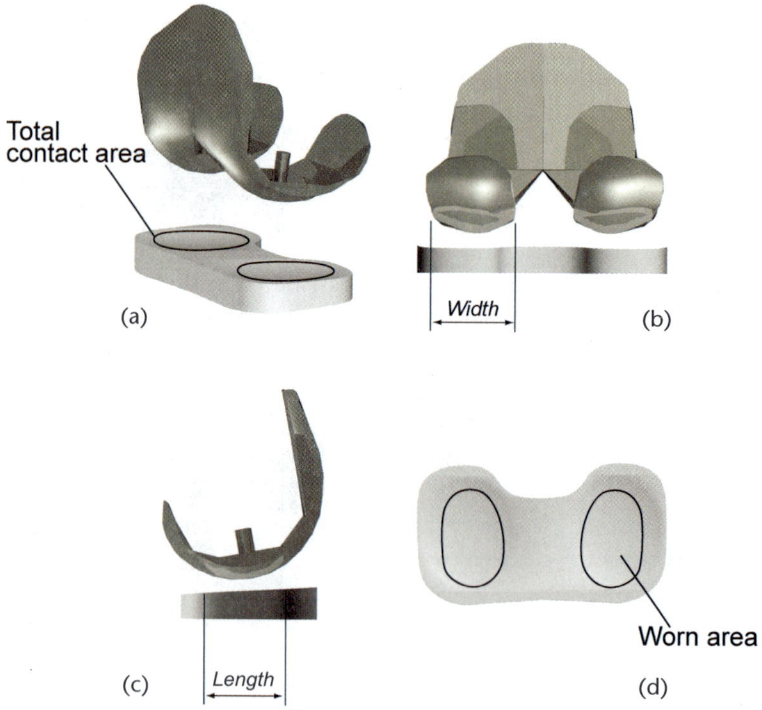

Figure 15.5 (a) Total contact area and (d) worn area for each condyle; (b) frontal view: shows the schematized area width; (c) lateral view: shows the schematized area length; (d) superior view: shows the final worn area marked with black lines.

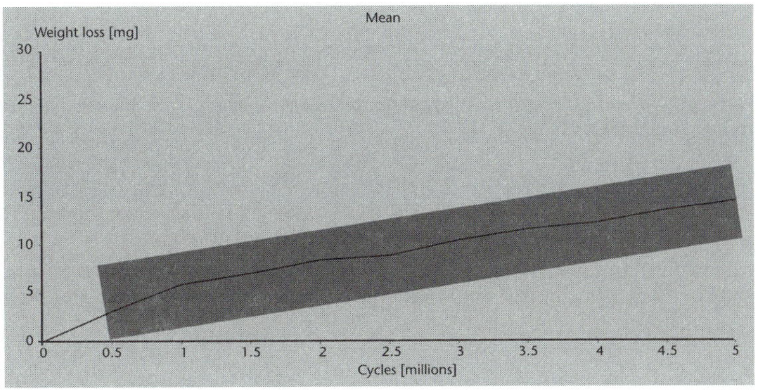

Figure 15.6 A box plot showing typical wear test results (SD in the dark gray box).

vitro experience and clinical retrievals proves acceptable qualitative matching of wear patterns demonstrating a good potential of such configurations to reproduce clinically interesting scenarios.

15.6 Discussion and Commentary

The wear patterns of the tibial inserts in this study were very similar to those seen in some retrievals. Although the wear tracks of the tibial inserts from the two sets of specimens in this study and those from a previous relevant in vitro study [16] were all similar (the percentage of worn area was 48% in this in vitro study and 52% in retrievals), being 50% (in a previous in vitro study), differences were seen in the tibial wear patterns and in the wear test results. Even though the tibial inserts used in this study and the previous one [16] were formed from the same UHMWPE starting powder (GUR 1050) and sterilized using the same method, they were classified as small and had similar thicknesses (13 and 14 mm). In a TKR, the tibial insert becomes molecularly oriented in the principal direction of sliding (A/P), producing a strain-hardened effect that increases the wear resistance [18]. Therefore, the component strain softens along the axis transverse to the sliding motion and exhibits less wear resistance in that direction. The presence of no physiological motions, such as internal/external rotation (which produces a friction force in the direction transverse to sliding), increases the polymer wear rate so that the greater cross shear, resulting from excessive rotation, accelerates the wear of the insert. Similarity in wear patterns observed on tibial inserts in this study and low wear rates measured from the present in vitro tests coupled with the results from other retrieval studies [19, 20] supports the efficacy of the new in vitro UHMWPE wear assessment method presented.

15.7 Application Notes

We have presented a new method for the in vitro assessment of the wear of the UHMWPE tibial insert in a TKR, which is based on fixing the femoral component to the distal end of a synthetic femur model (anatomic attachment). When a metal-block

Troubleshooting Table

Problem	Explanation	Potential Solutions
When using the analytical balance, the biggest difference among the three readings is bigger than 0.1 mg.	Inadequate stabilization time.	Continue taking readings in rotation until they are identical within 0.1 mg.
Preparing the synthetic bone by shaping the right geometry to host the femoral component by hand tooling.	Too rough control of cutting edges.	Use the CNC milling machine to achieve an appropriate control of the cutting edges.
The cement layer at the interface between the synthetic bone and the femoral component is too porous.	Wrong mixing composition of the cement.	Recement the femoral component by interposing a more liquid cement layer.

holder is used to attach the femoral component to the knee simulator, the wear pattern of the insert is different from that seen in retrieved inserts, and the wear magnitude referring to the insert is much higher than that of the anatomic attachment configuration, thereby indicating the significance of such a fixing method.

It is worth saying that the preliminary nature of this study is inevitably associated to some methodological limitations that need additional improvements. The small number of specimens (retrievals and in vitro ones) suggests caution in the considerations on the wear behavior of these tibial inserts. Before these results can be taken with sufficient confidence, it is necessary to improve the experimental protocol to reduce the interspecimens' variability. In particular, the method of attachment should be the only analyzed parameter in order to determine its quantitative influence on the setup. Moreover, a possible future target could be the development of a standardized protocol to determine the wear on retrieved inserts and carry out a quantitative comparison with in vitro specimens.

The presented innovative setup, coupled with current laboratory methods, allowed a high-accuracy evaluation of the elastic and permanent changes of position and orientation of the femoral components with respect to the host bone, and thus with respect to the loading directions. It was also possible to measure the wear induced by cyclic loading and motion while measuring changes of the relative position of the prosthetic components with respect to the loading directions [8].

15.8 Summary Points

1. We described a methodology through which it is possible to assess the wear of tibial inserts of total knee replacement (TKR).
2. This particular setup, while providing a more realistic biomechanical experimental simulation of the prosthesized knee joint, produces results comparable to standard tests and to retrieved implants.
3. The femoral holder provided room for the development of a more complex setup able to assess both wear and micromotions of the femoral component with respect to the host synthetic bone.

Acknowledgments

The authors would like to thank Luigi Lena for the illustrations and Paolo Erani for the support during the experiments.

References

[1] Merx, H., et al., "International Variation in Hip Replacement Rates," *Ann. Rheum. Dis.*, Vol. 62, No. 3, March 2003, pp. 222–226.

[2] Lachiewicz, M. P., and P. F. Lachiewicz, "Are the Relative Indications for Revision Total Knee Arthroplasty Changing?" *J. Surg. Orthop. Adv.*, Vol. 18, No. 2, 2009, pp. 74–76.

[3] Stea, S., et al., *Report of Orthopaedic Prosthetic Implantology. Overall Data: Hip and Knee Arthroplasty in Emilia Romagna Region: Istituti Ortopedici Rizzoli*, NIH Consensus Statement on Total Knee Replacement, No. 201, 2008, pp. 1–34.

[4] Miura, H., et al. "Prediction of Total Knee Arthroplasty Polyethylene Wear Using the Wear Index," *J. Arthroplasty*, Vol. 17, No. 6, September 2002, pp. 760–766.

[5] Naudie, D. D., et al., "Wear and Osteolysis Around Total Knee Arthroplasty," *J. Am. Acad. Orthop. Surg.*, Vol. 15, No. 1, January 2007, pp. 53–64.

[6] Schmalzried, T. P., and J. J. Callaghan, "Wear in Total Hip and Knee Replacements," *J. Bone Joint Surg. Am.*, Vol. 81, No. 1, January 1999, pp. 115–136.

[7] Martin, K. J., C. P. Neu, and M. L. Hull, "An MRI-Based Method to Align the Compressive Loading Axis for Human Cadaveric Knees," *J. Biomech. Eng.*, Vol. 129, No. 6, December 2007, pp. 855–862.

[8] Spinelli, M., et al., "Combined Wear Behaviour and Long-Term Implant-Bone Fixation of Total Knee Replacement: A Novel In-Vitro Set-Up," *Artificial Organs,* in press.

[9] Benson, L. C., J. D. DesJardins, and M. LaBerge, "Effects of In Vitro Wear of Machined and Molded UHMWPE Tibial Inserts on TKR Kinematics," *J. Biomed. Mater. Res.*, Vol. 58, No. 5, 2001, pp. 496–504.

[10] Cristofolini, L., et al., "Mechanical Validation of Whole Bone Composite Femur Models," *J. Biomech.*, Vol. 29, No. 4, April 1996, pp. 525–535.

[11] McEwen, H. M., et al., "The Influence of Design, Materials and Kinematics on the In Vitro Wear of Total Knee Replacements," *J. Biomech.*, Vol. 38, No. 2, February 2005, pp. 357–365.

[12] Muratoglu, O. K., et al., "Knee Simulator Wear of Polyethylene Tibias Articulating Against Explanted Rough Femoral Components," *Clin. Orthop. Relat. Res.*, Vol. 428, November 2004, pp. 108–113.

[13] Romero, J., et al., "Varus and Valgus Flexion Laxity of Total Knee Alignment Methods in Loaded Cadaveric Knees," *Clin. Orthop. Relat. Res.*, Vol. 394, January 2002, pp. 243–253.

[14] Cristofolini, L., et al., "Comparative In Vitro Study on the Long Term Performance of Cemented Hip Stems: Validation of a Protocol to Discriminate Between 'Good' and 'Bad' Designs," *J. Biomech.*, Vol. 36, No. 11, November 2003, pp. 1603–1615.

[15] Affatato, S., et al., "Fluid Absorption Study in Ultra-High Molecular Weight Polyethylene (UHMWPE) Sterilized and Unsterilized Acetabular Cups," *Proc. Inst. Mech. Eng. [H]*, Vol. 215, 2001, pp. 107–111.

[16] Affatato, S., et al., "Investigation on Wear of Knee Prostheses Under Fixed Kinematic Conditions," *Artif. Organs*, Vol. 32, No. 1, January 2008, pp. 13–18.

[17] Barnett, P. I., et al., "Investigation of Wear of Knee Prostheses in a New Displacement/Force-Controlled Simulator," *Proc. Inst. Mech. Eng. [H]*, Vol. 216, 2002, pp. 51–61.

[18] Wang, A., C. Stark, and J. H. Dumbleton, "Mechanistic and Morphological Origins of Ultra-High Molecular Weight Polyethylene Wear Debris in Total Joint Replacement Prostheses," *Proc. Inst. Mech. Eng. [H]*, Vol. 210, 1996, pp. 137–140.

[19] Ezzet, K. A., et al., "Oxidized Zirconium Femoral Components Reduce Polyethylene Wear in a Knee Wear Simulator," *Clin. Orthop. Relat. Res.*, Vol. 428, November 2004, pp. 120–124.

[20] Tsukamoto, R., et al., "Wear of Sequentially Enhanced 9-Mrad Polyethylene in 10 Million Cycle Knee Simulation Study," *J. Biomed. Mater. Res. B Appl. Biomater.*, Vol. 86, No. 1, July 2008, pp. 119–124.

About the Editors

Tim Maguire received a B.S. in chemical engineering from Rutgers University. His Ph.D., also from Rutgers and under the advisement of Martin Yarmush, was in the field of biomedical engineering and focused on applying novel polymer systems with embryonic stem cells to generate controlled differentiation systems. Tim also completed NIH-sponsored training programs, the Biotechnology Training Program, and an NSF-sponsored IGERT on microscale interfaces. Following the completion of his Ph.D., Tim went on to work in a formulation group at Merck for two years, where he helped implement new solid dispersion technologies, as well as establish a computational fluid dynamic program. He then went on to work at the Hurel Corporation where he was pursuing the creation of in vitro drug screening systems, integrating optimized human hepatocyte cultures with microfluidic systems. He is currently an associate research faculty member at Rutgers University where he is working on bioinformatics and microfluidics research. Collectively, Tim's work throughout his career has focused on engineering cells via culture processes, cues, or small molecules.

Eric Novik received a B.S. in biomedical engineering from Rutgers University. His Ph.D., also from Rutgers and under the advisement of Martin Yarmush, was in the field of biomedical engineering and focused on characterization and optimization of platforms for embryonic stem cell differentiation into liver lineage cells. Eric has been at the Hurel Corporation, a start-up biotechnology company, for over two years and is involved in all aspects of development and commercialization of patented microfluidic, cell-based platforms for use in drug discovery and development, consumer and industrial product testing, and related fields.

Index

A

Absorption, 19–38
  abstract, 19
  Caco-2, 28
  cultured cell system, 21–24
  data analysis, 25–26
  in dogs, 36–37
  in experimental animals, 35–37
  human, 19–38
  introduction, 20
  materials, 21
  methods, 21–25
  in monkeys, 37
  PAMPA system, 24–25, 29
  in rats, 35–36
  results and interpretation, 26–30
  summary, 38
  in vivo absorption measurements, 25
Absorption, distribution, metabolism, and elimination (ADME), 163
Absorption testing alternatives, 227–237
  abstract, 227
  biostatistics, 233–234
  chemicals and solutions, 229
  consumables, 229
  data evaluation, 232–233
  fluorescence, 233
  Franz diffusion cells, 229, 230–231, 234
  high-performance liquid chromatography (HPLC), 232
  introduction to, 228–229
  materials, 229–230
  methods, 230–234
  pitfalls, 236
  quality control, 231–232
  results, 234–236
  scintillation counting, 232
  skin absorption studies, 231
  skin penetration determination, 230–231
  skin permeation determination, 231
  skin preparation, 230
  summary, 236–237
  technical equipment, 229–230
  troubleshooting, 236
Aggregating brain cell cultures, 41–59
  abstract, 41
  analytical procedures, 53–54
  animals, 44–45
  application notes, 57–58
  cell isolation, 49–51
  cell type-specific immunostaining, 56
  culture preparation, 49–51
  data analysis, 54–55
  defined, 41
  discussion and commentary, 57
  equipment requirement, 43
  experimental design, 43–44, 57
  glassware washing/sterilizing, 49
  introduction to, 42–43
  maintenance, 51–52
  materials, 44–48
  media replenishment, 51
  modified DMEM serum-free culture medium, 48
  multiple endpoints, 57
  preparation, 57
  principle, 42–43
  reagents, 46–47
  replicate, harvesting, 53
  replicate, preparation and treatment, 52
  representative endpoints selection, 57
  serum-free culture medium, 58
  solutions/media preparation, 47–48
  special equipment, 45–46
  subdivision, 51–52
  summary, 58–59
  troubleshooting table, 58
Alcian blue staining, 217–218
Animals
  absorption and, 35–37
  aggregating brain cell cultures and, 44–45
  three-dimensional culture (chondrocytes, cartilage, bone/cartilage explants) and, 209–210

B

Benchmark concentrations (BMCs), 161
  determining, 164
  in vitro, 163–164, 165, 167
Benchmark dose (BMD) approach
  abstract, 159
  derivation of values, 163–164
  discussion and commentary, 167–168
  introduction to, 160–162
  materials and methods, 162–166
  summary, 167
  in vitro-in vivo correlation, 164–166
Benchmark doses (BMDs)
  derivation, 162
  determining, 164
  for developmental effects, 165
  plotting against BMCs, 164
  in vivo, 163, 166
  in vivo data quality, 167
Benchmark dose software (BMDS), 163
Benchmark responses (BMRs), 161
  choice of, 164
  defined, 163
  BMP-2 ELISA, 149
  DBM testing with, 154
  predictive power of a positive test (PPPT), 154
Bone morphogenetic proteins (BMPs), 148
Breast cancer resistant protein (BCRP), 32
Burst-forming units (BFUs), 124

C

Caco-2 cells/system
  calculation equations, 25
  compounds diffusing paracellularly and, 33
  discussion and commentary, 30–34
  experimental procedures, 23–24
  fraction summary, 30
  full automation, 30
  fully differentiated, 33
  with high-throughput LC/MS, 32
  PAMPA system with, 34
  permeability, 19, 27, 28, 34
  permeability measurement, 223–224
  seed and subculture, 22
  source, 21
  summary, 38
  use of, 30, 33–34
  validation, 26
  See also Absorption
Calcein-loaded liposomes
calcein separation, 104–106
  preparation of, 103–104
  See also Liposome assay
Canine uterine glands cell culture, 171–181
  application notes, 180
  cell culture, 173, 174–176
  discussion and commentary, 178–180
  electron microscopy, 174, 177–178
  imaging, 178
  introduction to, 172–173
  light microscopy, 173–174, 176–177
  limitation, 179
  materials, 173–174
  methods, 174–178
  results, 178
  summary points, 181
  troubleshooting guide, 181
CarcinoGENOMICS, 134, 144
Cell culture analog (CCA) system, 2
Cellular inflammation modeling, 67–73
  indirect response modeling (IDR), 68–69
  injury model, 68
  physiochemical cellular host response, 69–73
  proinflammation, 69
Cell uptake, in metabolic clearance, 14
Central nervous system (CNS), 63
Chondrocytes, isolated, 212–216
  2D culture, 214
  3D culture, 214–216
  histomorphological study, 217–218
  isolation, 212–214
  procedure illustration, 213
Clonogenic assays, 115–129
  abstract, 115
  application notes, 128
  colony counting criteria, 123–124
  colony scoring, 123
  data acquisition, 124–125
  discussion and commentary, 126–127
  experimental design, 116–117
  GM-CFU assay, 116
  GM-CFU test, 121–122
  human hematopoietic progenitors, 119–121
  human umbilical cord blood mononuclear cells (hu-UCB), 120–121
  incubator humidity test, 122–123
  introduction to, 116
  materials, 117–118
  methods, 118–124
  methylcellulose stock preparation, 118
  murine bone marrow cells (mu-BMC), 119
  murine hematopoietic progenitors, 118–119
  reagents, 118
  results and interpretation, 124–125

screening phase to IC determination phase, 122
statistical guidelines, 125
summary points, 128–129
trial layout, 117
troubleshooting table, 128

Colonies
burst-forming units (BFUs), 124
compact, 123
criteria for counting, 123–124
diffuse and spread, 123
morphology, 124
multicentric, 123
scoring, 123
See also Clonogenic assays

Confirmatory testing, 157
Control of Substances Hazardous to Health (COSHH) standards, 251
Corticosteroids, 87
Cortisol, 74, 75
Critical micelle concentration (CMC), 102, 109
Cultured cell system, 21–24
Caco-2 cell culture, 22–23
Caco-2 permeability measurement, 23–24
discussion and commentary, 30–35

D

Data acquisition
clonogenic assays, 124–125
hepatocyte suspension, 7–8
liposome assay, 108–109
physiologically based microfluidic system, 8
plated hepatocyte system, 7–8
stabilized primary hepatocyte cultures, 141

Data analysis
absorption, 25–26
aggregating brain cell cultures, 54–55

Demineralized bone matrix (DBM), 148, 150
Desmosomes, 197
Dogs
absorption in, 36–37
uterine glands cell culture, 171–181
See also Animals

Draize scores, 109, 111
Drug metabolism, 13
Drug metabolism and pharmacokinetic (DMPK) parameters, 19
Drug transporters, 35
Dulbecco's Modified Eagle's Medium (DMEM), 21, 22
Dye-loaded liposomes, 106–108

E

Electrocardiogram (ECG), 64
Electron microscopy, histological preparation
embedding and cutting, 177–178
fixation, 177
materials, 174
methods, 177–178
See also Three-dimensional culture (canine uterine glands)

Embryonic Stem Cell Test (EST), 160
Endotoxin
human response to, 73
intravenous administration, 65
neuroendocrine response, 75
rapid tolerance, 86
small dose administration, 86

Epidermis, 196
Epinephrine
acute infusion, 88
dynamics of, 74
immune function modulation, 74
plasma concentration, 64
signaling receptor, 74

Epithelial airway barrier model, 239–257
abstract, 239
anticipated results, 253–256
application notes, 257
discussion and commentary, 257
epithelial cell cultures, 241–242, 243–246
experimental design, 241
introduction to, 240
laser scanning microscopy (LSM) staining, 243, 250–251
materials, 241–243
methods, 243–253
monocyte-derived dendritic cells (MDDC), 242, 246–249
monocyte-derived macrophages (MDM), 242, 246–249
summary points, 257
transepithelial electrical resistance (TEER)
measurements, 42–43, 249–250
transmission electron microscopy (TEM), 243, 251–253
triple cell coculture, 242
troubleshooting table, 256

Epithelial cells
culturing, 241–242
currently used lines, 243–244
line cultures, 244–245
seeding, 245–246
thawing, 241
thawing protocol, 244

Extracellular matrix (ECM), 206, 207

F

Fetal bovine serum (FBS), 21
Franz diffusion cells, 229
  experiments, 234
  importance, 236
  in skin peneÜtration determination, 230–231
  in skin permeation determination, 231
  use of, 234
Free fraction, 14

G

GABAergic neurons (GAD), 55, 56
Gap junction intercellular communication (GJIC), 134
Glucocorticoids, 73
GM-CFU assay, 116, 121–122, 126–127

H

Hank's Balanced Salt Solution (HBSS), 21, 23
Hash value, 66
HDAC inhibitor TSA, 144
Heart rate variability (HRV), 63
  assessment model, 75–77
  autonomic dysfunction, 75
  measurement data, 75
  peripheral proinflammation effect on, 76
  prognostic significance, 75
Hematopoietic system, 126
Hematoxylin-eosin staining, 217
Hepatic clearances. *See* Intrinsic clearances
Hepatocytes
  cryopreserved, 13
  isolation from rat liver, 139–140
  isolation of, 135, 141
  plated culture, 4
  plated culture under flow condition, 4–6
  primary, cultivation, 141
  primary, dedifferentiation, 134
  rat, purification of, 140
  thawing, 3
  TSA-exposed, 143
Hepatocyte suspension system
  clearance study, 3–4
  data acquisition, results, and interpretation, 7–8
  discussion and commentary, 8–11
  as fully enclosed environment, 10
  intrinsic clearances, 9
  summary, 13
High performance liquid chromatography (HPLC), 167
  skin penetration, 232
  skin permeation, 232
Human endotoxin model, 64–65
Human hematopoietic progenitors, 119–121
Human umbilical cord blood mononuclear cells (hu-UCB), 120–121
  cryopreservation of, 120
  isolation of, 120
  thawing of, 120–121
  See also Clonogenic assays
Hypercortisolemia, 89
Hypothalamic-pituitary-adrenal axis (HPA), 63, 73

I

Immunofluorescent labeling, 192–193
Immunostaining, 56
Incubator humidity test, 122–123
Indirect response modeling (IDR), 68–69
Inflammation
  dynamics complexity, 65
  essential responses, 66
  reaction progression, 90
  response regulation mechanisms, 63
  stimulus dynamics, 69
  study of, 63
Inflammation response model, 61–92
  abstract, 61
  conclusions, 91
  data collection, 64–65
  dysregulation modes, 81–84
  elements of, 77
  expression motifs, 66–67
  human endotoxin model, 64–65
  increased insult implications, 81
  input/output data summary, 79
  introduction to, 62–64
  knockout in silico experiment simulation, 84
  lethal potentiation, 86
  materials, 64–65
  memory effects, 80, 84–87
  methods, 65–77
  modeling at cellular level, 67–73
  modeling at systemic level, 73–77
  parameter estimation, 77–80
  persistent response, 82
  qualitative assessment, 80–91
  results, 77–91
  self-limited response, 79
  stress hormone infusion evaluation, 87–91
  summary, 90–91
  temporal responses, 83
  topographical interactions, 78
  transcriptional dynamics and intrinsic responses, 65–67
  unresolved response, simulation, 82

Inflammatory and Host Response to Injury Large Scale Collaborative Project, 64
Influx transporters, 35
Intrinsic clearances, 1–15
abstract, 1
cell uptake role, 14
conclusion, 15
data acquisition, 7–8
hepatocyte suspension, 3–4, 9
hepatocyte thawing, 3
introduction, 2
materials, 3
methods, 3–7
"plasma shift," 13
plated hepatocyte culture, 4
plated hepatocyte culture under flow condition, 4–6
results and interpretation, 7–8
sample analysis with LC-MS/MS, 6–7
sampling, 6
summary, 13–15
In vitro skin substitute, 183–201
abstract, 183
antibodies, 187
anticipated results, 194–198
application notes, 199–200
bilayered, 197
culture media preparation, 188
desmosomes, 197
discussion and commentary, 198–199
epidermal differentiation markers, 195–196
experimental design, 185
fabrication, 189–191
fibroblast sheets assembly, 189–191
histological analysis, 193
histological modifications, 195
immunofluorescence preparation, 192
immunofluorescent labeling, 192–193
introduction, 184
involucrin marker, 195
keratinocyte differentiation process, 195
materials, 185–188
mechanical stability, 197–198
medium reconstitution, 189
methods, 188–193
proliferation markers, 184
reconstructed by self-assembly approach, 185–188
reconstructed skin assembly, 191
skin markers, 197
solutions/materials preparation, 188–189
statistical analysis, 193
summary points, 200–201
tissue-engineered skin maturation, 191
tissue preservation and sectioning, 191–192
transmission electron microscopy, 193
troubleshooting table, 200
ultrastructural analysis, 198
In vivo test reduction, 147–158
abstract, 147
confirmatory testing determination, 157
data analysis protocol, 152–154
decision analysis, 152
discussion and commentary, 154–158
experimental design, 149
introduction to, 148–149
materials, 149–150
methods, 150–154
recalibration frequency determination, 156–157
regulatory concerns, 156
results, 154
ROC analysis, 148–149, 155, 157
sample size determination, 155–156
secondary test selection, 154–155
statistical analysis, 157–158
summary points, 158
Involucrin, 195

K

Krebs-Henseleit buffer (KHB), 136

L

Laser scanning microscopy (LSM) staining, 243, 250–251
image restoration and, 250–251
materials, 243
methods, 250–251
triple cell coculture model, 254
LC-MS/MS, 6–7
Leukocytes, 67
L-glutamine penicillin-streptomycin (PEST), 21
Light microscopy, histological preparation
embedding and cutting, 177
fixation, 176–177
materials, 173–174
methods, 176–177
See also Three-dimensional culture (canine uterine glands)
Lipopolysaccharides (LPS), 63
binding of, 69
concentration, 81
pharmacodynamics model for, 69
in recognition process, 68
Liposome assay, 99–113
abstract, 99
application notes, 111

Liposome assa (continued)y
calcein-loaded liposome preparation, 103–104
calcein separation, 104–106
calibration, 101
critical micelle concentration (CMC), 102, 109
data acquisition, 108–109
discussion and commentary, 109–111
Draize scores, 109, 111
dye-loaded liposomes, 106–108
equipment, 103
experimental design, 101–102
introduction to, 100
materials, 102–103
reagents and supplies, 102–103
results and interpretation, 108–109
steps, 100
summary points, 112–113
troubleshooting table, 112
Liposomes
calcein-loaded, 103–106
diluted dispersion examples, 106
dye leaking, 108
dye-loaded, 106–108
extruded, 106
preparation of, 111

M

Madin-Darby canine kidney cells (MDCK), 33
Memory effects, 80
emergence of, 84–87
See also Inflammation response model
Microfluidic pump, 4–6
cell culture, 6
preparation, 4–5
MicroMass assay (MM), 160
Monkeys, absorption in, 37
Monocyte-derived macrophages (MDM) and dendritic cells (MDDC), 242, 246–249
air-liquid cultures, 249
isolation of, 242, 246–249
materials, 242
medium preparation, 247–248
methods, 246–249
triple cell coculture design, 248
Monoethyl hexyl phthalate (MEHP), 165
Multidrug resistant proteins (MRPs), 32
Murine bone marrow cells (mu-BMC), 119
Murine hematopoietic progenitors, 118–119

N

Neuroendocrine-immune system interaction modeling, 73–75
New chemical entities (NCEs)
hydrophobic, 33
lead optimization, 20
pharmacokinetic behavior, 2
NFkB, 69
nuclear activity, 71
protein inhibitor, 83
role, 69
signaling pathway, 70
Nonessential amino acids (NEAA), 21
No observed adverse effect levels (NOAELs), 159, 161

O

Ocular toxicity
area under the curve (AUC) and, 109
assay studies, 100
chemical compared to surfactant, 110
dye-loaded liposome experiments, 106–108
liposome assay, 99–113

P

PAMPA system, 19, 24–25
advantage, 34
Caco-2 cell model with, 34
calculation equations, 25
discussion and commentary, 34–35
experimental procedures, 24–25
permeability, 27, 29
transport buffer, 24
use of, 34
Permeability, 19–38
Caco-2, 19, 23–24, 27, 28, 34
PAMPA, 27, 29
P-gp, 32
Phosphate Buffered Saline (PBS), 21
Physicochemical cellular host response
model, 69–73
drug disposition, 71–72
kinase activity, 70, 71
LPS dynamics, 69
PK/PD model, 72
state variables, 71
TLR4 receptor, 70
Physiologically based microfluidic systems
data acquisition, results, and interpretation, 8
discussion and commentary, 12–13
schematic diagram, 12
summary, 14–15
Physiologically based pharmacokinetic (PBPK) model, 2
PK/PD model, 72
Plated hepatocyte system
biochip for cell seeding, 4

clearance studies, 4–6
data acquisition, results, and interpretation, 7–8
discussion and commentary, 11
microfluidic pump, 4–6
physiological basis, 11
setup illustration, 5
under flow condition, 4–6
Predictive power of a positive test (PPPT)
for BMP ELISA, 154
calculation, 158
Primary hepatocytes, 134
Proinflammatory response, 69, 85

R

Rats
absorption in, 35–36
hepatocytes isolation, 139–140
primary hepatocytes, cultivation, 141
Sprague-Dawley, 135
See also Animals
REACH (Registration Evaluation and Authorization of Chemicals), 134, 160
Reagents
aggregating brain cell cultures and, 46–47
clonogenic assays, 118
liposome assay, 102–103
stabilized primary hepatocyte cultures, 135–138
Recalibration
frequency determination, 156–157
sample size determination for, 155–156
See also In vivo test reduction
Receiver operating characteristic (ROC) analysis
computations, 149
curve, 149
curve examples, 157
defined, 148
plot illustration, 155
Reconstructed human epidermis (RHE), 228
Regulatory concerns, 156
Replicate cultures
harvesting, 53
preparation and treatment, 52
See also Aggregating brain cell cultures
Results and interpretation
absorption testing alternatives, 234–236
aggregating brain cell cultures, 55–56
clonogenic assays, 124–125
discussion and commentary, 178–180
epithelial airway barrier model, 253–256
hepatocyte suspension system, 7–8
liposome assay, 108–109
neurotoxicity tests, 55
physiologically based microfluidic system, 8
plated hepatocyte system, 7–8
stabilized primary hepatocyte cultures, 141–143
three-dimensional cell culture (canine uterine glands), 178
three-dimensional culture (chondrocytes, cartilage, bone/cartilage explants), 218–220
tibial inserts wear assessment, 268–269
in vitro skin substitute, 194–198
in vivo test reduction, 154
Rotary Cell Culture System (RCCS) bioreactor, 205
components, 209
defined, 209
illustrated, 210
operational conditions, 209, 218
Rotation-mediated reaggregation, 42

S

Sampling
analysis with LC-MS/MS, 6–7
for clearance study, 6
size, 155–156
Self-assembly approach, 183
advantages, 198–199
concept, 185
skin substitute reconstruction by, 185–188
in vitro fabrication by, 189–191
See also In vitro skin substitute
Sepsis, 62
Skin penetration
determination with Franz cell setup, 230–231
fluorescence, 233
HPLC, 232
See also Absorption testing alternatives
Skin permeation
determination with Franz cell setup, 231
fluorescence, 233
HPLC, 232
See also Absorption testing alternatives
Solid phase microextraction (SPME), 167
Stabilized primary hepatocyte cultures, 133–145
abstract, 133
application notes, 144
data acquisition, 141
discussion and commentary, 143–144
experimental design, 135
facilities/equipment, 139
introduction to, 134–135

Stabilized primary hepatocyte cultures (continued)
materials, 135–139
reagents, 135–138
results and interpretation, 141–143
summary points, 144–145
troubleshooting table, 141
TSA-optimized, 142
Stratum corneum substitutes (SCS), 235
Stress hormone infusion evaluation, 87–91
Systemic inflammation modeling, 73–77
neuroendocrine immune-system interactions, 73–75
reduced heart rate variability assessment, 75–77
See also Inflammation response model
Systemic inflammatory response (SIRS), 62

T

3R tests (reduction, replacement, refinement), 143, 222
Three-dimensional culture (canine uterine glands), 171–181
abstract, 171
application notes, 180
cell culture, 173, 174–176
electron microscopy, 174, 177–178
imaging, 178
introduction to, 172–173
light microscopy, 173–174, 176–177
limitation, 179
materials, 173–174
methods, 174–178
results, 178
summary points, 181
troubleshooting table, 181
Three-dimensional culture (chondrocytes, cartilage, bone/cartilage explants), 205–224
abstract, 205
animals, 209–210
anticipated results, 218–220
application notes, 222–223
cell/tissue culture equipment, 211
chemicals, 211–212
chondrocytes isolation, 212–214
culture models, 208
culture procedure summary, 219
discussion and commentary, 220–222
experimental design, 208–210
fragments of articular cartilage explants, 216–217
histomorphological study, 217–218
introduction to, 206–207
isolated chondrocytes (2D culture), 214
isolated chondrocytes (3D culture), 214–216
materials, 211–212
methods, 212–218
pitfalls, 220–221
RCCS bioreactor, 208–209
sample preparation equipment, 211
summary points, 223–224
tissue explants preparation, 212
troubleshooting table, 223
undissected proximal tibial epiphyses, 217
Tibial inserts wear assessment, 262–270
abstract, 261
anticipated results, 268–269
application notes, 269–270
cleaning procedure, 265
discussion and commentary, 269
experimental design, 263
gravimetric measurement procedure, 266
introduction to, 262–263
materials, 263–265
methods, 265–268
mounting procedure, 266
presoaking procedure, 265
specimens management, 265–266
summary points, 270
troubleshooting table, 270
wear test procedure, 266–267
worn surface examination, 267–268
TLR4 receptor, 70
Total knee replacement (TKR)
schematic, 262
tibial inserts wear assessment, 261–270
wear of, 262
Transepithelial electrical resistance (TEER)
measurements, 42–43, 249–250
materials, 242–243
methods, 249–250
Transmission electron microscopy (TEM), 193, 243
COSHH standards, 251
fixation and embedding of cells for, 251–253
of triple cell cocultures, 255
Trichostatin A (TSA), 135
cell cycle arrest, 135, 142
chemical structure, 136
effects on CYP450 protein, 142
HDAC inhibitor, 144
inductive effect of, 142
treatment assessment, 143
Troubleshooting tables
aggregating brain cell cultures, 58
clonogenic assays, 128
epithelial airway barrier model, 256

liposome assay, 112
stabilized primary hepatocyte cultures, 141
three-dimensional culture (canine uterine glands), 181
three-dimensional culture (chondrocytes, cartilage, bone/cartilage explants), 223
tibial inserts wear assessment, 270
in vitro skin substitute, 200

W

Whatman microfiber filter, 105
Whole Embryo Culture (WEC), 160